Electrifying Mexico

Joe R. and Teresa Lozano Long Series in Latin American and Latino Art and Culture

| DIANA J. MONTAÑO |

Electrifying Mexico

TECHNOLOGY AND THE TRANSFORMATION OF A MODERN CITY

University of Texas Press AUSTIN

Printed in the United States of America
First edition, 2021

Parts of chapter 4 appeared previously, in a different form, in "*Ladrones de Luz*: Policing Electricity in Mexico City, 1901–1918," *Hispanic American Historical Review* 101, no. 1 (2021): 35–72.

♾ The paper used in this book meets the minimum requirements of ANSI/NISO Z39.48-1992 (R1997) (Permanence of Paper).

LIBRARY OF CONGRESS CATALOGING-IN-PUBLICATION DATA

Names: Montaño, Diana, author.
Title: Electrifying Mexico : technology and the transformation of a modern city / Diana J. Montaño.
Description: First edition. | Austin : University of Texas Press, 2021. | Includes bibliographical references and index.
Identifiers:
LCCN 2021008822
ISBN 978-1-4773-2345-8 (cloth)
ISBN 978-1-4773-2346-5 (library ebook)
ISBN 978-1-4773-2347-2 (ebook)
Subjects: LCSH: Electrification—Mexico—History—19th century. | Electrification—Mexico—History—20th century. | Electrification—Social aspects—Mexico. | Electric power consumption—Mexico—History. | Mexico—Social conditions—20th century.
Classification: LCC HD9685.M62 M66 2021 | DDC 333.793/209720904—dc23
LC record available at https://lccn.loc.gov/2021008822

doi:10.7560/323458

Contents

Acknowledgments

Along an arduous pilgrimage, one accumulates unpayable debts. In conceptualizing, researching, analyzing, and writing this book, the author has found *sustento* (support/nourishment) from expected and unexpected quarters. I have struggled to do justice to the many individuals that have entered my life and have influenced me intellectually, academically, and personally.

It is only fair and necessary to begin by thanking William H. Beezley, whose mentorship has guided me for over fifteen years, a *quinceañera* in Mexican years. From supporting my ambitious master's thesis that took me from the state of Michoacán in western Mexico to East Los Angeles to drifting toward the field of history, to advocating for a semester abroad at El Colegio de México, to planting the seed for this book over thirteen years ago, the book in your hands bears the footprint of his strength as an academic mentor. He has been an engaged reader and commentator whose questions, concerns, and never-ending *quejas* made this book a better one. For that alone, he deserves a sack of *chitepines*. However, his enthusiasm for everything Mexico, for his incessant pursuit to do cultural history, I thank him the most. As many Latin Americanists trained at the University of Arizona, I was privileged to work with Kevin Gosner and Bert J. Barickman, whose intellectual rigor and dedication to teaching have become a guiding example. It pains that BJB left us too early. His brilliance and inquisitiveness have left an indelible mark, as has the singing of the hymn ao Senhor do Bonhim in the middle of the O Lugar Desprovido de Anchovas surrounded by strangers. I hold dearly the memory of him walking up and down the halls of a full auditorium around anxious undergrads immersed in their finals with my one-year-old son in arms exuded his kindness and effort of practicing inclusiveness.

I am especially indebted to Susie S. Porter, John Lear, and Andrea Friedman, whose intellectual kindness and generosity know no bounds. Susie championed the metamorphosing from a dissertation into the book. She commented perceptively on various drafts offering substantive feedback, and her work on working women in Mexico City encouraged me to cultivate grounded analysis. Her eagerness to workshop the manuscript along with John Lear at a critical juncture kindled its subsequent revision. John's work energized the visual analysis and a deeper dive into the electrical workers' union archives for chapter six. Having identified the engraving that tactfully introduces in the cover the argument of the following pages reaffirmed his title as the book's *padrino*. Andrea Friedman's incisive readings of numerous iterations were matched by her unwavering faith in this project. She has been a much-needed fountain of strength and courage—my heartfelt gratitude for her mentorship, caring and steady guidance, and friendship. Edward Beatty and Mikael Hård weighed in with comprehensive feedback, citations, and suggestions on several chapters. I am humbled by their generous and careful engagement with my work. Elisa Speckman-Guerra and Robert Alegre strengthened chapters four and six correspondingly through their insightful remarks and recommendations. William E. French offered invaluable help in the crafting of the book's architecture. Javier Duran deserves special recognition for his ongoing mentorship, critiques and for making me consider graduate school in the first place. *Mil gracias*.

I have been privileged to present pieces of this project to variegated audiences over the years. Individual chapters were workshopped thanks to William H. Beezley and David Pretel, who opened their graduate courses. Christiane Berth and Yovanna Pineda invited me onboard the workshop they organized at the University of Bern on the role of users in global technological history. Over two intensive days, feedback and conversations with Leida Fernández-Prieto, Mikael Hård, Clapperton Mavhunga, and Suzanne Moon provided ideas to improve chapter five. Mario Barbosa Cruz, Clara E. Lida, Diego Pulido, and María Dolores Lorenzo Río and participants in the Seminario Permanente de Historia Social of El Colegio de México's Centro de Estudios Históricos helped polish an early draft of chapter four. A later version of that chapter benefited from the sharpened eyes of David Pretel and Fabián Prieto-Ñañez.

Senior colleagues in the Department of History have injected enthusiasm for this project. Jean Allman, Monique Bedasse, Liz Borgwardt, Daniel Bornstein, Shefali Chandra, Alexander Dubé, Douglas Flowe, Christine Johnson, Peter Kastor, Sowande' Mustakeem, Christina Ramos, Nancy

Reynolds, Corinna Treitel, Anika Walke, and Lori Watt have all earned my heartfelt gratitude. I have felt incredibly energized by the collegiality of Andrea Friedman, Jean Allman, Shefali Chandra, and Nancy Reynolds, who have celebrated and engaged directly in the project's development. Liz Borgwardt also deserves an extra thanks for sharing her excitement about the book with Kathryn Lofton, who found its description "utterly fascinating" and put me in touch with the University of Texas Press. Across Washington University in St. Louis, I have found a stable intellectual community. I have benefited from the collegiality of William Acree, Mark Beirn, Adrienne Davis, Rene Esparza, Bret Gustafson, Mabel Moraña, Ignacio Sánchez Prado, Winnie Poster, Luis Salas, Akiko Tsuchiya, Miguel Valerio, Rebecca Wanzo, Helina Woldekiros, and Rafia Zafar. Special thanks to Sonia Lee, Andrea Friedman, Anika Walke, and Trevor Sangrey, who added chairs to their tables to welcome my family. My academic siblings Christina Ramos, Bahia Munem, and Miguel Valerio have collectively been a bedrock for the last few years. Thanks to each of you for the meals, laughs, and advice. I am incredibly thankful to Christina for entrusting me with her aunt's recipes, for her *sabor*, and for walking alongside as a *loyal hermana*.

Several institutions have funded my education and research. At the University of Arizona, the Tinker Foundation, the Social and Behavioral Science Research Institute, the Hanson Film Institute, and the Confluencenter for Creative Inquiry financed research at critical moments. My graduate education benefited greatly from the Graduate College Minority Fellowship, one of the many hard-fought accomplishments of the late María Teresa Velez, a role model for many women in academia and a fervent advocate for furthering diversity in higher education. A Greenleaf Visiting Library fellowship at the University of New Mexico granted me the opportunity to research cookbooks and culinary ephemera and present early findings to an engaged audience. At Washington University in St. Louis, a generous postdoctoral fellowship in the Department of History and a First Book Fellowship at the Center for the Humanities have provided the institutional and financial support to further my research agenda. Ignacio Infante, Barbara Liebmann, Tila Neguse, Shefali Chandra, and the other scholars in my cohort provided the ideal environment to polish the following pages' arguments and writing. The manuscript greatly benefited from their engaged reading and vibrant discussion. I appreciate their participation in the manuscript workshop and that of Deborah Cohen, Andrea Friedman, Winnie Poster, Christina Ramos, Nancy Reynolds, and Miguel Valerio. *Infinitas gracias* to Susie S. Porter and John Lear for leading us that day.

Over the years, I have found intellectual comrades along the way who have helped baptize ideas and solidify shared interests over endless cups of coffee and *un que otro mezcal*. At the University of Arizona, I was fortunate to cross paths with Marco Cabrera Geserick, Kelley Castro, Luis Coronado, Colin Deeds, Anabel Galindo, Rocio Gomez, Allison Huntley, Elena Jackson Albarrán, Amie Kiddle, John Kinglemann, Amanda López, Marco Macias, Max Mangraviti, mi boricua Zarah Moss, Lisa Munro, María Muñoz, Pete Soland, Ageeth Sluis, Aurea Toxqui, Cristina Urias, Jen Valenzuela-Sliger, and Robin Zenger. Sandra Aguilar, Steve Bunker, Justin Castro, Jethro Hernández, Christina Jiménez, Robert Jordan, Sam Martland, Christian Rocha, Anton Rosenthal, and Robert Weis provided insightful comments on numerous conference papers that made me revisit ideas and perspectives. Jeffrey Pilcher and William E. French must be doubly applauded for their generosity, mailing out a cookbook and a master's thesis early on in this project.

Archival research in Mexico City primarily benefited from the generous support of staff at the National Archives, the National Newspapers Archives, the National Library, Fundación Herdez, Miguel Lerdo de Tejada Library, Archivo Histórico del Distrito Federal, the Sindicato Mexicano de Electricistas (SME), and the numerous second-hand bookstores along Calle Donceles. A handful of individuals deserved my deepest gratitude for showing me the ropes to navigate the city's archives, principally Jen Boles, Carmen Nava Nava, Lucero Morelos, and Hayde Toledo. I am grateful to Sonia Robles and Aaron Shapiro, who welcomed me into their home to break bread. Special thanks to the SME *compañeros* who kept the archive and library open during their resistance and assisted me in 2012.

Writing a book is, in many ways, a collective effort and yet terribly isolating. The latter even more so amid a pandemic. In the last few months, Guillermo Cerón from the National Newspapers Digital Archives, Ana Carolina Abad and Gabriela Chávez from the Museo Nacional de la Estampa, art collector Ernesto Arnoux, and Fernando Amezcua Castillo, the SME's former Education, and Propaganda Secretary, kindly provided me high-resolution images and secured permissions for the reproduction of images. My gratitude to senior editor Kerry Webb, who expressed an unwavering commitment to this project and helped bring it to a successful conclusion. I am also grateful to the press's copy editor, Kathy Lewis, for her meticulous work to enhance this book's gracefulness and clarity.

A handful of colleagues have lessened this last *trecho* through moral and spiritual support. They have offered laughs, faith, and life lessons via creative and healing spaces through text messages, Facebook posts, phone calls, and zoom meetings. You have collectively nurtured me in countless ways. Trying to wrap up revisions and copyedits away from one's library, and with limited access to libraries, has been possible only with the support of Anabel Galindo and Marco Macias, who have earned lifetime *agradecimiento*. In particular, Anabel has galvanized me through endless conversations on scholarship, motherhood, homeschooling, disease, and loss. We have laughed and mourned repeatedly over the last year, and yet, I find glimpses of strength and encouragement that I treasure and take with me.

My family has been central to ensure this project reaches port. My parents, María Garcia de Montaño and Jorge Montaño have granted me an unsurpassable level of help and support in balancing academia, motherhood, and family life. Gracious storytellers, they have bestowed more knowledge, patience, and strength than I have been able to absorb. My siblings Sicilia, Jesús, Jorge, Rosa, Josefina, Luz, Lorenia, Dayan, and José, have encouraged and invigorated my professional goals. Special thanks to la tía Toty, whose strong presence in my children's life eased the completion of graduate studies. My biggest thanks go to José Román, who has tirelessly nurtured me to be freer, happier, and ever more determined. *Gracias por los sacrificios y los detalles*. I have fond memories of rushing back from the archives in Mexico City on Tuesdays for our weekly journeys to the street market. The stroller that doubled as a shoppingcart while Josecito sampled endless fruits and delicacies ran more miles than we care to count. Emiliana María rode out the dissertation with me—actively kicking—a constant reminder that there was little time to waste. Her sharpness, sense of humor, and inquisitive mind have made this journey to the finish line bearable. She and my *dos Joseses* welcomed the completion of this journey. *¡Provecho!*

Abbreviations

AGC	American Grocery Company
APHA	American Public Health Association
CADE	Compañía Argentina de Electricidad
CANACINTRA	Cámara Nacional de la Industria de la Transformación
CANAME	Cámara Nacional de Manufacturas Eléctricas
CEIMSA	Compañía Exportadora e Importadora Mexicana, SA
CFE	Comisión Federal de Electricidad
CGHC	Compañía de Gas Hidrógeno Carbonado
CGT	Confederación General de Trabajadores
CIA	Central Intelligence Agency
CME	Compañía Mexicana de Electricidad, SA
CMGLE	Compañía Mexicana de Gas y Luz Eléctrica
CMLE	Compañía Mexicana de Luz Eléctrica
CNDP	Comité Nacional de Defensa Proletaria
CNE	Cámara Nacional de Electricidad
CONCAMIN	Confederación de Cámaras Industriales de los Estados Unidos Mexicanos
CP	candlepower
CTM	Confederación de Trabajadores Mexicanos
DC	direct current
FNM	Ferrocarriles Nacionales de México

FUPDM	Frente Único Pro Derechos de la Mujer
GDP	gross domestic product
GE	General Electric Company
HP	horsepower
IEM	Industria Eléctrica de México
JWT	J. Walter Thompson
kW	kilowatt
kWh	kilowatt-hour
LEAR	Liga de Escritores y Artistas Revolucionarios
LTS	large technical systems
MACEC	Mexican American Commission for Economic Cooperation
MELC	Mexican Electric Lighting Company
METC	Mexico Electric Tramways Company
Mexlight	Mexican Light and Power Company
MGEC	Mexican General Electric Company
NAFINSA	Nacional Financiera
OIAA	Office of Inter-American Affairs
PRI	Partido Revolucionario Institucional
SCOP	Secretaría de Comunicaciones y Obras Públicas
SIDRO	Société Internationale d'Energie Hydro-Électrique
SME	Sindicato Mexicano de Electricistas
SOFINA	Société Financière de Transports et d'Entreprises Industrielles
STFRM	Sindicato de Trabajadores Ferrocarrileros de la República Mexicana
STS	science, technology, and society

Electrifying Mexico

Introduction

As she stepped out into the street, Doña Candelaria, a visitor to Mexico City, asked Señor Gutiérrez whether it was obligatory to hold a man's arm when out walking. It was not, he clarified, but it was the safest way, given the poor state of the streets. Suffering from eyestrain after years of sewing, Doña Candelaria accepted her guide's arm. At the Plaza del Seminario, their progress was stopped by her heartfelt cry: "Oh Jesus, Mary, and Joseph!" Confused, her escort asked what was the matter. She pointed to the lights and said, "The gas harms me . . . Look, what barbarity!" Señor Gutiérrez sought to explain that it was not gas but electric lighting, but the clarification made no difference to Doña Candelaria:

DOÑA CANDELARIA: What *elétrica*! It could make you go blind!
SEÑOR GUTIÉRREZ: It's a lovely light.
DOÑA CANDELARIA: How can you say that! How can it be beautiful if it's worse than an *ocote* [Montezuma pine] torch in your nose. Tonight, I certainly will be ill from my eyesight, Señor Gutiérrez . . .
SEÑOR GUTIÉRREZ: Don't fix your eyes on the *focos* [light bulbs].
DOÑA CANDELARIA: What light bulbs?
SEÑOR GUTIÉRREZ: In the lighting.
DOÑA CANDELARIA: Which ones are the light bulbs? You are a scientist, but I don't understand about light bulbs.
SEÑOR GUTIÉRREZ: Well, señora, don't look at the lamp.
DOÑA CANDELARIA: Even if I don't want to! Look at th— [Turning to her husband who joined them along the walk] Do you see it, Trinidad? I don't know how people put up with the, the what, Señor Gutiérrez?

SEÑOR GUTIÉRREZ: The light bulbs.
DOÑA CANDELARIA: The *focos elétricos*. Do you put up with them, Trinidad?
SEÑOR GUTIÉRREZ: I must confess, they are a little bit strong.
DOÑA CANDELARIA: Well, I see, and now I understand, Trinidad, my love, why there is a school for the blind in Mexico [City]. There's nothing like that in my town, and I now know why, because we do not have *elétricas*.[1]

José Tomás de Cuéllar, a prolific novelist critical of Mexico City's material progress, captured this scene in his 1883 novel *Los fuereños* (The provincials). Doña Candelaria's denunciation of electric lighting as a barbarity diverges from Señor Gutiérrez's attempted praise and demonstrates that personal reactions to modern industrial lighting were far from monolithic. Not everyone was impressed with what was new. Far from fictitious, Doña Candelaria's response echoed other documented reactions to electric bulbs in the early 1880s. The first forty arc lamps went up from the Zócalo, the city's main square, to the Reforma roundabout. *Capitalinos* (Mexico City residents) complained that pedestrians halfway along the lamp-lit blocks, dazzled by the radiant light, could not distinguish objects. Anyone in the radius of the light walked as if in a daze. Those directly underneath the lamps—the only ones who were spared being blinded—enjoyed an amusing sight. For them, the light perfectly illuminated figures from head to toe, in vivid colors, as these same people walked with eyes half-shut, comically trying to shade themselves from the harsh glare, so the light would not entirely blind them.

More than one hundred years after this spectacle, electricity has won people over in a way that its earliest incarnations did not. In our everyday lives, we spend little time thinking about electricity. Only major disruptions of service or high bills make us realize its presence.[2] We inhabit electrified spaces and even expect them wherever we go. All around us, high-voltage towers, transformers, posts, and power lines interrupt our horizon, weaving a web so familiar that we no longer recognize it. We live in a world where we hardly even think about electrification. Only when electricity is missing does it becomes a newsworthy event. But what was it like for those first experiencing electrification? How did it impact their lives, their ambitions? How did it register in their senses?

I answer these questions in the following pages by delving into the making of electrified spaces and how people navigated their way through them. I explore the experience of electricity through lighting in public celebrations, the social anxieties and legal complexities of streetcar accidents and power theft, the gendered and racialized scripts that framed the sale and consumption of electrical kitchen appliances, and the nationalization of the electrical industry. Since the late nineteenth century, technology has played a dominant role in modern life, with machines among "the most evident emblems and instruments of . . . modernity."[3] However, too often, technologies appear as "black-boxes" in popular discourse, as "fixed entities that irresistibly change society and culture," removed from the people and the ambitions that employed them.[4]

The fluid relationship between people and technology has escaped the attention of the traditional diffusionist model of the history of technology. This model was developed by George Basalla in 1967 to explain the simple transfer of "progressive" Western science to the non-European world.[5] Adopted for the study of technology, this model privileged innovation and dispersal at the expense of adaptation and use. In the histories of the ways technology has shaped modern life, too often technology becomes an all-powerful, unstoppable force. Not surprisingly, analyses that adopt the postulate of technological determinism to treat the history of technological modernity have produced "a single-stranded story of advancing, and indeed progressive, globalization."[6] This is particularly true for major technologies (such as railroads, telegraphs, and large-scale irrigation projects) introduced to Africa, South Asia, and Latin America. The model concedes no agency to local societies, which appear to be mere "perpetual consumers" of technology developed elsewhere.[7] Modernity, in these studies, always originates in an industrialized core.

Recent work has called this conviction into question, however, beginning with the very idea of what constitutes technology. As one seminal scholar has claimed, technology conceptualized as "abstract, reified, and universalistic . . . obscure[s] the significant differences between birth control and hydrogen bombs, and blind[s] us to the ways different groups and cultures have appropriated the same technology and used it to different ends." Far from dictating to people how they *must* be used, "technologies interact deeply with society and culture . . . [with these] interactions involv[ing] mutual resistance, accommodation, acceptance, and even enthusiasm." To untangle its connection to the way societies understood themselves and their reaction to modernity, we need to stop treating technology as a mass noun and instead, as Thomas Misa argues, "look more closely at individual technologies and inquire more carefully into social and cultural processes."[8]

This trajectory is evidenced in the history of electrification. The early twentieth century saw the earliest works on electrification, primarily by political economists, as a response to concerns over the power of monopolies in the United States and movements toward the nationalization of electric companies in Europe. The preeminence of the state and capital consolidated with works that predominantly centered on the technological, institutional, and economic aspects of electrification.[9] These have been the primary focus of the large technical systems (LTS) approach pushed by the seminal work of Thomas Hughes. Shunning technological determinist narratives, Hughes argued that a narrow focus on the development of technical devices (incandescent lamps, electrical generators, and transformers) could not explain the evolution of electric light and power systems. Instead, scholars had to look at the fundamental role of political cultures and system-building individuals (engineers and entrepreneurs).[10] Despite underlining the diversity of technical cultures, Hughes's demand-driven perspective left consumer demand for electricity unrecognized.[11]

Consumer and cultural perspectives revealing the social meanings of electricity emerged with the adoption of a social constructivist approach (social construction of technology: SCOT) that defines technological systems as "both socially constructed and society shaping."[12] After David E. Nye's work on electricity in the United States demonstrated the significance of this approach, electric grids are now recognized as "inescapably a social construction." The role of politics and capital has taken precedence, as demonstrated in works on South Africa, India, Palestine, and Latin America, where the footprints of the state and foreign capital loom large.[13] Concern has mainly centered on *how* electrification occurred rather than on the complex process of *whether* to electrify in the first place.[14]

In the last decade, scholars have questioned the inevitability of electrification by looking at the processes of domestication and energy transition. While Graeme Gooday, for instance, tracked the way Britons debated bringing electricity into their domestic spaces, Christopher Jones (privileging the environmental dimension of electrification) demonstrated how the construction of large energy transport systems in the nineteenth century set the United States on the path to fossil fuel dependency. Together these studies have shown how current electrified environments were "neither natural nor inevitable."[15]

Despite the level of sophistication reached by this historiography, it remains geographically unbalanced, disproportionately centered on the industrialized North Atlantic area.[16] Alexandra Hofmänner notes that Europe has remained the primary subject of electrification. The rest of the world, as Dipesh Chakrabarty writes, serves as "the historical residual against which difference is mapped."[17] The myth that electrification began in the North

Atlantic and then spread to other societies leaves intact "the conceptual confines of the assumptions that underlie the grand narrative of modernity."[18] The relativist approach recognizes that societies have embraced various technologies based upon local situations and needs, and even that the same technology may not be employed in the same way everywhere. While calling for multiple or alternative modernities, however, the relativist approach has struggled to produce "an alterity for the genealogy of modernity."[19]

Shalini Randeria underlines the urgency of overcoming the confines of a singular conception of modernity. The Western/non-Western binary defines non-Western societies "by negation." "Their historical and contemporary experience is then understood in such a framework not in terms of what it is but in terms of what it is not." She invites us to replace this with a "relational perspective [on modernity] which foregrounds processes of interaction and intermixture in the entangled histories of uneven modernities."[20]

The relation that a society develops with electricity and electrifiation must be treated as unique and discrete. This uniqueness does not preclude us from recognizing the patterns and trends that foreign capital and global electrification efforts established from the 1880s to the 1950s.[21] Through specific grounded analyses, we uncover the particular circumstances that shape and characterize different experiences of modernity and relationships to electricity. Mexico is no exception. This approach allows historians to bring new actors into focus. Discussion of people's doubts and fears about electricity, competition from other energy sources, legal complexities, and racial and class associations in the following pages helps us to see the adoption of electricity in the Mexican capital as neither inevitable nor linear. Electrification is entangled in other histories of the city, in the choices that capitalinos made in the transformation of their spaces, and in the transnational flows of capital, machinery, and technicians.

Most Mexicans in the 1880s saw modernity as something to be imported from abroad. Nothing signified the might of a modern Mexico more than electricity, with its brightness and foreignness. That did not mean that Mexicans agreed on what electricity was, its proper and potential use, and what a future electrified Mexico would look like. In different times and places, Mexicans could disagree vehemently. They may have imported a great deal of electrical machinery, along with scripts on how to use it, but, as I argue, these scripts were translated, edited, adapted, rewritten, or sometimes even entirely discarded. I go beyond official dictates and blueprints to look at what real people did every day.

The long regime of President Porfirio Díaz (1876–1880, 1884–1911), which became known as the Porfiriato, espoused technology as the guiding light to a modern, powerful Mexico. Until recently, however, scholarship on the history of technology has been limited. This lacuna is endemic to Latin America at large. The framing of the region as technologically backward has marginalized the interactions of Latin Americans with technology and, as in other areas of the Global South, has justified "giving greater attention to more industrialized areas of the world."[22] Latin American scholars have called for counteracting single-stranded narratives, for developing new narratives, for an "anti-grand narrative of Western progress," to strip away the vestiges of framing science and technology as imported magic.[23] Dynamics of technological change in Latin American countries remain underexplored. Too often historians slide back into the easy, unreflective position that the region's "contribution to patterns of technological change has been largely marginal, derivative, or mimetic" or that imported technologies "primarily served the interest of foreign investors and their local partners . . . [and thus] found little engagement with local society." This distorted image erases "dynamics of local invention, adaptation, and engagement with imported techniques."[24]

Over the last ten years, historians of technology in Mexico have worked to correct this image by documenting how technology was deployed as a tool for state-building and nation-building and in the pursuit of modernity. These studies have enriched our understanding of the environment in which technologies such as the railroad, radio, and sewing machine and technological systems and processes flourished. They reveal not only the institutional, national, and transnational networks of power that shaped such technologies, and how they entered discourses of modernity, revolution, and nationhood, but also the factors that facilitated and hindered local adaptation and innovation. Far from having a negligible footprint, new technological devices and systems became "deeply integrated into the social and cultural lives of many Mexicans."[25] By moving the lens to center on the local, Mexicans come into focus as agents of technological change. Concentrating on the country's capital, a handful of historians have followed the actions of politicians, social reformers, technocrats, merchants, industrialists, and ordinary citizens to transform the human environment and to make what Mauricio Tenorio-Trillo has called the "ciudad científica" (scientific city).[26] Informed by these studies, I take a user-centered approach to delve into the making of electrified spaces.

I challenge distorted images in the following chapters by exploring technological change in the Mexican capital. Capitalinos had to grapple with the quotidian opportunities and mishaps resulting from rapid urbanization, early industrialization, and technification of public and private spaces. In interrogating electrification of the city and grounding the process in everyday life, the inhabitants of the city emerge as agents who actively negotiated the extent and manner in which electricity entered their lives and lived spaces. In doing so, they established fluid relations with this new source of power and technology. My account moves away from elite actors in industry and government and privileges a broader and more diverse set of Mexicans (differing in race, class, and gender), thereby answering the call for historians of technology to shift "toward a more geographically inclusive understanding of technological development and change."[27]

The idea of a Mexican electricscape is central to a discussion of the capitalinos' relationship with electricity from the introduction of the first electric lights in the 1880s to nationalization of the industry in 1960. As in the case of an actual landscape, the electricscape offers a wide view of the physical and symbolic aspects of electrification. It captures not only how electrical production, the government, and consumers interact and how that interaction manifests itself tangibly in infrastructure, policy, and consumption but also the changing ideas, debates, and anxieties as well as hopes and dreams of electrified futures. Unlike the pristine views found in the works of the great nineteenth-century landscapist José María Velasco, the electricscape demands observation from multiple perspectives, placing equal emphasis on the individuals in the foreground and the expanse of urban scenery in the background. To capture the grittier aspects of the electricscape without losing a sense of perspective and depth, I acknowledge the value of granting equal footing to an inquiry of the human imagination.

According to the historian David Arnold, just as people (individually and collectively) can conceptualize modern nations only as imagined spaces, they imaginatively embrace particular technological goods, skills, and processes or, conversely, associate certain technologies with particular groups and seek to distance themselves from them. Electricity came to be seen as the engine that would propel the country into being an industrialized modern nation. Its power was not merely tangible. Sheila Jasanoff demonstrates how imagination, operating on an intersubjective plane, unites members of a social community in shared perceptions of futures that should either materialize or be avoided.[28] As the following pages show, the potential of the new technology loomed large in the minds of not only politicians, technocrats, and businesspeople but also ordinary people who debated and expressed anxiety and doubts of its application and those who sought to profit (even illegally) from its use. Visions of an electrified future were classed, racialized, and gendered. Participants came from all walks of life, from middle-class housewives to electrical workers. Capitalinos saw themselves and their city as becoming modern through electricity.

ELECTRICITY'S OFFICIAL WELCOME

Electricity was officially welcomed through the Electricity Exposition in 1886. El Duque Job, the favorite pen name of the writer Manuel Gutiérrez Nájera, captured the atmosphere on the streets of Mexico City when the exhibition opened in mid-December that year.[29] As darkness set in, joyful, rich young women—followed by their servants—hurried through cobblestone streets while crackling fireworks obscured the monotone prayers of a nearby procession. Illuminated shop windows on Plateros Street gleamed with trinkets for the *posadas* (traditional Christmas commemorations) and artistic objects for traditional New Year gifts.[30] The Electricity Exposition at the National Preparatory School housed in San Ildefonso, a sixteenth-century Jesuit building, drew in the crème de la crème of Porfirian society. Among the last to arrive, President Porfirio Díaz made a grand entrance, accompanied by his top ministers, a handful of military generals, and directors of institutions of higher education. They entered the colonial building as musicians of the Twenty-First Battalion played the arrival fanfare.

Celebrated at the time as the first scientific exhibition held in the country, the Electricity Exposition represented Díaz's hopes for stimulating the nation's economy. The organizers of the exhibition intended it to be the modernizing elite's introduction to electricity. Gutiérrez Nájera emerged from the exhibit at San Ildefonso impressed. He characterized electricity as a great magician and conjurer: through the telephone, telegraph, and electric lights, it demonstrated numerous marvels as it "illuminated, talked, and caught words in a net, throwing them like shuttlecocks over great distances."[31]

The emergence of these marvels of modern science in San Ildefonso, a former Jesuit boarding school confiscated in the period of liberal reform just two decades before, made the exhibit all the more powerful. The school, a bastion of Auguste Comte's positivist curriculum, stood as the center of scientific politics, training a generation of late nineteenth-century technocrats. The minister of education, Justo Sierra, trusted that—once they graduated—the nation's resources would be exploited and placed in global circulation under their supervision. If the Mexico of the past had been colonial and put itself in the hands of God, the Mexico of the Porfiriato would be independent and put its future in the hands of science and reason.

Electricity promised a way forward for energy-hungry Mexico.[32] In the inaugural address to the exhibition, Sierra surveyed science's achievements from antiquity to the present, all of which, in his estimation, had been leading up to electricity. He fervently believed that Mexico, as a land with limited fossil fuel resources, had everything to expect from this new technology.[33] After his address, the minister invited attendees to enter the exhibition. President Díaz pressed an electric switch: to everyone's amazement, the school's monumental door opened and powered the circuit, feeding lights and devices on display.[34] For the next fourteen days, scientists offered lectures to educate the public on the expanding subject of electricity.[35]

The marvels of science and the evidence of electricity's potential were there inside San Ildefonso for all to see. A dazzling banquet of light attested that science brought forth the visible out of the hidden.[36] Attendees first entered a small, lavishly lighted room where an elegant aquarium illuminated with Edison lights welcomed them.[37] The school's library shone with profuse lights, gracefully located. On one of the walls high above, the names of Benjamin Franklin, Alessandro Volta, André-Marie Ampère, Henri Becquerel, Auguste-Arthur de la Rive, and Claude Servais Pouillet appeared in golden letters, in recognition of their legacy in electrical science. Above the names, a painting of a stormy cloud with two lightning bolts served as the representation of nature's force. Dedicated salons displayed electrical devices by group: magnetism, static electricity, and the telegraph, among others. Students provided detailed explanations to the public as they operated the devices—proudly pointing out the ones that came from the workshops of Mexican inventors.

Spectators enthusiastically gathered in the telegraph room, where they witnessed the Hughes printing telegraph. Unlike earlier versions that used Samuel Morse's code, it employed a piano-like keyboard, with each of the twenty-six keys corresponding to a letter of the English alphabet. Operators typed out a message that was then "printed on paper tape at a receiving Hughes telegraph connected by wire to the sending telegraph."[38] After watching, many observers cheerfully walked away with a piece of printed paper as a souvenir.

Skeptics remained, however. Liberal intellectuals and political elites—and the press that catered to them—wasted no opportunity to lampoon conservatives as backward. Where liberals saw wonder, conservatives saw danger. Rumor had it that ghosts of Jesuit fathers, former teachers at the college before their 1767 expulsion from Spanish lands, lurked in the dark passages and former cells of the colonial building. The Jesuits became visible only because of the incandescent lights and made the sign of the cross when they saw them. Even the influential paper *El Partido Liberal* capitalized on such stories. It reported that the Jesuits (and by association deeply religious Mexicans more generally), unfamiliar with the wonders of science, saw electricity as the diabolical partner of Lucifer. Frightened by the big electric bulbs like fireflies, the fathers yelled, "Come back tomorrow, mob of ghosts!"[39] Forgotten among these rumors, of course, was the Jesuits' significant contribution to the advancement of science in the nation. It largely did not matter: despite the unfounded rumors circulating about Jesuits and complaints from those who had known only candlelight, the 1886 exposition at San Ildefonso announced that electricity was the path to the nation's destiny.

It was no surprise that the president was present for the exposition. The event not only educated the audience on its electrical future but also functioned as a public display of power: an announcement that Mexico was steadily moving along the path of progress, thanks to the Díaz administration. At the end of his speech (ostensibly about science and knowledge), Sierra thanked the regime for its tenacious national security campaign and financial reorganization, which he argued was an indispensable condition for the republic to raise its eyes from the present to plan for the future.

Like Sierra, Díaz's supporters felt that the nation had finally become stable enough to think about and even plan for a glorious future. The threat of revolt and unrest had subsided, and the regime's investments in education and technology made the future worth betting upon. Newspapers, many in league with the president, agreed. The leading financial journal, *El Economista Mexicano*, applauded the exposition as a scientific event of enormous proportions, which demonstrated the level of enlightenment that the nation would reach under Díaz's "empire of peace." The paper concluded by urging the nation to undertake another scientific exhibition soon.[40] Even the Catholic daily *El Tiempo* blessed the event, exhorting authorities to devote more funds to foster and protect scientific endeavors that would "earn the country a name for itself,"[41] through the example of Mexico City.

Mexico City is a particularly compelling case through which to study the tangible and symbolic deployment of electricity in technological modernity. During the Porfiriato, the combination of political stability, massive foreign investment, and consolidation of the export-based economy led not only to the physical transformation of the city but to a social, even attitudinal, transformation. Technological progress became a source of legitimization for the regime, with Mexico City as its showcase. Proud of the city's rapid urbanization, Díaz wanted nothing more than to transform the capital into a cosmopolitan center through the elixir of electricity. The city attracted capitalists from around the world, eager to profit from lighting and powering the capital's social and economic life. Díaz imagined that the vitality of the metropolis, like a dynamo, would energize provincial cities, which would mimic the city's electricscape by the turn of the century.[42] As went Mexico City, so went the nation: as the capital became an electrified industrial center, so did provincial industries.

No monolithic Mexico ever existed, however. In taking technology as an extension of human lives, I move beyond official blueprints to follow individuals on the ground in all their diversity. I examine how a staggering array of Mexicans—government officials, businesspeople, social commentators, inventors, doctors, electrical workers, domestic advisors, housewives, and ordinary citizens—became electrifying agents, who sold, bought, or even stole electricity to achieve their own ends. Their ambitions gave rise to a discourse on how electricity should be used and who had a claim to it.[43] That discussion was both concrete in its immediate application and symbolic in its hopes for Mexico's future. It was notably contentious, for hardly any electrifying agent could agree with another. Electrification was not inevitable: its development and speed resulted from decisions that facilitated the import of electrical machinery, the legislation that made the generation, transmission, and sale of electricity attractive to foreign capital, and the values associated with electric lighting and electrical devices that contributed to the material development of the industry.

My account favors perception and experience of everyday life. Informed by a user-centered approach, I treat users as agents of technological change, going beyond how technology changes society to examine how society impacts technology. In their user-centered studies, leading cultural and media studies scholars such as Roger Silverstone, Eric Hirsch, Metere Lie, and Knut Sørensen have reconceptualized the role of users in shaping technological objects by challenging the traditional distinction between production and consumption. In the process, they have breathed new life into Karl Marx's claim that consumption is production: as they make clear, consumption is "not complete until users have defined the uses, meanings, and significance of the technology."[44] In this reinterpretation of Marx, consumers are not passive agents and technology is not a universal standard. Instead of adopting a narrow understanding of how users interact with a given machine, which limits a priori how individuals interact, users and technology must be placed within in a broader set of social, cultural, and economic relations. To establish user–technology relations, we must understand how machines became integrated into everyday life. The concept of domestication of technology, which follows discretionary appropriation and incorporation of a new technology, is particularly valuable in the study of electricity, as demonstrated by the work of Graeme Gooday on Britain before World War I Britain.[45] We must see technological innovation through the eyes of those who witnessed it, including capitalinos adopting festive lighting or riding trams and home advisors integrating electric blenders to prepare pulque recipes.

Through the lens of everyday life, then, I document how electrical lighting, power, and transportation changed ideas of time and space: of bodies, self, others, and what it meant to be Mexican. Shifting the focus to the agency of individuals who confronted new electric technology allows a deeper understanding of processes of social change: how gender, race and class, and nationalism colored relationships between individuals and technology. Images of the electrified city—which celebrated human mastery of natural forces and resources, control, progress, and order—conceal the contingencies, complexity, and often violence that accompanied technological change. Electrocutions, tram accidents, and power theft were everyday realities that accompanied electrification and much as light bulbs and washing machines. Electricity brought with it debates on aesthetics as well as safety and legal issues. Looking at technological change through the lens of everyday life pulls us from the broad, abstract narratives of national "modernization" into individual personal quests for modernity: in the end, a "modern" Mexico could be defined only by actual Mexicans. To get at this, I explore a series of difficult, highly charged questions. What opportunities did Mexicans see in electricity? What dangers? And who, in the end, had a say in how to balance the two?

Historical research on the electric industry has dramatically expanded in the last two decades. In addition to general economic histories on its development, scholars have concentrated on foreign companies that dominated the industry from the late Porfiriato to its nationalization.[46] Indeed, historians have followed foreign investment in the electrical industry all the way down to the regional level. As has been the case with Latin American histories more broadly, private electric companies have received their due share of attention.[47]

Mexico, along with the countries in Latin America with the largest economies, joined the electric revolution early on. Mexican mines and refineries were the first in the region to introduce electricity, only a year after it was tested in US mines.[48] By 1889 the Batopilas mine in Chihuahua employed two hydroelectric turbines and two steam generators for crushing ore.[49] Most major mining centers were utilizing steam-generated electricity by the next decade.[50] The *Mexican Mining Journal*, the leading domestic trade publication, called electrical smelting one of the greatest moments in the country's mining history.[51] The journal noted the vast possibilities for iron production and the treatment of low-grade copper sulfide ores, which had remained unprofitable until then.

Within twenty years of the electrification of Batopilas, the application of electricity had reached the oldest and most important production centers.[52] Indeed, it had become the preferential energy source in Latin American mines, where it made the large-scale exploitation of low-grade ores profitable and reduced labor expenses. The introduction of electricity was one of the technological innovations that transformed the industry in the late nineteenth century. Despite extensive use of electricity in the various stages of mining activities, its most profound impact was on controlling the presence of underground water. Flooding of shafts had been a significant obstacle to the recovery of mining activity throughout the nineteenth century.[53] Electrical energy was either generated by the mining companies themselves or bought from commercial electric companies.[54]

Until the turn of the twentieth century, the generation, transmission, and distribution of electricity in Mexico remained isolated and uncoordinated. Over one hundred electric power companies were established between 1887 and 1911. All significant industries at least partly introduced the new energy source into their economic activities, hoping to increase production and save labor costs. *American Machinist* in 1904 recognized a widespread electrification drive in Mexico, with projects to harness the powerful waterfalls in the nation, with "every town and village of any pretensions demanding the introduction of electricity."[55] Textile, tobacco, paper, and glass factories were among the first industries that jumped on the electrical bandwagon. This led to the emergence of privately owned power plants that generated the energy for their machinery. Those with surplus energy often marketed it to municipal governments for public lighting.

I devote less space here to the armed struggle phase of the Mexican Revolution than to the postrevolutionary period, in which revolutionary rhetoric came to shape the electric industry. Unlike the railway industry, which suffered severe material damage caused by battles and sabotage aimed to disrupt supply lines, the electric sector emerged from the civil war without much physical damage.[56] Regions with critical power facilities did not experience military activity. Despite exaggerated coverage abroad, the conflict did not target foreign interests. Revolutionary governments were vigilant in protecting foreign property to prevent the likelihood of outside intervention.[57] For instance, at a formal banquet organized at the Necaxa hydroelectric plant in early 1913, President Francisco Madero assured attendees that Mexico welcomed foreign capital and that his government would protect their property from rebel bands.[58]

Hard data on Mexican electrification is uneven and limited for the first forty years of the industry. The Ministerio de Fomento (Minstry of Development) led the first efforts to document early applications in the late 1880s. Resulting publications traveled to world exhibitions as proof of the country's progress in the field of electricity. Economic assessments of the state of the industry emerged in the social reform phase of the revolution in the 1920s and 1930s. Studies dedicated to the industry's development and the challenges it faced became prominent with the renewed industrialization efforts led by the federal government in the 1940s and 1950s. In the post-1960s era, government officials and academics set out to evaluate the impact of the industry's nationalization.

Historical research on the electric industry has expanded in the last two decades. In addition to general economic histories on its development, the research agenda has concentrated on two aspects. The first is the foreign companies that dominated the industry during the late Porfiriato. Indeed, this has occurred at a regional level: foreign investment in the electrical industry (and in public utilities in general) in Latin America has received attention from economic historians. The other area of concentration includes studies that focus on an individual electric company. The application of electricity for power and lighting has also created a research niche. Additional work on the introduction of electricity by individual factories will certainly add a new dimension to our understanding of the context in which Mexicans secured electricity, subsequent transformations, and how workers manipulated it. Although economic and business histories and technical and policy studies of the industry have dominated such scholarship, a few works have ventured toward a more social understanding of the electrification process (including the educational training of engineering students who participated in constructing the Necaxa hydroelectric plant).

ELECTRIFYING AGENTS IN THE ELECTRICSCAPE

The French poet, essayist, and philosopher Paul Valéry reproached traditional historians for fixating upon politics and war while ignoring the real forces that shape the world and cited electricity's "conquest of the earth" as an exemplar of "those notable phenomena" that history neglects, even though it has "more meaning and greater possibilities of shaping our immediate future than all the political events combined."[59] While his critique was strong—and remains controversial—his take on electricity's importance is self-evident, especially when we consider that electricity, as a primary engine of a modern society, has permeated all aspects of life and weaves like a weft strand through the warp of city, transportation, labor, business, engineering, women, medicine, death, public celebrations, nightlife, advertising, literature, and architecture histories, to name only a few. Yet for all his radical insight—especially considering the state of history when he wrote his words—Valéry was wrong on one important fact: electricity did not achieve a "conquest." Rather, people took it, adapted it, manipulated it, negotiated over it, and in some cases rejected it. In short, they became what I refer to as electrifying agents. Electricity did not, like some cruel conqueror, hand people a set of demands. Everywhere the technology arrived, people incorporated it into their social and cultural norms and projected it into their dreams. That is, electricity served people; but in Mexico, as everywhere else in the world, the question remained: which people would it serve and in what ways?

Fortunately, a wealth of primary sources on the Mexican case help answer these questions. Government plans for electrical infrastructure and the records of private companies tell a story of how they wanted Mexicans to consume electricity, but it is incomplete. Newspapers, medical journals, novels, letters, court records, and popular sources complement the official perspective and help us to see how people became not just consumers of electricity but electrifying agents. Writers vented their frustrations at the inefficiencies of electrification in newspapers and in the penny press. Seemingly everyone had something to say about how electricity would change people's lives. Injured individuals, those suspected of theft, and electric companies went to the courts: along with lawyers and jurists, they debated and sought to shape the legislation of the new energy source. Even cookbooks, women's magazines, and other advice literature serve as windows into the way electricity, construed as the harbinger of modern living, was to transform homes into clean, efficient, and comfortable spaces. Families of trolley victims and those who defended the new transportation spoke their minds in the courts. Electricity theft cases, unknown a few years before, became a daily occurrences in these courts. Prosecutors attempting to defend powerful electrical companies in those cases ran head-on into injured individuals, suspects, and their lawyers, who all argued for a different perspective on what an electrified—and hence modern—Mexico meant.

No other group spilled more ink in debating and defending the place of electricity in the nation's future than the people who physically delivered it. Issues of *Lux*, the magazine of the Sindicato Mexicano de Electricistas (SME; Mexican Electricians Union), along with the syndicate's general-assembly minutes, reveal how the workers cast themselves as a new generation of revolutionaries in a struggle against a rapacious foreign company for the industry's Mexicanization.

I examine how Mexicans shaped the electrification process in the following pages, from electricity's first emergence in the 1880s until electrical appliances became truly ubiquitous—the smaller ones even unremarkable—in middle-class homes in the 1960s. The process of electrification did not follow a simple, linear process, so any accurate narrative about it cannot be linear. My account reflects that truth: though it moves along in a generally chronological direction, it also reaches across space and time to capture key moments of causation and contingency.

I interrogate the dreams and visions of an electrified future in part I. We must see that the history of electricity goes far beyond the material—the generators, transmission lines, transformers, dynamos, poles, and incandescent bulbs that made up electricity's physical presence—into the human imagination.[60] In the Mexican case, people were hardly bewitched by electricity, shrinking in fear and wonder, but almost immediately they began to imagine how the new technology could transform the nation, its industries, their neighborhoods, and, most commonly, their individual prospects.

I delve into the first manifestation of an electricscape in chapter 1. How did electricity, primarily electric lighting, redraw luminous boundaries within the capital? Lighting was one of the first and certainly the most widely encountered form of electricity. I ask how people understood and experienced the new illuminant as both tangible (lumen) and intangible (lux). I examine the transformation of public spaces and how the new illuminant both improved lives and reinforced inequality. I discuss the next most common demonstration of electricity's potential in chapter 2: exhibitions that aimed to document, display, and celebrate an electrified capital city. The grandest of these exhibits took place during the country's independence centennial celebration in 1910.

I shift from the introduction of electricity and the promises it held to an exploration of actual practices by electrifying agents in part II. A world driven by electricity, which had seemed like a dream just a generation before, became a growing part of everyday life during the late Porfiriato. With it came efforts to regulate the behavior of electric power users.

Chapters in this section build upon the work of previous historians who have found a "clear confrontation between an authoritarian regime's projects to reshape urban geography and the opposing forces of subordinate groups' use of the city, demographic growth, and technological change."[61] I try to put a human face upon these "subordinate groups" and show how they were not always as subordinate or powerless as they may appear. I transverse the city aboard electric trams to piece together how they altered daily life and the uses Mexicans made of them in chapter 3. In ways the regime did not predict, the *eléctricos* (electric streetcars) demonstrated the two sides to the imagined modern Porfirian city. On the one hand, they demonstrated great order, for they ran on tracks and were held to a regular schedule. On the other, that order came at the expense of horrific accidents, countless injuries, and everyday slights and insults.

Focusing on the debates surrounding accidents, I argue that, despite accidents, capitalinos did not advocate for the return of older modes of transportation. Instead, they categorized these quotidian misfortunes along with other lamented and frowned-upon events in city life. Along with public urinating, defecating, and drinking, "reckless" walking began to plague the city. Pedestrians faced calls to rationalize and standardize their behavior. The public discussion and debate about these accidents represented not a counternarrative of modernity but rather instructions on how everyday people had to adapt to the reality of technological change. Capitalinos would have to change their behavior in order to become modern. Discussions surrounding eléctricos indicated not only the growing anxieties of a society in a quest for modernity but also the certainty that the quest would continue despite these anxieties.

I deal with the first two decades of the twentieth century in chapter 4. Electric power moved from being a publicly consumed good to one that could be privately consumed during that period, and theft accompanied this change. I address who stole power and why, how the state and power companies responded, and the public debate over the regulation of proper use of electricity. Good money awaited active and former electrical workers as well as self-taught entrepreneurs who set up illegal connections to power lines or tampered with meters. The demand for their services spanned the social spectrum of Mexico City, including owners of prestigious theaters, small business owners, a town council member, the elderly, widows, and tenement residents. I follow electric company inspectors around the city as they sniffed out power theft and represented their company's interests in the resulting court cases. Their quarry was most commonly small business owners, who framed testimonies, defined the nature of electricity, and crafted clever arguments in court that challenged the very idea of private property and theft. Court records prove that many of the imaginative acts that shaped the electricscape arose from or were inspired by conflict.

After exploring the promise of electricity and how people actually brought electricity into their lives in parts I and II, I look at two very distinct, even wildly different, groups in Mexico in part III: middle-class housewives and workers in the electrical industry. They are, by design, an odd coupling. Yet, this may not be so unusual a pairing as it seems. In both cases, we can see how the experience of living with electricity filtered into a personal sense of identity and empowered people culturally and politically.

I examine how retailers promoted domestic gadgets not only as practical tools but also as ways to refashion the values of gender, class, race, and culinary nationalism in chapter 5. To understand the generational shift, I look across three decades (1930s–1950s), the period of the consolidation of the revolutionary regime and the economic miracle, exploring how middle-class women imagined themselves in an electrical future. Electrical appliances made housewives at once scientific, hygienic, and white.

Interestingly, gender and class also played a role in the nationalism of electrical workers. I examine the gendered language and imagery of the modern working-class man that permeated union-produced literature in chapter 6. These workers imagined themselves as a new generation of revolutionaries struggling against rapacious foreign companies. Their touchstone was the example of former president Lázaro Cárdenas and his oil expropriation; their imagined future was one where the state moved in to right the wrong of economic exploitation and fulfilled the promise of the revolution.

I conclude with the 1960 celebration of Mexico's nationalization of the electrical industry. The roughly eighty years that had passed since minister Justo Sierra promised his audience at San Ildefonso that Mexico's future lay in electricity had proven him more right than wrong. Mexico had electrified, and the electricscape increasingly spread out from the urban and industrial centers into all corners of the nation. Yet, at the same celebration, the stories being told about how Mexico became electrified did not reflect the real, historic past. In this first telling of Mexico's electrical history, the everyday people who did so much to expand and shape the electricscape began to disappear and be replaced by large, powerful actors—not least of which were electric companies and the government. I correct this imbalance by leaving the balcony of the Presidential Palace and taking readers back into the streets, businesses, and homes of Mexico City. It was there that the electricscape arose.

The Leyden jar was used in a demonstration of electricity popular in scientific gatherings during the nineteenth century. The demonstration, developed and perfected abroad, started with facilitators who dared attendees to witness an appealing and deceitful demonstration. First, they called for a group of people to stand in a circle holding hands, forming a human chain. At one end, an individual held a *botella de Leyden* (Leyden jar), a glass jar with its inner and outer surfaces coated with metal foil and sealed with a cork or wooden lid that had a copper wire running through it. An early form of batteries, Leyden jars had been essential for the study of electricity for nearly 150 years but were still a novelty in Mexico, as in much of the world.[62] Early on, jars were filled with water and some even with beer, as the liquid was thought to store electricity. The wire in the lid ran to an electrostatic generator or another source of electric charge; as it touched the inner lining of foil, it allowed energy to build up in the interior layer. The glass of the jar blocked electrical movement to the outer layer of foil, allowing it to remain grounded. For the stored energy to travel, it needed a pathway: a wire, rod, or human hand connected to the wire emerging from the lid served as a communication. A person at one end of the human chain would touch the top of a charged jar, while the person at the other end reached out to touch the foil-coated bottom of the jar with a free hand to complete the pathway.[63] Electricity would then flow through all members of the chain, each receiving the "gift of an electric shock."[64] Like the Leyden jar experiment, electricity would reach through all ranks of society when it came to Mexico, through the hands of capitalinos. Modernity would indeed bring shocks to all.

PART I

| ONE |

Sensing the Beautiful Stranger

Confessed lovers of their metropolis, Adolfo Prantl and José L. Groso were determined to correct unfavorable opinions of Mexico City penned by domestic and foreign writers. Their guide *La Ciudad de México: Novísima guía universal de la capital de la República Mexicana* (1901) promised to present the city just as it was, without exaggerating its natural beauties or hiding its urban defects. They painted two contrasting pictures of Mexico City after sunset, inviting readers to appreciate the capital's recent transformation. In the first scene, they conjured up dusk settling over the heart of the late viceregal capital of New Spain:

> Moments after sunset bells from all the temples rang . . . *Hail Marys* to announce the praying of the Night Prayer. Those on the streets regardless of caste or social class would stop, respectfully uncovering their heads, and almost in unison from their lips would emerge: The angel of the Lord declared unto Mary, and she conceived of the Holy Spirit. The city presented a dreadful aspect as dense shadows of the night covered the houses; the few calashes that ran during the day no longer circulated. Commerce had closed its doors. Noises had died down. The silence that reigned was interrupted only by the wind whistling or by a codger on his way to a friend's house for an intimate *tertulia* [get-together]. The entire city was a dark mass of ambivalent silhouettes: winding muddy streets, modest houses, and cold and gloomy convents through whose narrow windows the murmur of religious songs and prayers emerged. On every other alley, an oil lamp burned at a saint's niche . . . with its weak reflection struggling against the darkness. Everything was shadows, silence, [and] austerity.[1]

In the second scene, only one year into the new century, life in the city has changed dramatically:

> The melancholic voice of the bells could still be heard ringing the Angelus, but only here and there was the uncovered head of an old man or a lower-class individual seen, and far from indicating seclusion, prayer, calmness, [the ringing] announced the start of the racket of those having fun, the agitation of commercial life, the clamoring of the populous city. Hundreds of chimneys from countless factories, shops, laboratories, and furnaces spewed out thick puffs of smoke. Windmills sometimes spun steadily, sometimes vertiginously. [Flying over all was] an extensive and complicated network of wires that transmitted words from one corner of the Republic to the other or delivered electricity for lighting or the tramways.[2]

For these writers, the secularization and development of an industrial city were palpable—as was the birth of nightlife. They celebrated how the intimate, dour nocturnal activities of yesteryear yielded to smoky cafes, raucous bars, gaudy casinos, and squeals of glee from entertainment halls and theaters. Above all, people sought "to have fun, walk, laugh, and dance [and] they sing and cry, curse and bless, and everything happens like that, frivolously." It was a transformation that occurred without truce or respite and at a speed that astonished and troubled some. Prantl and Groso wrote about nineteenth-century Mexico City as though it had the modesty of a nun, but twentieth-century Mexico City was no longer so cloistered, undergoing a nightly metamorphosis. "Long after the sun rolls into the abyss of shadows, the city fastens diamond brooches to its splendid black velvet dress . . . resembl[ing] a pale and elegant lady ready for the opera, a soirée, or wherever avid entertainment awaits."[3] In both cases, they metaphorically feminized the city into a respectable woman: religious and modest in the nineteenth century, but wealthy, confident, and eager to enjoy life in the modern world.

At the center of these changes was light. Even a few years before 1901, when Prantl and Groso sang the city's praises, residents suffered in darkness, aided only by the oily flicker of small lamps in niches along alleyways. But by 1901 the dark night had become adorned with dazzling "diamond brooches": electric lighting, in less literary terms. Cultural historians of the twentieth century have dedicated many pages to the intersection of light, vision, and power in urban spaces, crafting a well-known and influential narrative of Western modernity as a "relentless expansion of illumination."

In this telling, the construction of modern illuminated spaces has fed two visual historical processes: the rise of surveillance and the development of spectacle. But is this true in Mexico City? In his seminal work on public lighting in Victorian Britain, Chris Otter encourages us to reject this paradigm, declaring that "the visual dimension of space was, in general, engineered with neither coercion nor seduction in mind." Heeding Otter's call to put aside reductive theoretical frameworks in favor of histories that remain "close to historical actors, their words, their eyes, and their physical environment" and render textured visual experience "in all its everyday richness and complexity," in this chapter I look at the introduction of electric light in Mexico.[4]

A cultural history of light in the city, as fascinating as it would be, is far beyond the scope of this chapter. Instead, I seek to understand both the diffusion of electric light across Porfirian Mexico City in its physical manifestations and patterns of dispersal and how capitalinos—from industrial leaders to common laborers—manipulated, sensed, and experienced lighting as they employed it to advance their social aspirations and economic interests.

THE PROBLEM OF LIGHT AND THE PROMISE OF THE ELECTRICSCAPE

Mexico City's experience with electricity began with a highly applauded public demonstration. On the night of November 2, 1850, a select party made its way to the Zócalo. Don Pedro Romero de Terreros, a successful businessman, had set up imported equipment to light the square. The equipment, likely a voltaic arc lamp, worked and left an impression on everyone there that night.[5] Referring to the trial as "strange works," *El Siglo Diez y Nueve* expressed the hope that it could be broadly employed.[6] A few days later, a second trial at the Palace of Mines brought even more prestigious guests, including President José Joaquín de Herrera, and generated even more enthusiastic coverage. One paper daringly compared the lamp's light to the light of *el astro rey* (the sun). Mexicans were aware of the importance of what they were witnessing. Just as the president was watching the electrical illumination, his office at the National Palace sent the first telegraph message in the country at the Palace of Mines. The success of both tests generated high expectations of electricity's potential. Mexico seemed on the verge of vanquishing time, distance, and darkness. Newspaper dailies confidently declared that the country would rapidly become modern through such innovations—and thereby wealthy and powerful.[7]

Light took on strikingly new meaning in Mexico City in the second half of the nineteenth century. Capitalinos born into a world of flickering flame suddenly confronted the dazzling—both metaphorically and literally—spectacle of electric light. Understanding how they experienced this shift between lighting regimes requires awareness of how electric light is sensed. Changes in lighting regimes (artificial lighting schemes) shaped distinctive experiences by producing different sensations. How did capitalinos experience the city under gaslights versus electric lights? Material culture studies have shown how "the sensation of things" as they come into contact with the body "is inseparable from the meanings and sociality of things and their production."[8] We do not come upon or sense something new as isolated beings living in a social or cultural vacuum; our minds bring with us ideas, notions, even physical and financial desires that directly relate to the way we experience the world.

Light, as an essential component of vision, plays a key role in how we sense and understand the world. Scholars have lamented that our earliest efforts to understand seeing led us to "effectively los[e] touch with the experience of light." This is not a trivial matter but a paramount loss, for "light is the experience of inhabiting the world of the visible, and its qualities—of brilliance and shade, tint and colour, and saturation—are variations upon this experience."[9] When we talk about seeing, we are really talking about experiencing light. Studies on light have long focused on either matter or vision. Scholars who focused on matter were engaged in understanding *lumen*: light as a physical phenomenon at once measurable and quantifiable. Studies on vision centered on light's "sensuous qualities formed as sight or vision: the *lux*."[10]

Beginning in the 1990s, the study of the perception of light attracted significant attention, but neurological determinists denied that culture could shape the way individuals "conceive and experience light and colour." For these positivist scientists, there was only one universal human response to light. This absolutist perspective, however, gave away to an avalanche of cultural studies, which proved not only that cultures could experience the same stimuli differently but also that different cultures give primacy to different senses. With this insight came the call to explore how light is and was perceived and experienced in culturally specific ways—that is, to document the social life of light. In recent years, shadows have emerged as a subfield in our scholarship of light. Scholars have come to see that any given environment offers unique lightscapes, changing landscapes of light, darkness, and shadow. As one seminal work declared, the world's appearance "is determined by the changing *lightscapes* cast by the shadows in the relationship between things, persons and light."[11]

Electric light would render astonishing transformations in the way capitalinos experienced light, but not immediately. Despite the successes of November 2, 1850, the lightscape of Mexico City did not change much for thirty years. The struggle between liberals and conservatives during that time, which led to the French Intervention (1861–1867), left little room for radical alterations of the nation's capital. With the consolidation of liberals at the national level and the implementation of liberal legislation, the cultural, financial, and spatial dominance of the Roman Catholic Church began to crack.[12] This was no minor achievement, for the church was the city's largest landowner, controlling fully half of the capital's buildings. The monumental ecclesiastical architecture, with soaring church towers and expansive monasteries and convents, dominated the cityscape until the 1870s, when the hand of state-led secularization allowed the emergence of commercial and residential architectural rivals. The introduction of the first electric arc lights thus coincided with a period of major changes in the city's built environment.

In the years after President Herrera's administration, the liberals would ascend to power, equipped with legislation that targeted conservative elements of society, including the church. Before the 1857 liberal Constitution was drafted, the passage of the legislation package known as the Laws of Reform curtailed the power of corporations. No corporation dominated Mexico City—or the nation—like the Roman Catholic Church. The Lerdo Law, which abolished the right of corporations to hold land, struck at the heart of the church's economic base as well as the livelihood of indigenous communities. The indigenous barrios encircling the city had lost their lands by the 1870s, with the exception for a few *chinampas* (floating gardens) and *huertas* (produce gardens).[13]

The defeat of conservatives and imperialists created the conditions for liberal authorities to reformulate the capital's urban space via the expropriation and nationalization of church property. With their victory in the War of Reform (1858–1861) and in the aftermath of the French Intervention, liberals were largely free to secularize the city and map a more rational design for its future, all while addressing the "need for breaking the paraphernalia of the Catholic Church's power." All three goals entailed the transformation of both physical and symbolic space: in addition to realigning streets to facilitate growth and development, the transformation was to cure the capital's spiritual and physical claustrophobia.[14]

Liberals and others sympathetic to a liberal ascendancy in the nation claimed that they had no anticlerical animus but were simply tearing down crumbling colonial buildings and realigning streets in the name of improved transit, but not everyone believed them. People took note that the liberals seem to target churches, monasteries, and convents for destruction. The widespread belief that the emergent political power had a strong antichurch bias was reflected in literature. Federico Gamboa brought it into his stories, as in the case of the character Don Isaac Cortijo, a former scribe of the Correctional Tribunal who loyally served under Maximilian's conservative regime and lived in fear of being executed by the liberals. Miserable and destitute, he lamented that his daughter Rosita might take holy orders and find herself suffering the "cruel persecution of those soulless *liberalotes* who seemed to have no other mission than emptying cloisters and demolishing convents."[15]

Nothing spoke to the power that dominated a place more than the way that power organized space. Liberal reformers of the mid-nineteenth century saw ecclesiastical fingerprints all over the city. Spanish urban design early in the colonial period privileged the layout of cities following a *traza* (plan), as far as the area's topography allowed. Linear streets met at right angles, resulting in a "grid pattern [that] represented their sense of perfect order and the neat placement of people within a well-defined space."[16] Outside the symmetrical traza the irregular layout of streets followed the earlier roads of Indian communities that once surrounded the city. Liberals called for efforts to recover the gridiron city, at least in the areas adjacent to the Zócalo and Alameda Park, the chosen promenade of the *gente decente* (members of respectable, cultured society).

While critics called for a realignment of streets in the heart of the city, a new breed of speculators set their sights on the relatively cheap farmland on its edge. Urban speculators and real-estate developers acquired agricultural lands in order to divide them to be sold as urban plots. The first colonias established the capital's growth toward the West: the early work of urban developer Francisco Somera showed future generations how to create wealth from "dust, little capital, ingenuity, and good relations."[17] During the French Intervention, Somera had led the construction of the Paseo del Emperador, the Parisian-inspired boulevard later renamed Paseo de la Reforma by the liberals, to connect the emperor's residence at the Castle of Chapultepec with the city center. Its construction was not part of a comprehensive urban design but rather a response to the emperor's immediate need for faster communication. This new axis shaped the capital's future: as Claudia Agostoni notes, "the expansion and direction of the city followed its course, and so did real estate speculation and construction."[18] The modern city rose along and across this cherished boulevard, serving as a stage for the middle and upper classes to perform their cultural and economic identity away from the city's underclasses.

The extension of urban land and the friendly environment for developers resulted in thirty-four new residential districts between 1850 and 1910, but rapid expansion occurred when the Díaz regime centralized the nation's power in the city. As the seat of the federal government, the capital not only amassed political and economic power but was widely promoted as "the showcase of political peace and social order."[19]

Keen to see a capitalist economy develop, Díaz used his growing power to push for lands to be taken out of the hands of traditional corporate holders like the church and indigenous communities and into private, more entrepreneurial hands. In the 1880s he furthered his goals by securing a series of laws that allowed individuals to claim vacant lands and surveying companies to retain up to one-third of surveyed lands.[20] Landless individuals fled to the cities in search of opportunities. This resulted in tremendous population growth in Mexico City, which grew from 250,000 residents in 1880 to 350,000 by 1900 and close to 500,000 by 1910. Its urban footprint increased from an area of 3.28 square miles (8.5 square kilometers) in 1858 to 15.63 square miles (40.5 square kilometers) in 1910. Díaz and his liberal allies celebrated this urban development as a necessary step for a modern economy, yet their calls for organized, rational growth went unheeded. People arrived in waves too large to be effectively controlled and settled wherever it best suited them. As a result, the downtown became densely populated—even overcrowded—while the city was ringed with low-density sprawl.

Critical of the developer's actions and the authorities' acquiescence, engineer and councilor Miguel Ángel De Quevedo, among others, stressed the irrationality of the capital's physical growth, with large, unpopulated tracts between the center and its multiplying tentacles.[21] De Quevedo, one of the country's early environmentalists, had been a fierce advocate of public green spaces for the lower classes, in contrast to the period's marked preference for French-style luxury gardens.[22]

One reason for this "irrationality" in Mexico City's growth is that a colonia could emerge in two different ways. According to the official process, businesspeople either individually or as a company approached municipal authorities and the Federal District with a project and a map to solicit permission to establish a colonia. In those cases, the project stipulated the party responsible for the provision of municipal services and the collection of taxes. The late codifying of official requirements for the formation of new colonias confirmed the Ayuntamiento's legal laxity toward development schemes.[23] "Spontaneous" colonias, in contrast, emerged without official permits and thus without a written agreement on who was responsible for the construction and distribution of services. This further accentuated differences between authorized colonias, which generally housed the upper-middle and upper classes, and those without official recognition, which housed the lower classes.

The city's enormous growth in the second half of the nineteenth century shattered patterns of habitation that stretched back to the seventeenth century. For generations, people's wealth or status had no clear correlation with the location where they lived. Colonial Mexico City was mixed and diverse. By the 1880s, however, a new phenomenon had begun to take hold: class-based neighborhoods. Urban growth in this period happened on a largely east-west axis in three stages, each building on and deepening class segregation in the city. The first stage (1870s–1883) was perhaps the most pronounced, as developers ushered in the colonias of Santa María and Guerrero.[24] Santa María was to house middle-class sectors, while the urban working class would reside in colonia Guerrero. Such deliberate segregation, blatantly departing from the previous pattern of residence in the capital, was strengthened during the second phase (1884–1899) with twelve new colonias. The largest ones in the northeast (Morelos, La Bolsa, Díaz de León, Rastro, Maza, and Valle Gómez) offered working-class housing near recently established industries (the city jail, the slaughterhouse, Hidalgo station, and the railroad tracks of the Guadalupe, Cintura, and Interoceánico lines). The colonias of San Rafael and Santa Julia in the east provided housing for the emergent middle class of clerks, teachers, government officials, and other white-collar workers. After the success of San Rafael and Santa Julia, developers quickly rushed to build additional colonias south of the city. Wealthy individuals (or, more rarely, partnerships) had financed the first two phases almost exclusively, but this changed in the third phase (1900–1910), as private companies, both foreign and domestic, eyed the undeveloped land to the city's west and southwest. The colonias of Juárez, Cuauhtémoc, Roma, and Condesa catered exclusively to the upper class, who would benefit from new laws and stricter requirements by 1903. By then, the city demanded public spaces like plazas and gardens and regulated the width and length for all streets. Every home in these swanky new developments would have access to water and sewer service. The liberal call for rational modern development finally came to fruition here, but only those who could afford it would benefit.

This pattern of socioeconomic segregation did not occur by chance:

it followed nature and the region's topography. Lake Texcoco perennially threatened the impoverished sectors in the east and northeast of the capital, while land in the west and southwest was drier and less prone to flooding. In an effort to make the city safe and sanitary, the government pursued three intertwined hydraulic projects: General Drainage of the Valley of Mexico, the Great Drainage Canal, and the Xochimilco waterworks. But rather than improve nature, the programs largely exacerbated existing problems.[25] Flooding and sewer waste were controlled only by directing them to the east and northeast. Users who squandered water in the west sent their wastewater east toward the low-lying working-class neighborhoods.

Public lighting, which also had inequalities in regard to location, at least helped everyone who came in contact with it. Lighting was first and foremost a practical necessity. Streetlights allowed individuals to traverse urban spaces by illuminating what lay ahead. This is paramount when considering the city's infrastructure. People out strolling at night ran the risk of running into people and things or falling into potholes and fracturing limbs. Even simply tripping and falling into the street was a fairly serious problem, not simply a matter of getting clothes dirty. The city lacked a proper infrastructure for sewage, potable water, and garbage disposal, so streets were filled with ever-present puddles of filth, waste, and dung. Increased traffic was also a problem in the growing metropolis. In the best of circumstances, pedestrians dodged carriages and wagons, which had to pay as much attention to other vehicles as they did to pedestrians. The danger grew astronomically at night. The Ayuntamiento deemed the dilemma so serious that it imposed a fine (1 to 10 pesos) on carriage owners for failing to carry *faroles* (lanterns) when traversing streets at night.[26]

The condition of Mexico City's roads was at best mixed. Some swanky areas of town enjoyed sidewalks and roadways that compared favorably to those anywhere in the world. Others were horrible by any measure. For years, *La Patria* ran a series of mordant illustrations. One of these, captioned "The unobstructed path through all streets," depicts a road, still partially dirt and poorly paved in some sections, with stacks of paving stones in others. A second illustration captures the ordeal of a contemporary street scene: a vendor with a broad basket over his head, a driver with a whip in hand and a team of mules hitched to a wagon, and a couple under an umbrella all attempt to cross a street flooded knee-high. The brief mocking caption reads: "Drain pumps at work."[27]

Writers chimed in with equally condemnations of the city's streets. If many real and fictional capitalinos walked the streets as a sign of their urbanity, others needed reliable lighting schedules to survive. In the area of

San Antonio Abad, the silk industrialist Hipólito Chambón demanded street lighting for his 150 factory workers (almost all of them women and children), whose shifts ended when the streets were immersed in complete darkness.[28] A bad fall for them could mean destitution.

With the church's influence removed, the city had changed drastically. Individuals and companies interested in change—literally invested in it—watched for what could come next. Where did future profits lie? One source of profits might include making their real estate and developments more attractive to buyers. Better amenities meant higher prices. Another was providing services for which they could charge customers. In both cases, that meant electricity.

Electricity burst upon—or, more appropriately, flickered through—the capital in 1881. The gaslight company in charge of public lighting—Compañía de Gas Hidrógeno Carbonado (CGHC; Carbon Hydrogen Gas Company)—installed the first forty lights on a trial basis to escort residents from the Zócalo to the Reforma roundabout housing the equestrian statue of Charles IV of Spain. *La Libertad* wrote of the event in glowing terms, claiming that the lights illuminated the area as much as daylight did. Despite its experimental nature, the city took pride that few other cities—and, indeed, no European city—could claim such an impressive area lit via electricity. Celebrated at the time, these forty lights were only one part of the city's *alumbrado* (street lighting) system. Until 1890, the city offered those walking its streets at night a "fascinating spectacle" with the juxtaposition of four different lighting technologies.[29]

A nocturnal bird's-eye view of the city in 1880 would have revealed a scintillating line south of the Zócalo, surrounded by opalescent gaslights throughout the downtown area, encircled by yellowish turpentine and oil lamps as well as ocote-filled earthenware pots scattered randomly. The first few electric lights in 1881 came into a city that had 3,153 other lamps: 2,090 gaslights, 1,041 turpentine lights, and 22 oil lamps. These numbers suggest that the addition of electric lighting to the city's existing lighting infrastructure allowed Mexico City to enter the "modern" world with one foot still firmly planted in the past. Indeed, Rafael R. Arizpe, chief of the Alumbrado (Public Lighting) Commission under the city council, who was tasked with supervising and maintaining the city's lights, described strolling from the downtown to the outskirts as "flipping through the city's streetlight history."[30]

That history reflected the realities of power in the city. Wealthy and politically well-connected residents secured electric lights for themselves to replace the gaslights first introduced in 1869. The gaslights, in turn, had been an improvement upon turpentine, which replaced the earliest form of lighting: oil lamps. The kind of lighting found in a neighborhood was a clear measure of the class of its inhabitants. As department chief Arizpe noticed, the farther from the influence of power of the Zócalo, the older and more inferior the lighting. The displacement of older systems meant much more than the relocation of brackets, gas pipelines, and lampposts. It redrew visible and invisible luminous borders within the city.

This history of public lighting began with gas, which was nonetheless a latecomer to the city. Although proposals to introduce the system dated back to the 1830s, it was not established until September 1869 along Plateros and San Francisco Streets. By the end of that year, over 258 posts and 480 brackets attached to structures held 813 lights, with some lamps housing several burners to intensify illumination in some wide-open regions. The little lighting that had previously existed had come from oil lamps or small candles in niches along the narrow streets of the city center or from ocote-filled earthenware pots scattered randomly throughout the city. The construction of gasworks and gaslamp infrastructure marked the arrival of industrial lighting. In 1868 the Ayuntamiento granted the concession to the CGHC, a company organized with British capital. It was represented in Mexico by the businessman Samuel Bishop Knight, whose interests until then had been limited to mining and commercial ventures. Before this, the municipal government had granted concessions to fuel and light oil lamps but retained ownership over the infrastructure. The concessionaires were not required to create or install new infrastructure or to expand the lit area—the Ayuntamiento saw to that.[31] Small-scale merchants thus had an incentive to be sparing with the allocated funds, which translated into an overall deficient oil lighting service. Short-lived concessions became the norm.

The CGHC's gaslights changed this and dominated the capital's lightscape until 1898.[32] By that time, the gas-lit zone had grown from San Cosme in the west to San Lázaro in the east and from San Lorenzo in the north to the Vizcaínas in the south. But gaslights required pipelines and more expensive installation—they were not going to be placed everywhere in the city, particularly in poorer neighborhoods. Areas such as La Merced, Plaza de San Juan, El Baratillo Garden, Tepito, and Candelaria de los Patos still lacked gaslighting and were collectively known as the city's "moles." These poorer communities continued to rely upon cheaper oil or turpentine lights.[33] Businesspeople who resided close to the gas pipeline network could donate streetlamps to the Ayuntamiento to be installed near their businesses or residences.[34] In those cases, CGHC would open up trenches, lay pipelines, and install lamps, but the donors paid for fuel.

Gaslights were supposed to offer a whiter, more intense, and more stable light than turpentine and oil lamps but the first results were not impressive. The CGHC had to figure out what height the lamps should be at and how they should be distributed, but one problem enraged capitalinos more than anything else: the service schedule. Service was supposed to begin at 7 P.M. during winter (8 during the summer) and end at 5 A.M., but the schedule was not always followed. In August 1882, for instance, a public lighting inspector asked for CGHC to end the service at 4:30 A.M., as "objects were already distinguishable" at that time.[35] The more routine problem for the general public was that the company sometimes decided by itself that moonlight was sufficient and did not light the lamps.[36]

Though they had high hopes that modern gaslighting would end the problem of weak and irregular service, capitalinos would quickly be disappointed. Criticism of service abounded. Dailies echoed the sentiment that the city lived *en tinieblas* (in darkness), with paltry street lighting that served only to make the darkness perceptible. *El Telégrafo* claimed that on gloomy nights, when clouds "inhibited [the moon goddess] Diana from giving her light, darkness flooded the city."[37] One problem for capitalinos was that the CGHC and the moon did not always agree upon how much light was needed.

Although the Mexico City Ayuntamiento stipulated the lighting schedule, the CGHC determined whether and when to begin service during moon-lit nights. The company and the public apparently disagreed on how much light was appropriate, which resulted in constant criticism. Although *La Voz de México* conceded that it was understandable for certain lights to remain unlit when the moon seemed unwilling to consent to the usurpation "of her titles and privileges as the queen of the night," the newspaper complained it was unacceptable to leave citizens in the darkness by *presuming* that the moon would shine.[38] *El Telégrafo* was more blunt and blamed the company's frugality for "exposing taxpayers to twisted noses or shins . . . whenever the moon refuse[d] to show her silver face."[39]

Despite the substantial funds assigned to the city's Alumbrado Commission, poor illumination and irregular service persisted. Gaslight service was one of the city's most important budget items: as early as 1882 the 7,000 pesos that the city spent on it exceeded the amount spent on public works and the police.[40] Still in a period of transition, turpentine lighting received a respectable sum of 5,757.10 pesos that month. It was an open question in Mexico City, however, what effect the money produced. In late 1884 an illustration in a national newspaper lampooned the costly gas lighting service. When a couple in European dress and a gentleman dressed in traditional charro garb meet, the male foreigner, pointing to a nearby gaslight hung from a bracket, asks, "What type of light is this?" The other man answers, "It is public lighting, which, despite its high cost, does not illuminate anything."[41] The foreigners, who likely hail from a country long accustomed to gas lighting, react in shock, reinforcing the public's common belief that the city's gaslights were not truly worth of the name.

Authorities were well aware of the service's inefficiency. In his 1882 report to the Ayuntamiento, the alumbrado inspector noted that he and other inspectors repeatedly notified the CGHC of the poor state of its pipework, but no repairs followed. As the inspector noted, the repairs were urgently needed in order to prevent "complaints from the police and residents." The Alumbrado Commission advised that the company needed to uphold the contract's stipulations, but it fell to the Ayuntamiento to verify whether the gas met prescribed quality standards and the light had the contracted photometric intensity.[42] Lacking the personnel to determine this, the Ayuntamiento commissioned professors Manuel María Contreras and Andrés Almaraz from the National Preparatory School to make the determination.[43] However, the two men did not have the equipment necessary to carry out the study and had to ask the CGHC to provide "the required instruments and apparatuses to confirm, compare, and verify the luminous power of hydrogen gas."[44] The Ayuntamiento records do not cite the results of the study.[45]

By the mid-1880s the CGHC pipework, burners, and lamps were in a dreadful condition. A cartoon satirizing New Year's gifts depicted a line of individuals carrying bundles of paper candles to an anthropomorphized gasometer. The caption read, "Some residents [give] paper candles to the gas company to increase public lighting."[46] This was just one instance of many that mockingly compared the luminosity of gaslights to older technologies. Journalist and writer Ireneo Paz bitingly claimed that yesteryear's large lanterns, with their thick tallow candles, surpassed gaslights that "certain English companies serviced during this period of modern civilization."[47]

FUMBLING INTO THE ELECTRICSCAPE

With the introduction of electric lights in 1881, the phrase "fascinating spectacle" in *La Voz de México* succinctly encapsulated the city's changing lightscape.[48] Forty arc lights of the Brush System graced San Francisco Street to commemorate Independence Day and add a new element to the luminous festivities, which included the traditional fireworks. Charles Brush had designed a direct-current generator to feed arc lights, an innovation that made electric arc lighting practical for commercialization.[49] Twelve tall wooden columns, higher than the balconies of the adjacent houses, held the lights in place. As people's first impressions attest, however, the immediate assessment was not favorable, for the lights cast otherworldly shadows. According to the Catholic newspaper *La Voz de México*, the *reflectores eléctricos* (electric reflectors) brightly illuminated a section of the street but cast equally vivid, nearly supernatural shadows of passing pedestrians.[50] Within its radius, arc light perfectly illuminated figures from head to toe in vibrant colors. Passersby wandered by, dazzled by the radiant light, barely able to distinguish objects. They walked with their eyes half-shut so that the light would not harm them.

Those standing at the foot of the columns enjoyed the spectacle, watching others stumble past with their eyes closed, barely avoiding each other. Farther out, critics claimed, Brush arc lights made people look like pale ghosts with undefined colors and ill-defined contours. Waves of shadows in the middle of the road seemed to push one another. The journalist Enrique Chávarri, who wrote under the pseudonym "Juvenal," described the phantasmagorical spectacle: the lights cast unnaturally sharp shadows—in some cases, three *varas* long—ahead of the pedestrians who passed under them, like ghosts of some sort: "black, rigid, and fateful." The powerful lights shone out so far that enormous shadows traversed the street as if "a world of black giants commanded another world of pygmies illuminated by pale, azure reflections emanated by the moon of a drawn-out cemetery." He described the gloomy spectacle, "out of a Hoffmann story, a light under which fantastic heroines of Scandinavian legends should walk, flap, and dance."[51] *La Patria Ilustrada* caricatured the experience in an amusing cartoon in which people walking right under a lamppost are so blinded that they cannot use their eyes and must guide themselves with extended hands to prevent running into others (fig. 1.1).[52]

This new illumination was not part of a fantasy world: real people—real bodies—passed under the bulbs. Their glare held these people up to uncomfortable, even invasive, inspection. "Juvenal" remarked that "beautiful dandizettes with half-face vividly illuminated in our boulevard" could be scrutinized so closely that observers could discern bismuth goldenrod powder on one side of their faces, while the other side, left in the shadows,

FIGURE 1.1. *"Public lighting has greatly improved in this capital."* La Patria Ilustrada, *May 19, 1884. Hemeroteca Nacional Digital de México.*

was bewitchingly indistinct. The scene painted by the light, he said, made Rembrandt van Rijn's use of shadow look muddled.[53]

Manuel Gutiérrez Nájera, the capital's most prolific chronicler, who later confessed his desire to capture Rembrandt's chiaroscuro through words, echoed this criticism of the intrusive light. An article on the disadvantages of electric light exposed the new lights' ruthless unkindness to passersby. Under the harsh light, dandies confronted their bald spots and the discolored areas of their frock coats, while prominent women suspected that "the atoms of rice powder were enlarged when the ray, emitted by the large lamps, strike them." With a touch of gentle humor, Nájera baptized it *luz byroniana* (Byronic light). The new light, like its namesake, was powerful but ultimately flawed. Women in makeup, he declared less gently, looked like deceased dancing nuns from the French opera *Robert le diable*. He advised them "[t]o sensibly abstain from crossing the streets illuminated by electric lights."[54] Extravagantly dressed women on the streets, in search of a boyfriend, according to Nájera, who *se empenachan* (adorned themselves with feathers), appeared to be at "a shop counter under the electric light." Despite Nájera's snide commentary, women did not hesitate to come out into these lighted spaces to enjoy the city's nightlife.[55]

Advertisers and entrepreneurs were only too happy to capitalize on people's fears of the way electric light presented them. The skin toner El Agua de Juvencio (which claimed to whiten and preserve youthful skin, protect it from sunlight, and add an afterglow) had been widely used before electric lights went up. Like many other cosmetics, however, it had not been designed with the glare of electric lights in mind. According to one satirical poem, after the toner had boasted of besting the sun, electric light responded by threatening to burn female beauty in its "crucible."[56] Electric light won the ensuing battle, leaving a woman's beautiful face badly battered. However, the war was not over, as *El Telégrafo* noted. After criticizing electric lighting, the chemist Francisco Patiño, the toner's inventor, went back to the lab to rework his formula to work well in electric light. Were he to achieve this, the newspaper concluded, "women would likely erect a pedestal in his honor":

EL AGUA DE JUVENCIO Y LA LUZ ELÉCTRICA

A burla del mismo sol	Mocking the sun
Dijo el agua, ha de brillar	The water said, it will bring out
Mi magnifico arrebol	My magnificent afterglow
Y la luz: he de quemar	And the electric light: I will burn
Tu belleza en mi crisol.	Your beauty in my crucible.
La luz al agua encontró	The light encountered the water
En un rostro satisfecho	In a satisfied face
Y la lucha se entabló;	And the battle began;
Pero en la lucha quedó	But the struggle left
El bello rostro maltrecho.	The beautiful face badly battered.[57]

Whether or not the new light was injurious to female beauty, fashion commentators noted that it exalted colors, fabrics, and textures. Electric light, some claimed, transformed a humble blue skirt into a wave on a German lake, rippling under moonlight. Silk acquired marvelous shades and gleams unavailable in ordinary sunlight.[58] Lace gained visual density, resembling dragonfly wings. Satin dresses, particularly those in blue and white, gave "a truly marvelous effect under the influence of electric light."[59] Goose feathers, which commonly decorated elite women's hats, resembled even more regal pheasant feathers. Of course, as Nájera noted, the light was ultimately "Byronic, and in the sudden, stark shadows these same feathered hats acquired a gloomy quality similar to enormous ravens looming over a dead body." Men, dandies in particular, also were not exempt from transformation under the new light. Their emblematic long frock coats, "with collars as tight as a noose ready to strangle," cast shadows like those from the tall and slender eucalyptus trees.[60]

The Brush System's curse was that it provided *too much* light. While the technology behind the system was old, it had only recently been put into use when introduced in the Mexican capital.[61] In 1809 Sir Humphrey Davy, a British chemist and director of the Royal Institution, dazzled a London audience with the precursor to the Brush System. Having previously supervised the construction of a voltaic pile—the first electrical battery, invented a few years earlier by Italian physicist Alessandro Volta—in the basement laboratory, Davy stood before his audience holding two thin charcoal sticks. He connected one stick to the voltaic pile then, as electricity began to flow through it, put it in contact with the second stick, which produced a brilliant spark. The entranced audience gasped as the chemist pulled the sticks farther apart, enlarging the spark: "electricity traveled in a dazzling arc of light between the two slender carbon rods . . . [a] bright blue-white light would burn until either the carbons or the electricity ran out."[62]

So the arc lamp was born, along with the race to commercialize it. But for decades the high costs of the carbon arc delayed successful commercial application. Despite improvements in the 1840s, the main obstacle remained: the expensive batteries that supplied the power. However, by the 1870s, the advent of the dynamo had solved this difficulty, enabling the development of the Brush System, which allowed several lights to run in series. The door to commercial electric street lighting was now open.

The output of the Brush System, which was at the cutting edge of technology in the early 1880s, could be measured in thousands of candlepower (instead of dozens for gaslights) and had a dazzling effect on the eye.[63] Indeed, the staggering amount of light produced by the system made sense only in ample, open spaces such as railway stations, factories, and mills, but even here the system drew complaints. A British trade journal noted that electric light was disagreeable even when applied in large public spaces. "Overcharged with actinism," the property of radiant energy, the electric light was "blue in colour, and injurious to the eyes." The author of the article attested that after an experiment in Paris "I could scarcely see my way out of the place. I was so dazzled, and was made so uncomfortable, that it was some minutes before I regained my proper vision."[64]

These complaints were widespread, not just the grumblings of traditionalists. Perhaps the simplest explanation was that real physiological adjustment was necessary when moving through spaces lit by arc lights, which had a "spectrum similar to that of daylight."[65] Hence eyes observing arc lighting behaved as they did during daylight, using the same retinal cones, which, of course, was problematic, as the lights were on only at night. However, in gas-lit spaces the eyes used the same retinal rods as they did at night. Therefore, going from the darkness of night or from even a gas-lit space into a space lit by electricity required an uncomfortable adjustment.

That adjustment could happen literally instantaneously, as in the case of a scene in Alameda Park. A few months after lights went up to San Francisco Street in 1881, the city erected a scene around the park's fountain designed to resemble a forest just after a snowfall. Mirrors surrounded the fountain, reflecting the water and, in the darkness of night, making it look like a lake. Around these mirrors stood a portico made of tree trunks wrapped in vines, with plants filling in the space between. Behind it all stood a painted background, which the designers illuminated with a series of small Edison lamps of various colors. Their light bounced off the trunks, the fountain, and the mirrors, creating a truly enchanting scene, with a magical effect. The portico portion, however, was lit by four arc lights and sixteen bronze gas chandeliers, which should have been more than enough light and yet failed to illuminate the wooded rotunda. The problem, as *La Voz de México* reported, was that the inequality of the lights proved unsettling to the eyes: "One was forced to go from [a space with] great clarity to shadows." This abrupt transition caused what was beginning to be referred to as *deslumbramiento* (blinding). Instead of evoking a wintery pastoral scene, the display became a "contest of shadows" that looked like the staging of a cemetery scene from a Zorrilla drama.[66]

After the acknowledged failure of this initial attempt, designers of the fountain spectacle made minor adjustments that did little to correct the problem. Finally, they shut it down and made more substantial adjustments. When they turned the power back on and lit up the plaza at nightfall on December 11, the conflict between the two technologies remained. The sharp contrast between intense light thrown off by the arc lights and the depth of dense shadows created by gaslight once again ruined the effort. A journalist from *La Voz de México* was on hand to witness the unveiling, noting that the struggle between clarity and darkness gave the plaza a melancholic aspect, like those "fields of the polar region where weak sunrays appear and retire without twilight, brusquely invading certain sites and leaving others in complete shadows."[67]

During the hour-long trial test, the electrical system suffered two total failures. The first lasted seven minutes, while the last ultimately ended the test and forced the use of only gaslights. When the arc lights were not flickering and failing, *La Voz de México* reported, they had a tendency to shift in color from reddish to violet to whitish, which bothered the eyes of everyone and even caused some to faint. The paper concluded that electric lighting would need many improvements before it could effectively replace gas.

PUBLIC AND PRIVATE HEALTH IN THE ELECTRICSCAPE

Doña Candelaria may have been fictional, but her anxieties over electrical lighting were real and were shared by government and professional circles. Before permanently establishing electric lights and considering extending them to the barrios, town council president Pedro Rincón Gallardo convened an advisory commission to listen to expert opinion on how—and even if—the existing electric light system should expand. One person who spoke before the commission was Juan María Rodríguez, a physician who considered municipal lighting a public health matter meriting expert discussion. In his view, the Ayuntamiento had to devise a *zona de luz* (light zone) capable of sufficiently illuminating the city for smooth traffic and the protection of pedestrians' life, honor, and interests. He began with the widely known limitations of the Brush System: the light was excessively intense and unbearable to the eye, given the flickering and great quantity of violet and ultraviolet rays. In addition, it created inky shadows that contrasted sharply with the lit-up objects next to them.

None of this was particularly novel; however, Rodríguez's concern for the light's physiological effects was a new concern. He explained that the body, including the *tutamina oculi* (protective appendages of the eye: eyelids and eyelashes), reacted to the lights. As the natural defense failed to protect sight, individuals on the street resorted to the brim of their hat, the full collar of their cape, and even their hand as a makeshift visor to avoid being blinded and slamming into others on the street. Arc lights, the doctor explained, simply gave too much light, a problem never faced by previous lighting technologies. Their intensity called for further innovation. He noted that among those answering the call was Thomas Edison, who had not sought to develop lights of higher luminous intensity but rather "light bulbs suitable to the eye and uniformly distributed to achieve an appealing comfort, beauty, and safety." He cited the example of the minister of development, who had installed the first Edison bulbs in his house, which, he claimed, everyone agreed was a "very favorable" example.[68] Rodríguez admitted that eye diseases lay outside his expertise as an obstetrician, but this did not prevent him from having convictions and expounding upon them. Certain parts of the ocular apparatus, he explained to the commission, such as the cornea, crystalline lens, and the retina, were known to undergo a molecular shock when encountering electric lights.

Medical officials had recently identified *fluorescencia* (fluorescence) as an origin of diseases that, in the short and long term, produced amaurosis (partial or total blindness). How Rodríguez arrived at these convictions is unclear, but he likely consulted Jules Regnauld, whose influential work "Phosphorescence in the Eye" revealed that "in man and other mammals, the cornea and the crystalline humour are endowed with a manifest phosphorescence, and that the retina possesses it also, but in a less degree." The presence of phosphorescence—which was a harbinger of eye disease—was attributable to the prolonged action of intense light, particularly the shorter, more intense wavelength light at the violet end of the visible spectrum. As early as 1860 Regnauld warned that the introduction of electric lighting on a large scale would "become injurious unless a way [could] be devised for cutting off the extreme violet rays."[69] Much of this was largely speculative at that time, yet the Frenchman advised further research before electric light became common.

Twenty years later, Rodríguez agreed enthusiastically. He even introduced a motion at the October 3, 1883, meeting of the National Academy of Medicine, stating that electric light was not merely an issue of public safety and the city's embellishment but was directly related to public and private health. Citing Regnauld, he explained that once a physical agent came out of the exclusive dominion of science and was applied industrially, "it was the doctor's responsibility to foresee its useful and harmful consequences."[70]

Based on this, Rodríguez called for a detailed study on the issue and solicited commentary from the president and secretary of the academy's board, Agustín Andrade and Manuel Domínguez, respectively, who both had completed specific studies on eye diseases. Although Andrade felt unprepared to comment officially, he diplomatically said that not enough experiments had taken place in Mexico. He added he was aware of particular amblyopic distress in individuals employed in establishments illuminated by electric lighting. Board member Manuel Ortega Reyes concurred, citing the trial installation of arc lighting at the capital's post office. Employees could not tolerate the illumination, making it necessary to bring back the older system. He suggested "possible visual distress in people exposed to [the] intense rays." However, he countered, this might not be the case when it came to lighting in public spaces. Despite initial complaints after the introduction of electric streetlights, criticism markedly declined.[71]

Rodríguez remained unconvinced and said that any public toleration of electric light was deceptive. The Brush lights at the Zócalo and the city center, he argued, were of lower intensity than those first introduced, as could be seen in the intensity of the shadows that passersby cast onto sidewalks, walls, and the pavement. Despite this modification, he still distrusted the fundamental technology of the arc lights and advised the adoption of Edison or Maxim lights made with carbon filaments so that the air would not receive excessive quantities of carbonic acid like that produced by Brush lights. Instead of producing an intense and intermittent light with abundant violet and ultraviolet rays (unbearable to the human eye), the lamps would emit a more bearable light, similar to sunlight with its yellow-reddish color. Rodríguez exhorted the board to address this important matter in the future.

To the academy's credit, it took up the issue. In one meeting José Bandera admitted that he had not diagnosed any eye disease linked to electric lighting in his extensive practice as an ophthalmologist since the establishment of the new lighting technology in the city.[72] He agreed with everyone else that the light was far too intense in its illumination and the blackness of its shadows but confessed that he lacked data proving that electric lighting was harmful to health. Doctor Andrade concurred, citing his numerous observations at the Valdivielso hospital, which specialized in eye diseases. As a final nail in the coffin, Ortega Reyes supported his colleagues' appraisals. None could see an objective reason to consider electric lighting a health hazard. Undeterred, Rodríguez adamantly responded that research on the subject raised doubts. Works on public hygiene, he insisted, recognized that electric light created a "certain shock on the outer layers of the eye," which in his view could not be "totally inoffensive." He urged the doctor and engineer Antonio Peñafiel to continue research on the effects of the new lighting technology.[73]

With that, the matter rested for the next two years. When Mexican doctors revisited the effects of electric lighting in late 1885, they were not concerned with its effects upon humans but on much smaller animals: mosquitos. The capital witnessed a "prodigious multiplication" of the mosquito population that year, and the Pedro Escobedo Medical Society put together a commission to research the possible factors.[74] The society asked physician Antonio Peñafiel, an earlier witness to Rodríguez's attack on arc lights, to review the hygienic aspects of the infestation. He responded with a very uncontroversial call to drain away raw sewage that served as an ideal nesting site for in incubation of mosquito eggs, to which the Mexican medical establishment readily agreed. Much more debate emerged over his second finding: that electric lights encouraged mosquito reproduction. Since their establishment, electric lights had appeared to the common observer to attract large quantities of insects. This might have been true, for electric lights, especially the blazing Brush System lights, stood out from a great distance and could well have lured the mosquitos more than gas or turpentine lights did.

To Peñafiel, this was enough to raise suspicions and warrant research that might lead to a condemnation of electric lighting as demonstrably unhealthy—as opposed to merely physically uncomfortable. One researcher, however, was not convinced. Jesús Sánchez, a zoologist and recent member of the academy, agreed with Peñafiel's assessment of sewage and draining stagnant water but was skeptical about electricity's influence. Citing several other natural phenomena that, cumulatively, might explain the jump in the mosquito population (such as the decline of insectivorous aquatic birds), he saw no reason to think that electric lights made the insect problem worse. He offered the case of New Orleans, a city surrounded by swamps, which also had electric lighting but nevertheless did not have an excessive quantity of mosquitos. If mosquitos in fact followed electric lighting, he reasoned, then there should be no mosquitos—or at least fewer—in areas without electric lights. Peñafiel conceded that swarms of mosquitos were found in the Santa María and Guerrero districts had been sighted back in February, where there was no electric lighting.

Peñafiel was not a Luddite, eager to push back against what was new. During the earlier Rodríguez-led debates on electric light's health consequences for the human eye, he had remained open to accepting evidence. As long as the impact of this new lighting technology remained unknown, however, fear would find fertile ground. Even trained doctors were not immune to it when enough people voiced concern. When confronted with evidence, Peñafiel surrendered his position, to his credit, but not everyone was so open-minded when it came to electric lighting. Dr. Rodríguez kept up the drumbeat against electric lights, shifting his opposition to defend the flora of the city. He believed that electric lights, among other factors, were harming the *ahuehuete* trees (Montezuma cypress) in the Zócalo.[75] The good doctor claimed that the lights, rich in chemical rays, "kept those plants in perpetual wakefulness."[76]

LIVING IN LIGHT AND SHADOW

While the debate about the physical effects of public lighting was slowly coming to a consensus, Mexicans' understanding of the emotional consequences of light was unformed. It would fall upon Mexico's artists and not its scientists to define this realm. One who took up the call was Federico Gamboa, who, in 1889, wrote "¡Vendía cerillos!" (He sold matches!), a story that follows the short and tragic life of Luis, a homeless teenager who sold matches to pedestrians on the capital's streets. Heartbroken and resolved to take his own life, the boy contemplated the means and location. He looked toward a canal but, knowing he could swim, thought suicide by drowning might be too difficult. He looked around the growing city but found few choices.

Finally, after much scouting, Luis found himself at the Paseo de la Reforma, built after 1900 as a showpiece of the nation. By the time Luis trod its sidewalks, it had been illuminated with electric lights, but the trees along the paseo sculpted those lights, stretching out the distance between them until, as Gamboa wrote, they appeared interminable. At the *plazoleta* (small square) housing the Christopher Columbus monument, the boy felt fear of the harshly lit statue of the explorer and the accompanying friars. They seemed to both know and condemn what he was about to do. Nervous and panting, he hoped that someone might come out of one of the homes along the avenue, but no one did. When he collected himself, the figure of Columbus no longer frightened him but instead "sympathized with his torment and, almost guessing his state of mind, with its extended right hand, it seemed to indicate the only refuge for the disinherited ones: the afterlife."[77] Still, the place did not suit him; it felt too big and too bright. He ran to Cuauhtémoc's plazoleta, a site more fitting: isolated and—without any lighting—darker. The Paseo de la Reforma was abandoned, he felt. People did not come to unlit squares anymore, so no one would find his body, at least not until the following day.

Where and how to take their own lives were choices through which individuals attempted, ironically, to construct a sense of themselves.[78] Luis's calculated decision novelized the decision-making process for young individuals looking to take their lives. Spaces such as parks, plazas, and streets—which the city's youth claimed for courting, socializing, and, for the city's homeless like Luis, sleeping—also became spaces for committing suicide. For Gamboa's protagonist, the darkness and shadows of an isolated plazoleta afforded sought-after privacy.

For those in the expanding middle class and the elite, to see and be seen were aspects of life that Porfirians held dearly. This could occur in many places, both public and private, but the two most important and most scrutinized venues were the city's theaters and department stores. Theaters were the far more obviously public arena where lighting was everything. The very popular nightly performances depended on artificial lighting, and audiences immediately sensed any changes. As late as 1880, the Teatro Nacional remained illuminated with oil lamps, but by 1889 people expected better: theaters were at the cutting edge. In that year, Guadalajara's Degollado Theater had its own steam plant with three dynamos and an impressive regular supply of electric lighting. Embarrassingly, however, Mexico City's Teatro Nacional lacked a reliable power supply. The same dynamos that powered street lighting fed its modest lighting scheme. Each night, only for performances, the electric company redirected power so that the theater could illuminate four arc lamps along the stage and four additional ones high above in the auditorium. The theater still had to rely upon gaslight to complement its modest scheme.

The installation of electric lights, all else being equal, was a signal that the site being lit was an important, modern space. In 1885 the Ayuntamiento approved public money so that the Teatro Nacional would have electric lights for the debut of Mexican artist Manuel Múgica in an Italian opera.[79] Despite the Ayuntamiento's high hopes, the results were disappointing. The intense lights distorted faces and bodies and disrupted the spectators' imagination by revealing the flying system, cables, and other stage equipment.[80] Others wrote about similar disasters. During a theater scene in one of Gamboa's novels, the individual responsible for the lights turned the wrong key, "giving torrents of light, simultaneously illuminating halls and high and lower balconies."[81] It was a comic error that ruined the atmosphere of the performance and was topped only when a few doves fell fluttering at the end of the show, dazzled by the light and frightened by the applause.

Ten years after the Ayuntamiento's failed attempt to electrify the Teatro Nacional, the newspaper *El Siglo XIX* blasted the 1895 installation of two large electric lights on the stage of the Teatro Nacional. The lights' intensity, the paper claimed, dazzled attendees in the balcony, injuring their eyesight and preventing them from seeing the play. Later, during a performance of Giuseppe Verdi's *Rigoletto*, attendees enthusiastically applauded when the lights were turned off. They were glad that the "*annoying light stopped pestering for a few seconds*, only to return with greater intensity."[82] As the preeminent place to see and been seen, the theater catered to the elite and the aspirational. If this section of the capital city could not find satisfaction under the new lighting regime, then who could?

While too much lighting posed consistent problems in a theater, this was less true in the windows of department stores, whose owners were eager for rich and poor alike to see and dream. Among the prestigious central city department stores that lined Plateros Avenue, the opportunity for electric lighting was great. With the mid-century development of large sheets of glass, the fronts of stores became "a glassed-in stage on which an advertising show was presented."[83] Mexican firms began importing glass sheets only in the late nineteenth century. Mass production halved prices, which meant that the glass was expensive but still within reach. With the installation of these showcases came window shopping, which Steven Bunker has called "one the most popular urban entertainments of the Porfiriato."[84]

When window displays arrived, out went the plain stacks of goods and large written advertisements. As the visual took precedence, so too, by necessity, did lighting. An editorial in *Municipio Libre* discussed the importance of this new form of commercial representation. After briefly dismissing the traditional monotonous display of wares, the writer exclaimed: "[H]ow many times a humble object has awakened our desire, given the artistic taste when displayed inside a shop window!" In Paris, the recognized leader in the field, stores paid high salaries to shop assistants, whose responsibilities centered on artistically exhibiting fabrics, jewels, and flowers behind glass windows. The paper advised Mexico City merchants to follow Paris's example and respect the importance of this new space. Stores that ignored this advice and failed to engage the customer's eye with fanciful preparation and artistic ensembles risked losing out to competing stores—especially European ones that did embrace a celebration of luxury. Their customers would instead continue to frequent European and other department stores, which "we are almost obliged to enter, given their great charm for the eyes."[85] Although national merchants complained that only foreign retailers made fortunes in Mexico and that people spurned their products, *Municipio Libre* argued that the issue was fundamentally not nationality but aesthetics.

Store light has a long history in Mexico, but few were prepared for the saturation of arc and Edison lights that would come within a decade. The contrast between those areas of the city that embraced light and commerce and those that did not would become stark. At mid-century, *quinqués* (oil lamps) installed by businesses helped pedestrians navigate streets. Even with their glow, Mexico City's nights were still dark, as Francisco Zarco attested. A generation later, in the late 1880s, the darkness was losing ground: an array of bright arc lights shone out from the windows of businesses.[86] In *Metamorfosis*, Gamboa wrote about Fray Paulino leaving his church one night in 1899 and walking down San Francisco Street captivated by the pleasant spectacle of the city's principal avenue. Illuminated stores, cantinas, and cafes spilled streams of light over the stone pavement, where it mixed with light from the unsteady headlamps of carriages. Electric streetlights swept out and across the streets and even into the side streets, providing a calm, almost lunar illumination. The electric lights strung up above across side streets looked like "large opal strawberries about to be plucked from an antediluvian and invisible tree." The lights' brightness transformed the appearance of whatever its rays touched. Signs, windowpanes, and merchandise "acquired the exaggerated brilliance of an opera stage."[87] At the corners where the lights of a store shot out in multiple directions, beams of silver illuminated the interiors of coaches instantly. Just alike an ill-mannered customs officer, Gamboa wrote, the vulgar rays revealed to the world the carriage's female occupants and drew inappropriate attention to the curvature of the women's décolletage.

As Fray Paulino reached the Zócalo, he stopped to compare the colonial buildings around him with the artery he had just passed through, a space that seemed "to go up in flames and throb with its lights and noises." Just north of the Zócalo, however, the ugly and gloomy black walls of the Academy of Medicine, housed in the former Inquisition building, unlike the vibrant commercial district that he had just passed through, seemed to absorb and swallow the few lonely electric lights at the corner. A block south of the academy, the Old Customs Building sat in shadows along a street plunged in abandonment and silence.[88] After the riotous animation of San Francisco Street, the buildings around the Zócalo gave the impression of a city bewitched or dead. Their dark stone walls along shadowed streets made them look, in Gamboa's telling, like forgotten bodies, rigid in their beds without someone to bury or pray for them.

No one would ever confuse Mexico City's vibrant commercial district with the yawning silence of the Zócalo. At the intersection of Capuchinas and 2da de la Monterilla Streets stood the modern four-story building of El Puerto de Veracruz, one of the city's most prestigious clothing stores. The building dominated the intersection of the streets with enormous street-level windows on both sides. The light of a nearby arc lamp flung over wires flooded the building's *escaparates* (shop windows), which were themselves illuminated with Edison incandescent light bulbs. J. Figueroa Doménech's *Guía general descriptiva*, the premier guide to the city in the late nineteenth century, described the store's bright, artistically decorated shop windows with recently arrived goods from Europe as a constant attraction.[89] El Puerto de Veracruz was challenged only in 1899, when El Centro Mercantil became the city's largest department store, with a four-story building occupying an entire block of Tlapaleros Street on the southwest corner of the Zócalo. Its prime location, combined with twelve-foot-high storefront windows, made the store an obligatory stop for nocturnal strollers. On its inauguration day, arc lights profusely illuminated the exterior of the building. Doménech noted that the lights gave the building's decorations—as well as the elegant ladies crowding around them—a sense of lavishness that would remain indelible in the memory of all attendees. As the crowds moved inside, incandescent bulbs showed why they were becoming a preferential lighting method for commercial lighting. Edison lights charmingly lit glass display cases, while an electric elevator moved people and merchandise, to the enchantment of all.[90]

After electricity's arrival, department stores became a showcase of lavish modernity for capitalinos in no time at all. The new lighting technology highlighted fabrics, colors, metals, and jewels, at times with a surreal effect. Doménech's guide described La Perla jewelry's showroom on San Francisco Street: "hundreds of small electric lights dotted the coffered ceiling, cascading down on the pristine glass displays and striking off bright sparks from the jewels," a scene that could be understood only after seeing it firsthand.[91]

With spectacles like this, it is no wonder that captivated passersby spent precious time inspecting, commenting on, and dreaming about the goods behind the glass. Once again, it fell to Gamboa, the chronicler of life during the Porfiriato, to give voice to everyday reactions. For Javier and Elisa, trapped in the spiral of a failing marriage, a walk down Plateros Street (the two block-section of San Francisco Street closest to the Zócalo) offers an unexpected respite.[92] Elisa is shocked when she realizes that they have been holding hands for some time without realizing it. They have forgotten themselves—and their conflicts—amid the wonders of the stores. Even better, their irreconcilable differences have faded away—if only momentarily—as they inspected every single shop window and rediscovered that they shared an identical taste that harked back to their honeymoon.

Though this dramatic scene is fiction, Javier and Elisa were not alone. Store windows became meeting points where crowds gathered to take part in the emerging consumer culture. The bigger, more artistic, and flashier the presentations—all of which typically called for more electric lights—the larger the crowd. Intersections like Plateros and Palma drew crowds attracted by the lavish shop windows like those of Le Paris Charmant and Sorpresa y Primavera Unidas. Women, in particular, the Doménech guide noted, became entranced while contemplating displays of Parisian novelties.[93] The displays at the escaparates offered rich and varied items. In her report on Mexico City for the US publication *Harper's Weekly*, Helen S. Conant predicted that Plateros Street would remain "a brilliant thoroughfare, where jewels glisten in the shop windows, where peaceful processors parade in gorgeous array, and where crowds of fair dames go shopping or drive in handsome carriages on their way to the afternoon promenade."[94]

For the window shoppers ambling down San Francisco Street at night, the mass of lights blazing all around was not just a wonder but entertainment eagerly consumed. They could rest assured that the illumination was a clear signal of Mexico City's emergence into the technologically and culturally modern world. For those hurrying home toward Santiago Tlatelolco, the experience was different. Light for them was not only a utility—as it was on a fundamental level on San Francisco Street—but also a necessity. The metaphorical aspects of electric lighting gave way to practical issues of security and safety.

People had to cover their eyes in 1881 during the first demonstrations of the Brush System above San Francisco Street, but in the late 1890s envious and aspiring capitalinos peered reverently into store windows. None of the people in the store thought or wrote of themselves as agents of modernity, yet they did as much as any other historical actors to make electric light not just useful but luxurious and amazing. Shop windows, aided by an ever-growing number of luminous advertisements, marked the city's nightscape. Their proliferation and significance were evident by the late Porfiriato. A problem at the Necaxa hydroelectric plant left sections of the city without power in 1909. In contrast to the lights usually disseminated throughout the downtown area, which "made merchandise sparkle and nocturnal walks so pleasant and attractive," *El Tiempo Ilustrado* described the scene: the "escaparates were in the dark and inside stores salespersons walked back and forth with weak candle lights over the counter." The stores had the "gloomy aspect of a wake."[95]

This description shows how accustomed capitalinos had grown to the electric mantle that covered familiar spaces. Although the curious scene seemed to come from a half-century ago, the daily noted, it would have been horrifying were it not for the electric power that fed light and power to the rest of the city, factories, and tramways. A sudden paralysis of work would bring "the starving ones to the streets, who, taking advantage of darkness, would satisfy their bellies even in lieu of scandal and crime."[96] The dreaded scene, normal fifty years before, had become a backdrop for lawbreaking by the late Porfiriato.

THE FAILINGS OF THE MASSES AND THE PROMISE OF THE ELECTRICSCAPE

Electrical lighting was a symbol of modern scientific achievement by the late nineteenth century. Celebrated in international exhibitions, it had extended the workday in workshops and factories and was transforming urban landscapes. This last aspect, which seemed to serve the public good so clearly, was an especially important talisman of modernity. After all, electric lights illuminated the streets and parks in the newly developed areas of the city. Along with the completion of the extensive drainage works that aimed to secure the city from flooding, electric light promised that life would be safe and comfortable, two of the most important characteristics of material modernity, showing that nature might not be the threat that people had long supposed. *El Imparcial* echoed this view when it held that civilization had seemingly trapped lightning, exploiting its brightness to meet its needs. Like their counterparts in America and Europe, those in Mexico who preached the gospel of modernity associated lighting with public safety and progress. Unlike drainage projects, electric lighting was a public works project that transformed spaces nightly, but this physical transformation was only half of the equation. The second, arguably even more important task, was peopling these spaces with modern individuals who espoused bourgeoisie values (thrift, order, temperance, prudence, and punctuality). With these citizens, Mexico City would have the foundation for a capitalist society.

The Alumbrado Commission retained its colonial association as a guarantor of both security and moral behavior while becoming associated with hygiene during the Porfiriato. A note in *El Centinela Español* in late 1881 applauded beautification projects such as the construction of gardens in the central plazas, the alumbrado, and the erection of statues but could not remain indifferent to the Ayuntamiento's neglect of sanitation in the barrios, which were "the eternal focal point of endemic diseases that flagellate the population."[97] Here the pathologizing of poverty was evident. The author noted that those who passed by those nests of misery were overwhelmed at the sight of unfortunate individuals who spent their lives as specters, inhabiting nauseating shacks, breathing a contaminated atmosphere, and crossing streets by night under the weak light of an oil lamp inside a dusty and broken lantern. Their abandonment by the Ayuntamiento was censurable. The barrios had no electric light or garden or statues or even a pure atmosphere.[98]

The Porfirian ruling class fretted over what would become of the city without light, however, especially as poorly lit neighborhoods expanded its borders. As the city grew, along with a menacing mass of anonymous lower-class citizens, its commercial, political, and cultural leaders wondered what would happen: would their city rise into modernity or sink into chaos? Public lighting seemed to be one easy, universally celebrated way to assure the former. *El Mundo Ilustrado* elucidated this in its report applauding the Siemens and Halske electric installation. Mexico, the paper declared, "is now well lit and has taken one of the most critical steps toward progress in terms of public safety." As the weekly newspaper succinctly concluded, the absence of light did not mean merely poorly illuminated spaces: "without light, no hygiene, public morality, *policía* (public order), or safety was possible."[99]

Bringing in reliable, effective lighting was tantamount to ushering in order. Light, according the logic of time, drew moral geographies, scaring away thieves, moderating the intemperate, and reining in the depraved. In the same way that public health officials had worried that electric lighting attracted and encouraged mosquito populations, they now worried that a lack of light did the same for criminal activity. The writer and journalist Ignacio M. Altamirano, known for his unforgiving opinions, shed all delicacy and described the outskirts of Mexico City in the early 1880s as "unknown regions [housing] savagery, nakedness, infected houses where the squalid and starving population huddle together, entire families of sickened and beggars. Proletarianism in its most repugnant expression." In keeping with the times, he naturally had to bring in the problem of lighting. "[T]he municipality barely hangs an oil lamp at night," he wrote, and "the police send gendarmes to stalk rather than to safeguard."[100] *El Telégrafo* seconded this popular opinion, proclaiming—in notably more moderate tones—that poor lighting exposed residents to physical harm or "to be[ing] robbed by one of the many pickpockets that stalk us without fear of the police or pistols."[101]

The association between public lighting and *policía* harked back to the late eighteenth century, when the colonial viceregal capital of New Spain underwent demographic and territorial expansion and the establishment of the first lighting system was construed as a tool to instill public order. Juan Vicente de Güemes Pacheco de Padilla, 2nd Count of Revillagigedo, led a comprehensive urban renewal program during his tenure as viceroy of New Spain (1789–1794) to address the "continuing issue of street flooding, an inadequate sewage system, and disorganized public spaces."[102] Facing a wave of urban growth prompted by an agricultural crisis in bordering areas, colonial officials perceived the "need to monitor and control the poor," who accounted for 85 percent of the city's population.[103] Officials drew up a three-tiered approach to deal with nocturnal safety in 1790. Besides adding a ninth municipal court, Revillagigedo installed oil lanterns as street lighting system, which "not only enhanced the capital aesthetically but prevented the public sins that darkness facilitated and covered." A corpus of ninety-two guards called *guardafaroles* or *serenos* (night watchmen) would also safeguard and maintain the city's investment.[104] Although each individual watchman was exclusively responsible for looking after twelve lamps, collectively their jurisdiction was citywide. They carried arms and could apprehend offenders who would subsequently be tried by the new tribunal.

Serenos were the tangible manifestation of the Bourbons' new interventionist vision of power, which would not condone any questionable behavior in public spaces in its efforts to attain urban moral regeneration. Streets and alleyways lacking lighting allegedly provided "a perfect context conducive to crime and disorderly conduct."[105] It is in this context that lighting became associated with functions of public order and safety. Colonial authorities described serenos as those "who watch over during the night . . . preventing the frequent thefts, assaults, homicides, and other crimes that darkness gives room to."[106] To help the serenos achieve this, Revillagigedo believed, he would need to lessen the dark, for the poor, lacking internal regulation, "would be transformed if . . . visible" to the upper classes and the authorities.[107] Because his fight against sin and public immorality could not go far enough, Revillagigedo also demanded that bars, which many officials believed were epicenters of crime and disorder, be well lit after 9 P.M.

Lighting was thus not solely a struggle against darkness as an aesthetic issue but against real and, perhaps more often, imagined dangers that lurked in the spaces where it reigned. Therefore the presence of lighting and serenos every night "had significant symbolic weight to safeguard residents." The policing power of 1,128 oil lamps placed at intervals of 35 meters guarded by 100 serenos supposedly secured the possibility of strolling streets at night without danger of becoming a victim.[108]

This notion that fighting darkness equated to dampening disorder and insecurity was not particular to Mexico City. In large metropolises around the world, particularly in the influential circles of Paris and London, elites shared a fear of the masses and the crime that they imagined was leaching out from their neighborhoods. By the end of the century, the association between the two had even gained a scholarly endorsement with the rise of positivist criminology. According to Italian criminologist Enrico Ferri, lighting was a social factor that influenced the occurrence of crime.[109] As a pioneer in the field of criminology who studied under the founder of the Italian school of criminology, Cesare Lombroso, Ferri focused on the social and economic influences rather than biological ones in understanding criminal activity. He believed that private and public lighting had a positive effect on the individual; likewise, the lack of light was unavoidably detrimental.

In an 1898 lecture at the University of Naples, reproduced by the Mexican legal journal *El Foro*, Ferri described "the revolting hovels in the [Italian] city where entire families lived without air, without light, and without personal space, curling up one against another." Under those obscene circumstances, he argued, how could society be surprised by the poor's constant violation of modesty? How could society punish when it consented to such horrors against humans and allowed people to become ruined? Rather than inflicting punishment, he argued, the conditions of social life needed to be changed: "give them air, given them light, regenerate the blood of those miserable people and crime would be decimated."[110] If this were true for the individual, it was no less true for society. For a city suffering from repetitive nocturnal armed robberies, Ferri asserted, abundant street lighting was cheaper and more effective than hiring an army of guards to persecute robbers. Plentiful light, unlike the guards, both deterred crime and addressed one of the fundamental factors that caused it.

As influential as his ideas were globally, Ferri's reception in Mexico led to little action by the government. Instead, as the city grew during the Porfiriato (with both planned, legal neighborhoods and spontaneous, unofficial neighborhoods), capitalinos could only wring their hands in worry over the problem of darkness in the barrios. And the problem was not just the newest additions at the edge of the city. The newspaper *La Patria* pleaded for better lighting on behalf of citizens living on the streets north of the city toward Santiago Tlatelolco, where "the horrible service left the honest and peaceful residents at the mercy of crooks ready to ambush them."[111] Too little, the paper concluded, had been done to save this barrio and others from danger.

Another paper, *El Diario del Hogar*, frustrated that electric lights were not coming out to the barrios quickly enough, demanded that the city transplant decommissioned gaslights and use them there: some light, no matter how weak, was better than none at all.[112] Residents from the colonia Valle Gómez followed the paper's suggestion in 1898, asking for the turpentine lights or gaslights freed up by an upcoming extension of the electric system in a more fortunate area. The colonia's petition to the city read, "[W]ithout being pretentious, as other colonias have, we do not ask for electric lights; we will be satisfied with *una poca de luz* [a little light] to guarantee our property and, even more valuable, our lives."[113] This request, deferential in tone, showed that the residents recognized the unspoken ranking of colonias and how lighting was concomitant with their status. As inhabitants of a more humble neighborhood, they did not expect electric lights but would take what they could get.

Lighting was a powerful point of leverage that the city had over not only its own population but even towns and neighborhoods outside its jurisdiction. For essentially the same reasons expressed by the capital's less-fortunate, ill-lit neighborhoods, other towns sought to secure the older gaslight technology displaced by electric lights. Ayuntamientos within the Federal District such as Coyoacán, Santa Fe, Iztapalapa, Cuajimalpa, and others as far away as Cuernavaca in the neighboring state of Morelos, petitioned the Mexico City municipal government for faroles. Demands from many quarters for gas and turpentine lights peppered the Ayuntamiento meeting minutes from the 1880s and 1890s. While a few were denied, the council often approved the requests. Whenever lanterns were not immediately available, a request for them at least was filed away for a future opportunity. Sometimes a failed request would be granted later because the suggested project was so modest. In 1897, for example, the Alumbrado Commission notified Bernardino Salazar, who had a pending petition, that the turpentine lights freed up by the installation of electric service would go up on 2da de Carpio, 3ra de Olivo, 3ra de la Rosa, and 6ta del Chopo Streets.[114]

Until its dissolution in 1903, the Ayuntamiento mediated calls for more light. As the authority channeling the demands of residents and provisioning by companies, the institution administered material realities and dealt with social and cultural expectations. Decisions on which neighborhoods were blessed with electric lights and which continued in darkness or in the muddled light of older technology all came down to the Ayuntamiento, which understood the power vested in it and wielded it accordingly. With the establishment of new colonias, the Ayuntamiento was called on to illuminate their streets. In their petition to the Ayuntamiento for electric lights, Severo Monterrubio and his *vecinos* (neighbors) in the colonia Valle Gómez also tried to influence the location and number of lights.[115] The Ayuntamiento brushed them away, claiming that the number of houses and daily movement dictated the distribution of lights. The council went on to reproach the colonia with a reminder that the allocation of street lighting was "a special benefaction of the Ayuntamiento" in their case: it could be removed at will, because the colonia had yet to be legally recognized. The council had approved the expansion of electric lighting to the working-class colonia but was not willing to allow its residents to determine the location of the lights—they had to know their place and the limits of their power.

The Ayuntamiento's gentle reprimand of Valle Gómez was not an anomaly. The power relationship between the city government and the citizens shifted between the late 1880s and the early twentieth century. One measure of that change was the preparation of requests for street lighting. Light had become an obvious marker of difference between neighborhoods in this period. Whether a neighborhood had flood control or sewer service was difficult to discern throughout the year, but everyone could see who had electric lights and who did not. Consequently, neighborhoods supported petitions to secure lighting. The sizable number of signers was startling, particularly in a society that was still mostly illiterate, even in this period. The vecinos of Calzada de La Verónica had 256 names on their petition, a demonstration that the ad hoc organization, although informal, was a real neighborhood movement, willing to make demands on the authorities. Perhaps more crucially, there was now what one historian has labeled a "revolution of expectations." The publicity that came along with the introduction of new technologies, followed by immense—sometimes critical—public discussion, captured all capitalinos' attention: people knew the improvements that were taking place around them. The increased number of petitions is proof that people recognized the shift that had taken place and wanted to keep up with it. Those with oil lighting wanted gas, and those with nothing wanted whatever was available. After the official signing of the 1886 CGHC contract, six petitions registered 398 signers; after the 1896 contract, a total of fourteen petitions gathered 248 signatures. Besides this increase in the number of demands in the aftermath of improvements, people now demanded things they had not previously asked for.[116]

Despite the earlier power of the Ayuntamiento, after the implementation of the 1903 Municipal Law it became an advisory board for the country's president, thereby losing its governance over the city and its role as advocate for neighborhoods. That same year, only two alumbrado petitions came before the Ayuntamiento, a substantial decrease from those filed in the 1880s and 1890s. It appears that members of the public went silent because they had lost their natural interlocutor within the Federal District.[117]

COMPETING VISIONS OF THE ELECTRICSCAPE

While many capitalinos craved electric lighting, not everyone in the city cheered its growth. Gas lighting companies saw a threat in the expansion of electric lighting, but CGHC would not dim its lights without a fight. In addition to modifying its name to Compañía Mexicana de Gas y Luz Eléctrica (CMGLE; Mexican Gas and Electric Light Company) in 1883, the company integrated horizontally and introduced electric service in addition to gaslights. People began to suspect that the company's interest in electric lighting came at the expense of its gaslight service. In 1881 *La Voz de México* opined that the intensity of gaslight had diminished to the point that on certain nights the city had the same dim appearance that it had in the eighteenth century.[118]

Despite criticism of the 1881 forty-light installation, the Ayuntamiento applauded the convenience of electric lighting and moved to extend it. In doing so, it considered the benefits of combining arc lights with tower lighting. These towers were posts that rose 160 feet into the air with enormous lights. Enterprising individuals had promoted the use of tower lights long before they were technically feasible. These projects shared a belief that a handful of enormous lights could flood entire cities with light so that "darkness would be totally banished." Whereas such schemes remained in the realm of utopian literature in Europe, they materialized in the United States. Detroit's system, for instance, had 122 towers that illuminated 21 square miles.[119] Perhaps inspired by its American counterparts, the CMGLE entered conversations with the Ayuntamiento for a new contract in 1883. The draft for a six-year contract stipulated the combination of 600 arc lights (2,000 CP) of the Brush System and ten tower lights in the suburbs with each tower producing a total of 16,000 CP.[120] Given their novelty, one tower light was installed as a test-run: the city council would decide within six months whether to adopt it. At the end of the six-year contract, the Ayuntamiento reserved the right to purchase the installation, but this never happened; the contract, for unknown circumstance, failed to materialize.[121] Instead, the two parties entered a much less aggressive three-year contract for a hundred arc lights in 1886.

Over the next two years, the CMGLE's monopoly over electric lighting in the city was on shaky ground. Julio Labadie, a French citizen and owner of La Profesa Drugstore, exhorted the Ayuntamiento in 1886 as representative of the Compañía Mexicana de Luz Eléctrica (CMLE; Mexican Electric Lighting Company) to open up to public auction a three-year contract for a hundred electric lights that the council was considering at the time.[122] Labadie promised better-quality public lighting at a lower price than the CMGLE and offered to subject his company's service to both photometer and dynamometer testing. His pitch seemed to work: the Ayuntamiento contracted his company for twenty-five electric lights in the Colonia de Guerrero, but the company failed to execute the contract, which is not surprising, given how new it was. Just a year earlier, journalist José F. Godoy, who like Labadie also served as a CMLE representative, requested a two-year tax exemption and permission to erect posts, extend lines, and place "other accessories" to offer incandescent lighting to commercial and residential homes.[123]

With the failure of CMLE, the Ayuntamiento and the CMGLE went back to the drawing board in 1887 to expand the earlier hundred-light contract to 300 lights. The expansion was to take place within a six-month period, but the Ayuntamiento hinted at further expansions in the near future. The revised contract included a stipulation that allowed the council to request expansions of 60 or 120 lights to be completed within six months. Alderman Ignacio Cejudo, frustrated and enraged by incomplete and inadequate work, advocated for the inclusion of penalty clauses for failing to install the 300 lights within the timeframe established: "since 1885 the establishment of electric lighting had been agreed upon, but the company had not even imported the generators."[124] Alderman Manuel González Cosío sympathized with his colleague's frustration and acknowledged that his proposal seemed attractive but contended that the CMGLE was not at fault in the case of the 1884 contract for six hundred lights and ten tower lights. The Ayuntamiento itself was at fault for not covering the large expense to expand electric lighting.

The Aguirre Hermanos Company presented the first serious threat to the CMGLE's monopoly over the provisioning of electric lighting. In 1887 it requested permission to install posts on the streets that would hold its power lines. The company requested that posts be established not only "on the streets listed, which for the most part were not already crossed by electric lines," but also "in streets where the good judgment of the Ayuntamiento considered it useful."[125] The request triggered the CMGLE's alarm bells, for it appeared the company aimed to stretch into areas that the CMGLE already controlled.

The Public Works Commission's recommendation of the Aguirre Hermanos proposal crystallized the menace to CMGLE. The commission, which looked after changes to the city's infrastructure, argued that it was the Ayuntamiento's duty to seek improvements that would benefit the public, including comfortable, bright, abundant, and safe light.[126] The commission believed that it had found all these qualities in the Edison System. Commercial houses and even residences would promptly request the system because it was simple to transport and install and relatively cheap to use. Hydrogen gaslights were expensive and gave off poor light; arc lights were viewed as tiring and harmful to the eyes. But Edison lights offered distinct advantages that made them the favorite choice of great European and American cities. Incandescent lights, the commission explained, offered good coloring that neither harmed nor strained the eye, were free from fire hazards, and did not corrupt the air.

The commission's findings would also introduce competition among electrical providers. Aguirre Hermanos and the Edison System would run headlong into CMGLE and the existing Brush System. This raised questions, not least of which was the Ayuntamiento's responsibility in regulating companies offering the same service. Councilman Miguel Iglesias noted that Aguirre Hermanos had carried out several successful exhibitions of Edison lights. Similar results around the world had prompted the city of Berlin to choose them for street lighting. For Iglesias, the matter was clear: the Ayuntamiento "ought not to close the door to progress over unworthy excuses." Alderman Pedro Ordoñez chimed in, noting that Aguirre Hermanos was a Mexican company, which merited consideration. The argument that Edison lights were an improvement carried no weight for councilor González-Cosío, who had previously defended CMGLE against criticism, particularly for private installations, where the Ayuntamiento had no say over what he deemed was purely the "scientific and exclusive interest of private citizens." Deciding what was "better" was either a purely subjective issue or else an esoteric problem for scientists, not the concern of the Ayuntamiento, which simply had to provide illumination on the best terms for the city.[127]

Other ethical and logistical questions arose from the issue of having multiple companies provisioning electric lighting in the city. Councilman Manuel Gargollo demanded that the Ayuntamiento resolve the question of what to do with companies that had previously requested to offer such service. For him, to allow Aguirre Hermanos to enter the market and provide light and sell electricity amounted to an injustice to those who had been previously approved. Furthermore, granting concessions to more companies would fill the streets with posts, an injustice to the city's residents. Iglesias disagreed. It was clear that the Ayuntamiento was not forced to approve all permits. In his view, once street lighting and electric service were satisfactory in an area, there was no need for new concessions or more infrastructure. Alderman Ignacio Cejudo supported Iglesias, explaining that concessions made sense only once an improvement or innovation merited them. The council had no obligation to bring in more and more companies simply because they asked. But, in the end, the questions were not answered to the satisfaction of all. The petition of Aguirre Hermanos was voted down 3 to 8.

The company did not give up but revised and resubmitted its proposal in August 1887.[128] This new version solicited authorization only to replace turpentine and naphtha gas lamps with Edison incandescent lights. In other words, the company would not threaten areas where electric lights already existed. The rate for each incandescent light would remain fixed in exchange for a ten-year contract. The Ayuntamiento approved the substitution of up to three hundred turpentine and naphtha lamps on streets jointly selected on a trial basis. Fifty lights were to be installed by Independence Day and the rest by the end of October. At that point, the council would assess whether to adopt the system permanently at the end of the year.

The years between 1886 and 1890 were a period of profound change in the lighting of the city. Given the attention that electricity received—and deserved—several different lighting technologies cohabitated in Mexico City. The arrival of electric lighting did not remove its competitors. Instead, companies that had offered gas lighting before expanded into electrical service; new competitors, keen to expand the new technology, still offered older ones. Even new technologies, like *solarina* fuel, had much more in common with gaslights than with electric lights. Numbers for 1886 reveal the shifting lightscapes: 116 electric lamps of 2,000 CP fought against darkness alongside 1,273 gaslights, 860 turpentine lamps, and 182 oil lanterns. The Aguirre Hermanos Company, following in the footsteps of the CMGLE, became a gas *and* electric provider by introducing lamps fueled by solarina, a fuel that produced a whiter light with an intensity equal to that of turpentine.[129]

The move to introduce new fuels while also entering the electric lighting market spoke of electricity's uphill race to become the city's sole illuminant. A sizable addition of 500 of solarina lights in 1886 transformed the capital's nightscape, but this was the last innovation before electricity came to dominate the city's lighting. By 1890 companies were no longer servicing oil lanterns, but oil lanterns did not disappear entirely just because they were being phased out. They would continue for years, especially in the poorer areas of town. By contrast, the city recorded 298 and 135 electric lamps of 2,000 and 1,200 CP, respectively, that same year. Both gas and turpentine lights declined, from 1,273 and 860 to 408 and 461, respectively. Meanwhile, solarina lights increased to 1,169 lamps.

FIGURE 1.2. *"Reforma Avenue's public lighting."* La Patria Ilustrada, *January 5, 1885. Hemeroteca Nacional Digital de México.*

While it is true that diverse technologies cohabited throughout the city for decades, they were not all equal. Electric public lighting spread out from the city center toward the northwest and west, along with the districts undergoing fast development, such as colonias Guerrero, Santa María, San Cosme, San Rafael, Arquitectos, Reforma, and Bucareli. Meanwhile, gas lighting remained entrenched toward the east and south while turpentine and oil lamps, except a few here and there in the interior, "formed a crown" around the capital.[130] As the numbers revealed, this was no minor crown.

No matter how impressive the growth in electrical lighting was between 1886 and 1890 in terms of lamps or money spent, it was not enough for the public. The demand was so great that only a light on every street would have made capitalinos happy. They complained that the expansion was unnecessarily slow, even delayed, and half-hearted. A series of illustrations titled "Obras municipales" (Municipal public works) in 1889 mocked the council's poor performance in addressing the capital's infrastructural needs. These systematic failures included alumbrado. In one illustration, a line of tall trees directs the viewer's eye toward the Cuauhtémoc statue in the background center (fig. 1.2). Unveiled two years earlier, the monument celebrates the last Mexica ruler of Tenochtitlan, a powerful national symbol. The sculpture anchored the Paseo de la Reforma (Reforma for short)—

a broad, modern avenue lined with beautiful tall trees, designed to be a favored promenade for the well-to-do. However, in the image of the statue at night, Cuauhtémoc sits encircled in darkness, lost among towering trees, with a laughably small luminous globe at his feet. Hardly the striking visual display that others were sculpting with light at the same time, it looks rather like someone with a flashlight standing next to the monument. The municipal government, according to this illustrator, had left the capital in the dark.

SAFETY AND FEAR IN THE ELECTRICSCAPE

On the night of February 14, 1898, the Siemens and Halske Company of Berlin fitted "a splendid choker of luminous pearls" around Mexico City's neck.[131] Capitalinos had long awaited the arrival of a permanent electric lighting installation, which unexpectedly came with underground cable networks and elegant electric lamps powered by a central station, the use of alternating current (AC), and transformers distributed throughout the city. A total of 480 high-intensity (2,000 CP) lights hung above intersections, gardens, and other expansive public areas. Downtown streets whose length and importance merited it held an additional 120 lights of lower intensity (1,200 CP). The National Palace alone was graced with 100 incandescent lights (though only 48 shone initially). The weekly magazine *El Mundo Ilustrado* applauded the partial tests carried out a few days earlier, extolling the company's pearlescent light, simultaneously soft and intense, which did not harm the eyes, offering complete steadiness and a tremendous luminous power. The Brush System's lights had been clear globes, but the Siemens egg-shaped lights were of an opaline shade, which softened the unpleasant impression of the light radiating down from directly overhead. Nothing, added the weekly, was more attractive "than the interminable lines of opaline globes along our avenues, suspended like aerostats in space."[132] Much ink was spilled praising the latest installation. As *El Imparcial* proclaimed, "We have electric lighting, it has entranced us. We sing anthems to it."[133]

The new installation had been long in the making. The contract between the city council and Siemens and Halske had begun two years earlier, at the height of the Ayuntamiento's role as an interlocutor in the area of lighting. That year, the Ayuntamiento sent out a call for bids that various Mexican consuls around the world translated and distributed—notably, along with a map of the city. In the end, two firms, Siemens and Halske of Germany and the Mexican company Francisco Espinoza y Com-

FIGURE 1.3. *Group picture of Siemens and Halske workers in Mexico City. Courtesy of the Siemens Historical Institute.*

pañía, tendered offers for the contract, but the Ayuntamiento rejected the Mexican company for failing to meet the invitation's guidelines. Guillermo Brockmann signed an eight-year contract on behalf of the German firm. The Ayuntamiento gave the company fourteen months to build a central station, lay underground and aerial lines, and install the required posts and lamps.

The new electric system arrived in a cacophony of machinery, tools, and shouts. Siemens and Halske brought an invading army of 15,000 workers with state-of-the-art "unknown tools" with which they set about breaking sidewalks and pavement, opening trenches, laying down wires, and erecting posts and candelabras.[134] A group picture of the company's employees, both native and foreign, behind an imposing generator inside the central station showed the importance of the enterprise (fig. 1.3). *Semana Mercantil* reported that the installation cost upward of 3.5 million pesos—1 million alone was spent on 124 miles of cable.[135] While the workers, under the supervision of German engineer Francisco Neugebauer, received accolades for the work's order and precision, the trenches that they dug and the underground wires that they laid still generated anxiety.[136] As rainwater filled the trenches, they were mistaken for simple puddles and swallowed the drivers of two carriages. This was particularly worrisome at night. *El Popular* recommended the installation of a *farolito* (small lamp) to indicate danger.[137]

The transformation of Mexico City's lightscape had been years in the making, with older lighting technologies refusing, in the proverbial sense, to go quietly into the night. Like other cities around the world, in the 1880s Mexico City experimented widely with different public lighting regimes. Not every trial ended well. A mix of high hopes, much skepticism, and critical assessments characterizes these early installations, but the arrival of the Berlin company began to change that. Social critics interpreted the capital's luminous borders through a combination of religious and bellicose metaphors of light battling darkness. *El Imparcial* held that before the Siemens contract the city moved like "a somnambulist stumbling in the shadows of a sleepy town."[138] In the presence of modern technology, older lights became like fantastical heads whose macabre winks and ironic blinking were rocked by gusts. Colonial-era lanterns hanging from ropes brought to mind the silhouette of a person hanged, but modern Edison System lights tore away the dark shadows and saved the people from these specters. As the city's overseer, the Ayuntamiento could take credit for redeeming the capital from the darkness that seemingly haunted it.

While the Siemens and Halske contract brought the novelty of underground wires, most wires in Mexico City still resided high overhead. The danger of these overhead cables was well known.[139] Not only were arc lights, strings of incandescent bulbs, and power lines classed as dangerous in themselves, but other wires that crisscrossed the city could potentially become perilous if they came into contact with them. In 1884 the Compañía Telefónica Mexicana (Mexican Telephone Company) asked the CMGLE to place "an insulating cover" over its wires to prevent tragedies.[140] Later that year, when the Fourth Court Tribunal inquired whether or not the CMGLE was obliged to use insulated wires, the Alumbrado Commission responded that no such stipulations had been written into the permit for electric street lighting but promised that any new contract would add them. With this promise the commission signaled its intent to make power safer as power lines continue to proliferate.

In addition to complaining about the danger posed by overhead lines, capitalinos sneered at the lines in aesthetic terms. The Public Works Committee conceded that the expansive series of posts and cables were *molesto a la vista* (unsightly) but insisted that the welfare and advantages that the

Edison incandescent lights would provide to all residents superseded the issue of ornamentation and aesthetics.[141] Alderman Manuel Gargollo worried that, if authorized, companies would fill streets with posts. They would, in his odd analogy, "become like the multitude of arrows from Xerxes' army and block out the sun, and capitalinos would have to walk in the shade." The sense that the modern city would suffer from interrupted views had become commonplace by 1889. That year, a Mexico City paper published a political cartoon of a street forested with towering telephone posts over miniature pedestrians, with an ironic caption: "The telephone's little posts give such a beautiful vista."[142] Alderman Agustín Lazo summed up the mixed feelings of capitalinos about their electric infrastructure: although telephones provided excellent service to commerce and the public, the poles disturbed traffic and blocked views.[143]

As they multiplied, the hazards of posts and overhead lines only grew, while developments in insulation were simultaneously making underground wires more feasible. A preference for cabling—running wires and cables underground—had become established by the mid-1890s. The Siemens and Halske 1896 contract for electric public lighting stipulated a mixed installation of overhead and underground cables. The underground cables would power 475 arc lamps and 150 incandescent lamps, while aerial wires would feed 125 arc lamps.[144] Yet underground cables carried their own controversies, which were only amplified by the lack of specifications by the Ayuntamiento. Clause 10 of the 1896 contract called for cables to be laid at a depth of seven decimeters and the cable to be the model invented by Siemens and Halske, but nowhere did the contract give any detailed instructions on where precisely to run the cable. Regulations for underground and aerial wires passed the following year included only general precautions on resistivity and insulation. The underground cable must "resist the likely chemical or mechanical actions given the subsoil's nature and conditions."[145] A decade earlier, in 1887, the Ayuntamiento held that allowing the CMGLE to use underground wires in the permanently damp ground would condemn the city to constant interruptions and expensive repairs due to the inevitable destruction of metal wires.[146]

Once underground, overhead cables were out of sight but not out of mind. By 1897 dailies such as *La Patria*, *El Universal*, and *El Nacional* noted that elsewhere in the world utilities buried cables toward the middle of the street, not under the sidewalks, as in the Siemens and Halske installation. (The company had apparently decided to undertake cabling under sidewalks after receiving a high estimate of what repairing damages to the street's new asphalt pavement would cost.) This difference could pose real danger, according to *La Patria*, which looked to New York City for proof. The city council there had appointed engineers to audit the "insulating companies" after reports emerged of horses shocked by defective underground cables.[147] Mexico City had not even taken the precaution of insulating the cables; because its lines were under the sidewalks, the city had shifted the risk from horses to people.

Every broken sidewalk served to remind capitalinos that Siemens had decided to protect its profits and not the lives of those walking the city's streets. Every crack was an opportunity for a current to escape the wires and *trastornar* (disturb) pedestrians. Allies of electric companies marshaled scientists and experts to launch a counterattack upon those who questioned the safety of buried cables. *El Imparcial*, in particular, condemned the alarmist reports of danger under people's feet. The regime-friendly daily, an unwavering ally of industrialization and technological innovation, prided itself on being unlike the newspapers of yesteryear, which "took joy in alarming the ignorant with announcements of the end of the world." The paper boasted that its modern form of journalism was founded upon facts and research. An early article on the subject of the "cable-issue" transcribed an interview between one of the paper's reporters and a renowned electrician, who declared, "In my opinion, my most intimate conviction is that not even in the most remote cases is there such risk [from underground cables]." He congratulated the paper, saying that he had little to add to the articles from a scientific perspective already published on the topic. It was obvious to him that whoever was behind the alarmist articles "knew nothing about electricity."[148] For this electrician, it would have been a different case if the reports had set out to demonstrate how underground installations caused more frequent *desgracias* (tragedies) than overhead lines. In underground lines 1,500 volts pushed alternating current (AC) via resistant and insulted wires under a 27-inch layer of soil, whereas overhead lines carried the same loads with virtually no insulation. They were more dangerous for those working with the lines, but this was not true for the general public.

The anonymous engineer said that he was available to debate with an *hombre de ciencia* (scientist) in order to demonstrate, in a technical discussion, the errors behind the panic and would put his name to the article. In the meantime, he asked for anonymity, judging it sufficient for now to destroy *temores* (fears) with useful data and time-tested experiences. Numerous cities had buried cables, he added, including Paris, but no data supported the claim that they presented higher risks than aerial wires. No doubt accidents occurred, but they had explanations. Isolated incidents should not serve as the basis to anathematize the procedure. It was foolish, the engineer concluded, to deduce from one or more separate accidents that wires represented imminent risks; experience and theory demonstrated that the city was safe.

For all the engineer's confidence, he still demanded anonymity. Though he may have had many reasons for this, one may have been that the science behind his assurance was not as certain as he claimed. In an 1893 speech at the World's Electrical Congress in Chicago, D. C. Jackson, a renowned electrical engineer, admitted that there were many unknowns regarding the two methods to install wires underground (built-in systems and conduit systems).[149] For instance, no consensus existed as to the proper depth at which conduits should be laid or, equally important, their material. The durability of clay pipes depended on where and how they were laid. Mexico City was constantly growing and building, creating considerable danger of clay pipes being cracked by picks and other tools during excavations. Furthermore, if the soil around clay pipes was replaced incorrectly, the entire site could settle awkwardly, causing considerable damage. Iron pipes fared little better. In this case, water ruining underground cables was the most significant cause of concern but had not yet been fully studied, as iron pipes had only recently been invented. Treated-wood conduits, being neither air-tight nor water-tight, engineers acknowledged, were the worst of the three choices. Despite some successes, none of these conduit systems had been in use long enough to inspire confidence. The success or failure depended upon on local conditions like soil, terrain, and drainage, which created more than enough ambiguity to allow the debate to rage for years.

In short, people had real reasons to question the safety and durability of underground cables. Readers of the pages of *El Imparcial* might not be aware that this was the case. The newspaper found the alarm so ridiculous that it could afford to mock it with satirical headlines, such as "Capital Residents in Danger!" It is unclear what effect the paper's efforts may have produced, as an uninformed reader may well have missed the irony and become more alarmed.[150] Perhaps that is why the newspaper simultaneously kept presenting experts to deflate the panic. An "intelligent electrical engineer" explained the science behind the insulation that people fretted over, noting that it had become axiomatic within the field that the air provided the best insulation, which, of course, was not feasible when power lines were underground.[151] Another material would have to act as insulation.

The growing demand for lighting and traction power translated into higher voltages supplied, so the choice of insulation conduit was critical. The article explained that notable studies on the insulating properties of different materials and compounds had been carried out around the world. Without listing or providing details on any of the studies, the paper presented a list of materials with their corresponding resistivity. Siemens used a high-voltage cable that, the "intelligent" engineer claimed, Paris had adopted in 1891 with no recorded accidents despite the various companies that ran their gas and other utility pipes nearby. For fourteen years Berlin had yearly dug up and inspected its underground installations and had found no deterioration of the wires.[152] Perhaps taking his cue from the Germans, the minister of the interior had responded to the outcry over cabling by demanding personally to inspect the cable being employed. Having done so, he declared that it met resistance and insulation requirements.

Another paper was not ready to give up the fight. *El Universal* responded to *El Imparcial*'s accusations of alarmist reporting by focusing not upon the power lines' composition or insulation but on their location, thus undermining *El Imparcial*'s so-called scientific take on the issue. *El Imparcial* had prided itself on being an outlet aimed at publishing fashionable scientific ideas of the period, but *El Universal* rejected the idea that criticism of underground cables was ill-informed superstition and published accounts of accidents abroad.[153] *El Imparcial*'s pronouncements were not to be blindly accepted just because of the paper's pretentious tone but were to be subjected to reason and demonstration just like everybody else's.[154] For its part, *El Universal* reasoned that both overhead and underground wires presented dangers, but cabling under sidewalks augmented this by placing pedestrians at risk. *El Universal* further undermined *El Imparcial*'s claim of being scientific and fully informed by pointing out that the paper had not even gotten its terms correct when it reported that underground cables in most European cities carried a *fuerza* (force) of 3,000 volts. Volts measured not force but rather electric voltage. Moreover, most underground power lines under streets used direct current (DC), with a voltage never exceeding 500 volts in extreme cases, with the median between 100 and 300.

Wise to the journalistic tricks of newspapers, *El Universal* called into question both *El Imparcial*'s experts and the account of the minister of the interior. Invented correspondence and interviews, especially with self-proclaimed "experts," was a well-worn ploy. The tell-tale sign, according to *El Universal*, was a demand for anonymity by the "expert" interviewee. Likewise, the paper questioned the minister's judgment on electrical matters. He could have looked at the cable as a "strange thing, ugly or beautiful, whose resistivity and insulation he measured *a ojo de buen cubero* [by the rule of thumb] or because someone told him."[155]

The two papers locked horns because people were genuinely worried about the dangers of running power lines throughout the city. By the late nineteenth century, constant reporting of railroad wrecks and urban fires had captured the attention and incited the imagination of capitalinos. Newspapers, the penny press, and popular outlets such as broadsheets and corridos (popular ballads) vividly narrated and illustrated accident scenes.[156] Journalists and engravers skillfully mined the devastating consequences of train wrecks, producing detailed descriptions of mangled iron and maimed bodies that together cast a horrifying shadow upon the utopian promises of a technified future.

Industrial accidents were hardly unique to the Mexican experience, but President Diaz's encouragement of foreign investment meant that railroad networks and other utilities were owned and operated by foreigners, primarily Americans. Fears about new technology thus mixed with a potent antiforeigner animosity. Crackling like electricity through the debates over underground power lines were the shockingly common allegations that authorities privileged foreign interests over the safety of the country's citizens. As I discuss in chapter 3, perhaps nowhere was this resentment more pronounced than in the foreign ownership of the city's electric tramways. The tram accidents that frequently injured members of the public gave rise to the notion that foreign investors mattered more than the lives of capitalinos.

THE ELECTRICSCAPE: FROM A FLICKER TO A SHINE

There was no comparison between the older Brush System and the new Siemens System in either cost or performance measures. A report in 1898 by the alumbrado supervisor calculated that the Siemens and Halske lamps were stronger in terms of luminosity. In addition, the cost per candlelight of the CMGLE reached 0.0642 pesos, while the Berliner was only 0.0634.[157] The difference, multiplied by the yearly 3,600 contracted hours, translated into 31,788 pesos in savings. The superiority of the new system did not stop there. *El Nacional* agreed that the new service was demonstrably better not only in terms of intensity and fixity but also in the aesthetics of its infrastructure. Earlier lights had required ugly wooden posts and webs of lines overhead, but the new system hid all that.[158] Residents had long expressed distaste for wooden posts. During a city council session a few years before, two aldermen voted against a permit for a company that wanted to install posts in the middle of the street. One of the men, Jesús F. Contreras, pointed out that the permit was unacceptable: in addition to the narrowness of downtown streets, the posts would be hideous.[159]

Yet the new system, for all the praise it garnered, did not deliver—at first. *El Popular* observed that many lamps were still missing and sarcastically added that residents were yet to see whether Siemens or the moon would illuminate the city that night.[160] Problems continued in the following days, as noted by a February 25 report from the police inspection office that registered "deficiencies" for the nights of February 17, 18, 20, 21, 22, and 23. Although it was not specified, a deficiency was an interruption in the contracted service, whether in the number of lights, luminosity, or number of service hours. Subsequent reports added nearly another week of nights—February 24, 25, 26, 27, and 28—to the list. *El Popular*'s assertion that no one could be certain where the nighttime light would come from in the first few weeks after the new system came into operation was not outlandish. Within a month of the new service, the Alumbrado Commission had to coordinate with the Inspector General de Policía to economize on lighting during moonlit nights in order to prolong or improve service whenever necessary.[161]

A few months after the Siemens and Halske System's inauguration, with the initial hiccups worked out, the Ayuntamiento judged its service a success. It even considered extending the area lit, claiming "an urgent need" to expand not only to neighborhoods that lacked lighting altogether but even to those currently served by naphtha and turpentine.[162]

The CMGLE could do little to stop the entrance of Siemens or its success, but it was not for lack of trying. In 1896, when the Ayuntamiento informed CMGLE of the Siemens contract, the company's representatives, Braniff and Knight, moved to exercise their *derecho de tanto* (right of first refusal). Until then, the CMGLE insisted, that right had been respected. The company interpreted this right "within an ancien regime logic, insisting that in its relationship with the city and the Ayuntamiento the rights acquired through seniority outweighed proper contractual rights." In 1896, for instance, when its 1886–1887 contracts were about to expire, the company argued that these contracts were obligatorily extendable because CMGLE had priority whenever another party proposed a contract.[163] The company also objected to calls to remove its posts from the city's streets. Braniff and Knight demanded that the Ayuntamiento "declare that the company had the right to keep the posts on the streets for the distribution of electric lighting to residential homes." The Ayuntamiento disagreed, arguing that the concession CMGLE enjoyed was limited only to the time and duration of the contracts.[164]

Knight, for one, was not done. As the CMGLE's long claimed "genealogical preference" lost currency, he tried other strategies to prevent the entrance of Siemens. In a new 1897 complaint to the Ayuntamiento, he argued that Siemens had installed an iron lamp post in the 1st Street of Bucareli that made contact with eleven of the CMGLE cables. Such a situation would inevitably cause disasters. A few days later, Knight informed the Ayuntamiento of his petition to the government of the Federal District for police assistance to prevent damages.[165] Engineer Alberto Best, chief of the Alumbrado Commission, accompanied the governor to verify whether the post made contact with the lines. Although Siemens had placed pieces of wood to insulate their cables from CMGLE's and the alumbrado service had not been affected till then, Best demanded that Siemens fix the issue. The engineer shut down the power in the area until workers installed safe new insulators. Even though he waved the CMGLE's fines for the delay in the lighting for that night, he reproached the company for not notifying them sooner.

Unbowed, Knight still tried to harry Siemens, notifying the Ayuntamiento a short while later of a similar situation near the Ciudadela. In this case, the Siemens post had made contact with their company's lines, burning the insulating cover. He demanded the council's intervention, but Francisco Neugebauer, as the representative of Siemens, countered, asking the Ayuntamiento to take measures to prevent CMGLE from complicating the installation of his company's posts and lampposts. In the end, Knight could only delay, not prevent, the Siemens victory in public lighting.

Having lost its long-held alumbrado contract, the CMGLE went after the market for residential lighting by expanding its services to include developments in gas lighting. In 1897 it ran an advertisement for acetylene gas, a new lighting source and a product of calcium carbide.[166] The company's hopes to compete with electricity with gas was not entirely off base. As late as 1908, chemistry professor George Gilbert Pond claimed in a treatise on calcium carbide and acetylene that electric illumination had given gas illumination real competition for over twenty-five years, but its cost had prevented the destruction of the gas industry. Gas, including acetylene gas, was still cheaper, despite its limitations.[167] As if to rewrite its institutional history, the ad noted that the new illuminant was the company's latest effort in its long effort to introduce improvements in the area of lighting, stretching all the way back to 1868, when it pioneered carbonated gas.

CONCLUSION

Encouraged by Siemens and Halske's initial efforts, on March 1, 1899, the Ayuntamiento authorized a second contract between the Berlin company and the city to expand public electric lighting to the entire city.[168] In his address to Congress later that year, President Díaz applauded the achievement as essential and sang the praises of Siemens, but it would not be the Berlin company that would fulfill the second contract. Siemens had transferred both its street lighting contract and the right to install and run posts and power lines to the Compañía Mexicana de Electricidad, SA (CME; Mexican Electric Works, Ltd.), a free-standing company organized in London. Siemens and Halske retained one-quarter of the CME's ownership along with Indelec, its Swiss holding company, while the Dresdner Bank owned the other three-quarters.[169]

Mexico City concluded the century with a citywide electric mantle, with the CME firmly in control. It dominated the new public lighting market and had an advantageous position to enter into the home and business electrical market. Before signing the second contract, the CME requested the extension of its first contract, arguing that the eight-year period was too short to recoup the capital invested in the installations.[170] The Ayuntamiento agreed, extending the contract to twelve years on the condition that the company renounce its right to charge a premium for lights fed by underground cables outside the zone it was obliged to service. The Ayuntamiento could also request additional lights beyond the previously imposed 218 yards limit from the existing service area. Although the Ayuntamiento extended the CME's monopoly, it secured some flexibility. By 1900 the CME controlled 287 miles of underground cables that powered 480 arc lamps of various amps (900–2,000), over 10,000 incandescent lights, and 144 miles of aerial lines where cabling was not demanded by the Ayuntamiento. These were impressive numbers, considering the company had started less than a decade before. Only two other companies offered CME any real competition for residential and commercial lighting. The CMGLE fed 209 arc lights (700–2,000 amps) and 9,190 incandescent lights across the city, while the Compañía Nacional de Luz Eléctrica fed a mere 60 arc lamps and 6,650 incandescent lights. Yet, although the CME had the full backing of the Ayuntamiento and the Dresdner Bank's deep pockets, it would not last long.

In 1900 most capitalinos imagined that these three companies would dominate the electrical market for the near future. When they walked home under streetlights or gazed at the curiosities in department store windows, they had one of these three companies to thank. When they talked around the dinner table about bringing electricity into their homes, lighting up rooms, or plugging in one the growing number of appliances that (the advertisements promised) would change their lives, they knew that they would be customers of one of these three businesses. What no one could anticipate was that at this moment a new player had arrived in Mexico that would turn all these expectations upside down. In a few short years the Canadian-owned Mexican Light and Power Company (Mexlight) would introduce vertical integration to the nation's electrical market, offering customers everything from the generation of power to bulbs and appliances. It would also horizontally integrate and gobble up its competitors. The result was a company that offered Mexicans more than they had ever dreamed of. It also created an economic and political power so strong and centralized that it became a nightmare.

| TWO |

Exhibiting the Electric City

William Axford, accompanied by his wife and son, arrived in the port of Veracruz aboard the steamship *City of Merida* in January 1882.[1] Among their precious possessions was a wooden crate with a delicate passenger: a *gallina de vidrio* (glass hen). After a short stint at the port and the nearby town of Orizaba, the Axfords and the chicken began a winding 420-kilometer trip along the recently established Veracruz–Mexico City rail line. Inaugurated nine years earlier and celebrated as an engineering marvel, the line journeyed through the mountains, up steep grades, around sharp curves, through tunnels, and along nerve-racking canyons before finally arriving at the nation's capital.[2]

CAPITALINOS, CHICKS, AND THE ELECTRICSCAPE

Before the railroad came, the rough terrain made the journey from the country's largest port to its largest city difficult and memorable chiefly for its dangers. But the Veracruz–Mexico City line represented prosperity for Axford and hundreds of Americans and other foreigners who came before and after him. Axford, who had already made a name for himself in the United States with innovations in artificial hatching, looked to Mexico as an unparalleled opportunity for market expansion.[3] He thought Mexico was a ripe market for a marvelous glass chicken that laid bare the miracle of life. The glass hen made it to the capital in one piece and commanded a place of honor, sitting for weeks under a tent right next to the Presidential Palace. Curious capitalinos could inspect it for five minutes for twenty-five cents, half price for children.[4] The hen was both simple and amazing. Essentially a four-foot cylinder with a one-foot radius, it looked only roughly like a chicken. Axford had added embellishments that mimicked what a chicken should look like, but people did not pay to see the glass hen as a work of art. They came to see what was inside. Axford had installed an incubator near the bottom of the gallina and covered sections of its iron frame in glass, allowing visitors to see hundreds of chicken eggs incubating inside its temperature-controlled chambers.

Even in Mexico City, watching a chick be born was hardly a novelty. What made the glass hen something that every capitalino had to see was not only the number (800 chicks at a time) but also how electricity made it all possible.[5] The incubator's innovations included an electric thermostat, a turntable that held the trays of eggs, a special egg-turning arrangement, and a combined heater and electric current that regulated the temperature.[6] They all worked together to incubate the eggs and nurture the newly hatched chicks. Axford, a clever entrepreneur, had made sure to stagger the incubation so that births would take place every five to ten minutes: every paying customer could be sure to witness the cracking of shells. From there every newborn chick went to an electrically heated chamber where electrical bells announced when the temperature dropped so that the chicks required more heat from a lamp.[7] Another bell rang when the temperature was too high or when fresh air was needed. Guests could complete their visit by buying a three-day-old chick for five cents.

Capitalinos simply could not get enough of Axford's device. For months the glass hen attracted some of the largest crowds in the city, while other entertainment venues could only stew in envy. Managers at El Principal theater watched as a stream of people headed to the Zócalo, where they could see "the curious exhibition of the glass hen brooding thousands of eggs through heaters and electrical currents."[8] Theater owners may have hated it, but capitalinos loved the gallina. When they peered inside the hen, it was like looking into their own future. Would they too be warmed by electric lights? Would bells announce to them the change in temperature or the quality of their air? Only chickens enjoyed these luxuries in 1882, but who knew how soon in the future people might enjoy them?

In the exhibitionary culture of the late nineteenth century, Axford's glass chicken was only a small entry in the long history of artificial incubators. Some questioned its contribution to science, while others praised it. *El Monitor Republicano* recommended that the poultry industry take advantage of the glass hen. The city's English-language newspaper *Two Republics* considered the glass hen a clear example of the triumph of science over matter: it "completely dispensed with the role of the mother hen."[9]

For the next three decades, capitalinos harnessed the formula that made

Axford's device a success: the combination of electrical technology and the invitation to witness and experience events as never before.[10] Instead of seeing a chick hatch under a reflector, capitalinos witnessed the city, patriotic symbols, and civic rituals transformed under the shine of thousands of electric lights. No longer confined to a shabby tent next to the National Palace, the power of electricity moved beyond a few select demonstrations, through the city streets, and into a broader electric culture. The electricscape would come to mean more than the mere utilitarian aspects of lighting streets and workshops. In this period from the 1880s to the 1910s, electricity took on a larger meaning not only for capitalinos but for Mexicans generally, as well as for those looking in from abroad.

Twenty-eight years after Axford's chicken, electricity had clearly come into its own and offered a spectacle worthy of notice. The Metropolitan Cathedral's bell towers stood stoically on the north side of the Zócalo, dominating the cityscape.[11] A nighttime photograph taken in 1910 (fig. 2.1) captured them wonderfully: the photographer structured the scene so that the viewer's eye travels to the upper half of the image, where the two bell towers, profusely illuminated by incandescent lights, stand out against the night. The trees in the bottom half of the picture blend in with the black background, producing a phantasmagoric vision of luminous, floating towers.[12] Taken during the massive celebration staged for the Centenario (the centennial anniversary of the nation's independence), the photograph exemplified the Porfirian desire to showcase how much technology, investment, and modernization had improved the country, through highly theatrical public displays. This wish to glorify the changes, shared by government officials, intellectuals, and wealthy citizens, focused on the man they all credited for the change: President Díaz. Mexico City, as the seat of the federal government and self-proclaimed social and cultural center of the nation, was the natural focus of this patriotic drama. The city had to demonstrate, often in highly symbolic terms, the nation's commitment to order and progress—that is, Porfirian technocrats' notion of modernity. At the turn of the century, electricity was one the clearest demonstrations of progress globally.

Editors of the American journal *Illuminating Engineer* chose the photograph of the haunting bell towers for its November 1910 cover, an acknowledgment of Díaz's and capitalinos' mastery in portraying the Mexican capital as modern.[13] Mexicans must have prized the accompanying article's recognition of the city as one of the best-lit municipalities in the continent. The city's electric mantle, almost three decades in the making, placed it alongside cosmopolitan centers renowned for their broad tree-lined ave-

FIGURE 2.1. *"The Great Cathedral, Mexico City."* Illuminating Engineer, *November 1910.*

nues, belle époque architecture, and active nightlife. The Porfirian regime wanted desperately for Mexico to be seen as the cultural equal to any other nation in the world. While electric lights alone did not reach those heady heights, the lights did signify the nation's commitment to achieving such success. Throughout his tenure, President Díaz would keep sending out these signals. No advance would be too small to announce to the world. After all, Díaz wanted people like William Axford to take an interest, bring their ideas and capital, and help him usher Mexico into the arena of powerful nations.

In this chapter, I examine the exhibitionary culture that developed in Mexico City as it became one of the best-lit cities in the Americas. Government officials, business leaders, and ordinary citizens became versed in the importance not only of displaying profuse, festive electric lighting designs but of outdoing previous ones. Spectacular lighting—episodic in the early 1880s and dramatically tied to Independence Day celebrations—had become a regular, dazzling occurrence by the time of the Centenario. The evolution of artificial light served the practical purpose of extending commemorations after dark and aesthetically enhancing public spaces but also as visual evidence of the country's material progress.[14] As I demonstrate in the following discussion, this liturgy was not limited to civic ceremonies or official events. After the Centenario and its impressive displays, lighting would be more regular, reliable, and refined. Indeed, it would become common enough that capitalinos—and Mexicans more broadly—began to see it as an indicator of progress.

EXHIBITING THE CITY ELECTRIC: PRAISE AND DOUBT

Axford's glass hen was immersed in and capitalized on the visual culture of the second half of the nineteenth century. The cultural studies scholar Tony Bennett argues that exhibition institutions (museums, international exhibitions, arcades, and department stores) formed a complex of disciplinary and power relations during this period. Embodied in this exhibitionary complex was a rhetoric of power that manifested itself "not in its ability to inflict pain but by its ability to organize and coordinate an order of things and to produce a place for the people." Expositions, at home and abroad, thus transformed "displays of machinery and industrial processes, of finished products and *objets d'art*, into material signifiers of progress—but of progress as a collective national achievement with capital as the great co-ordinator."[15] This exhibitionary complex existed in Mexico as much as anywhere else in the world and was powered by electricity.

World fairs were the quintessential sites where the power of nations and capital manifested itself, compared itself to other venues, and secured the public's acceptance of the promises of science. For peripheral countries such as Mexico, the historian Mauricio Tenorio-Trillo notes exhibitions served as opportunities to form part of "the admirable group of countries that, sharing ideals, ambitions, and trends, advance together, led by progress."[16] Functioning less for the technical education of the working classes, fairgrounds became what Walter Benjamin described as "places of pilgrimage to the fetish Commodity," where the emphasis shifted from the process of production to the products themselves. Consequently, Benjamin argued, these developments transferred "the rhetoric of progress from the relations between stages of production to the relations between races and nations . . . [with]

the achievements of the metropolitan powers, invariably most impressively displayed in the pavilions of the host country." As promissory notes, the displays in these fairs vowed to improve social conditions like health, sanitation, education, and welfare and encouraged a vision of the future in which "the engines of progress would be harnessed for the general good."[17] Fairgrounds sought to inscribe and broadcast a universal truth: if life were to get any better or easier, it would happen only through science and industry.

Electricity lent itself to exhibitions. A cadre of men recognized this at an early stage and crafted their careers and reputations on it. Michael Faraday, for instance, made himself at home at the Royal Institution and the Royal Society in London, where he amazed the prestigious clientele with spectacular electrical wonders. Such demonstrations revealed not only the raw power of electricity but how it could be tamed and assimilated for consumption. The rise of international exhibitions provided them with an enormous stage.[18]

Electricity's profuse presence and symbolic value matured and, to a certain extent, coalesced on fairgrounds across the globe in the last quarter of the nineteenth century. Each new exhibition aimed to surpass its predecessor in terms of the number of incandescent lamps installed, lighting schemes employed, or the power of its machinery. World's fairs offered nations an unmatchable stage and an audience that came with high expectations: visions of a fully electrified world first emerged on the fairgrounds. Electrical exhibits served to expose, astonish, and often inspire individuals by the 1890s, making them aware of the potential of the invisible current.[19] Buildings dedicated to electricity, machinery, and manufacturing displayed a plethora of devices, introducing new applications of electricity. Spectacular electrical lighting at fairs at night produced wonder and awe. Thousands of incandescent lamps illuminated the grounds, adorned with electric fountains that transfixed visitors.

Hailed as the most magnificent world's fair of all time, the 1893 World's Columbian Exposition in Chicago embodied the promise of the Machine Age: electricity transformed the entire fairgrounds every night into a magical scene.[20] The monumental Edison Electric Tower, one of the main attractions at the center of the Electricity Building, rose high over visitors, while the electric intramural railway shuttled millions of visitors around the fair. Visitors saw an astonishing demonstration of electricity power and potential: every building was outlined in white bulbs, and giant searchlights swept the fairgrounds.[21] Those who spent a day and evening at the fair walked off with "a lasting impression, if not faith that there was an electrical device for every purpose."[22]

Closer to home, municipal authorities, business owners, social critics, and ordinary citizens contributed to, exploited, and sometimes questioned this exhibitionary culture. Electricity was at the heart of visions of progress at the turn of the century, so in questioning these exhibitions they implicitly questioned the use—and abuse—of electricity. Take, for example, Axford's glass hen. Capital periodicals criticized its residency in the city center, claiming that it risked becoming a mere spectacle and not a source of scientific education. A local agricultural and veterinary journal complained that the *vulgo* (uneducated masses) "cried out 'a miracle!' A miracle when what they had before their eyes was nothing more than the natural effect resulting from the actions of nature applied to live matter." Electrical bells gave the public the impression that electricity bred chicks, a misapprehension that Axford did nothing to correct. Not content to let the mistake go unchallenged, the editors felt compelled to run an article on "a mere artificial incubator that many were erroneously calling *máquina de sacar pollos* [a machine to produce chicks]."[23] To make a real contribution to the science of chicken incubation, the author noted, Axford needed a truly automatic device that eliminated human intervention every time the chamber's temperature rose or fell. Instead of just ringing an electric bell, the apparatus could close the heat vent through a diaphragm and open it when the temperature dropped, instead of having a human do so. A better artificial incubator, the article concluded, would be a real benefit for countries like Mexico, with its high egg consumption: chickens could then invest their energy in producing more of them.[24]

The city council was not listening to the reservations and outright criticisms expressed by some journals—especially the more scholarly ones. Its Markets and Expositions Commission recommended approving Axford's petition to exhibit his invention on the Zócalo, judging it "an undertaking worthy of support [given its purpose] to entertain and instruct the public."[25] Yet for all this boosterism and acclaim, Axford's exhibit did raise questions that were rarely spoken out loud but still haunted the spectacle. What was being replaced? What did this mean for the cycle of life? Between the nineteenth and twenty-first day of residency inside the glass chicken, the chicks could be seen cracking their eggshells and tumbling down, flapping and taking their first shaky steps in search of a safe refuge, but there was no brood chicken to run to. *El Monitor Republicano* achingly detailed how the chicks grouped together within hours of their birth, rubbing their frail bodies against each other, with many leaning their heads as if unable to support their weight. Visitors could "barely hear them cheep.

One would think chicks needed the mother's wing warmth, that they felt abandoned, orphaned."[26]

La Patria chided that the chicks "did not have the joy of hiding under the mother's loving wings." It was pitiful—and touching—to see the newly hatched chicks gather together, cheeping timidly under the crystal's heater. The reporter could not see past the chicks' orphanhood: despite the rice and breadcrumbs provided and the pleasant temperature produced by the heater, the chicks saw "no chicken clucking with hackles raised from maternal pride and eyes illuminated by the fire inspired by the love of a family."[27] No doubt science and electricity could do some things more efficiently, but the central question raised in these reports was whether that was an improvement.[28] Perhaps industry and science were not as wondrous as the fairs and exhibits made them out to be.

EXHIBITIONS OF INFORMATION

As exhibits rendered possible futures visible, nations became obsessed with measuring where they stood vis-à-vis other nations. Beginning in the late nineteenth century, the gathering of statistical data became imperative for governments. The gathering of statistical data relating to electricity became a significant endeavor, a barometer of the country's material progress. The minister of development, General Carlos Pacheco, an influential figure in the establishment of the Dirección General de Estadística (National Directorate for Statistics) and the deployment of the science of statistics, asked Alberto Best in 1882 to join the commission overseeing Mexico's contribution to the 1889 Paris Exhibition in the field of electricity.[29]

Best had graduated in 1886 from the Escuela Nacional de Ingenieros (National School of Engineering), specializing as a topographer and hydrographic surveyor, and would later serve as the chief of the Alumbrado Commission.[30] Upon graduation, along with classmate Rafael R. Arizpe, Best undertook an eighteen-month government-financed tour to inspect the most modern electrical installations in Chicago, New York, Berlin, Paris, and other cities.[31] The Siemens Company in Berlin markedly impressed them both, in particular its cable factory, which manufactured submarine cables and underground cables for electric lighting (then still a novel method of routing cable throughout the city streets). Open trenches around the city offered them an unobstructed view of the work that installing cables entailed. In Paris they not only took theoretical and practical courses on the subject of electricity but attended scientific conferences and joined the International Society of Electricians.

As a local expert on the subject, Best was an ideal choice to carry out an assessment of Mexico's electricscape. Arizpe remained in Paris as part of the Mexican delegation to the fair (and to attend the International Congress of Electricians, a meeting offering the latest development in the still new field of applied electricity).[32] Within months, Best published *Noticia sobre las aplicaciones de la electricidad en la República Mexicana*, which he unassumingly described as a compilation of "statistical data to give an idea of our progress in the area." With firsthand knowledge of the state of the field abroad, he informed readers that Mexico was likely to present "little or nothing . . . in the theoretical or in applied electricity" but "in terms of consumption, we are at par with average consumers." Throughout the book, he underlined electricity's foothold in Mexico: the countryside was covered with telegraph and telephone networks, electric-lit cities, factories, mills, and railroad stations as well as growing applications in mining, electrotherapy, galvanoplasty, and many others. Together, he concluded, "this evinces our country's excellent disposition to accept advances and the full confidence in a future of peace and progress."[33]

Best's celebratory tone would remain characteristic of the promotional tracts that the government published and distributed among fairgoers during the late nineteenth century, which painted the country "as a legal and political haven for foreign entrepreneurs."[34] A consistent and important participant in international fairs, the Porfirian regime understood early on that a precondition for economic development entailed "reforming Mexico's image before the world." As would be the case in future boosterish tracts, Best had no qualms about specifying his intended audience—business leaders and industrialists in the field—and his intention of positioning Mexico as an open market for European capital and goods to rival American interests. His assessment statistically demonstrated to all who received a copy at the Paris world fair that the Porfirian regime had steered Mexico "on the straightforward road to progress . . . and had positioned the country, in terms of material developments, at par with truly-civilized nations."[35]

Mexico returned to Paris eleven years later for the 1900 exposition with not one but two statistical compilations, differing in scale and perspective, that attested to the country's electricscape transformation. This time the task of assembling them fell to Rafael Arizpe, Best's colleague who had followed in his steps and joined the rising class of government bureaucrats. As a class, they had enjoyed a growing influence since at least the late 1870s. As the Porfirian regime centralized power, it needed people like Arizpe to keep track of what was transpiring throughout the nation. The simplest, most expedient, and—to a man like Arizpe—most rational way to do that was via statistical aggregation. He would be joined by others, like Best, who used statistical descriptions to paint a picture of electricity's penetration into to the nation.

Yet, as in all paintings, a certain degree of artistic license was involved. Best and Arizpe used these statistics to define what was "advancing" and what was "failing" in the nation's quest for modernity. These two men, sarcastically anointed as the "wizards of progress" by the historian Tenorio-Trillo, ultimately defined what modernity was in the nation. Tenorio-Trillo argued that members of the Porfirian exhibition team quickly recognized that "modernity . . . was a diverse and comprehensive set of techniques to be mastered and . . . a means of showing that their interests coincided both with those of the nation and with those of the modern civilized world."[36]

The use of statistics was perhaps the most essential of those techniques. One of the two documents that Arizpe produced, *Estadística de las aplicaciones de la electricidad* (Statistics of electrical applications), was a study of electricity's penetration on a national scale. Arizpe interpreted his role as a mere interlocutor who gathered and ordered data. Steeped in the positivist naivety of the late nineteenth century, which was repackaged by post-revolutionary technocrats, Arizpe imagined that his work—and especially his role in this work—was completely objective: he had simply compiled the responses to a questionnaire sent to state governments and private companies along with reports from the ministries of development and communications. But his fingerprints and a picture of the future that he wanted to see were visible throughout the report. To help readers, Arizpe offered several tables that showed the demand for electrical applications through "the simple and eloquent language of numbers" and how these numbers compared to other countries.[37]

Four chapters dedicated to the telegraph, telephone, street lighting, and electric power occupied the bulk of the report. Arizpe highlighted the rapid expansion in the areas of communications (telegraph and telephone) and street lighting, but he saved his most rapturous tones for the potential to harness the electricity in nature, citing the power held by "the unruly currents that plunge over our mountains."[38] This was no longer the modest electricscape drawn by Best. Arizpe cleverly emphasized not only the growth and established strength of Mexico's electric market but also its untapped potential. A few years later his words led to the construction of the Necaxa hydroelectric plant.

Despite the decade that separated Best and Arizpe's endeavors, the two shared a central goal: to show Mexico's modernity through the expansion of electrical applications and attract investment in that field. They were far from alone in recognizing electricity's importance. Displays at the Paris fairground in 1900 were classified in groups (such as education, raw and manufactured products, and agriculture). Electrical development was so important that French fair organizers granted it the status of an entire group (Group V) instead of a small class within a group, as it had been the case in 1889. For Arizpe, tasked with directing the Group V commission, Mexico had little to present in the area of advances in electricity compared to other nations, but it would send to Paris a contingent with a great interest in the innovations of others.[39] The nation would present itself in 1900 as an eager consumer of electrical investment with a promising future. The numerous waterfalls dotting the country, along with its overall growing economy, promised strong returns to foreign investors.

To ensure that wealthy foreigners got the message, Arizpe wrote a history of Mexico City's street lighting that combined historical narrative with statistical analysis.[40] He crafted a triumphant tale of the nation's dedication to electrical consumption even when Mexico itself had little to invest in the new technology. An introductory chapter on the global history of artificial lighting preceded a decade-by-decade analysis of lighting technology in the capital: oil and turpentine (1500–1899), gas (1830–1898), and electric (1879–1898). The 200-page study also included copies of the 1896 Siemens and Halske contract and its 1897 reform that extended electric lighting to the entire city limits. The account concluded with the argument that few cities of comparable size, particularly in Europe, had such rich and uniform visual design of arc and incandescent street lighting as Mexico City's, which, according to Arizpe, spoke highly of the capital's culture.

THE SHOCK OF FOREIGNNESS?

The railway had just passed a long path of colossal trees in the hacienda of Chapingo, on the outskirts of Mexico City, when Harriett W. Sherratt looked out the window and saw the glimmering lights of the capital in late 1898. The gleam intensified until it became a blaze as the train pulled into San Lázaro station; after miles of nearly complete darkness passing fields of maguey, Sherratt was mesmerized by the city's electric lamps. She was in the city only briefly before leaving for the rugged simplicity of the Teotihuacan archaeological site, where she spent the day inspecting the wonders of Mexico's past. At the end of the day, she and her traveling companions settled into the comfortable cushions of the carriage that returned them to the city. As she entered the brightly lit metropolis, she could feel herself leaving behind the Aztecs, Quetzalcoatl, and other gods and heroes of the past. She and her friends enjoyed a well-cooked dinner in one of the city's French restaurants. As they ate and talked, they broke out in gasps of relief. They had come to Mexico expecting rustic comforts, but, as "the children of steam and electricity" who enjoyed modern newspapers and cooking, they were surprisingly at home in the capital.[41]

Like Sherratt, foreign travelers to Porfirian Mexico City marveled at the contradictory levels of social development they encountered, especially in and around the country's capital. As "the children of steam and electricity," they were attentive to signs of material progress. Foreigners often encountered the capital's railways, streetcars, electric lights, and elevators—among other technologies—with surprise, describing them as elements out of context. These technologies were second nature to them back home, but here in "Old Mexico," as Americans in particular were fond of saying, they seemed out of place. This was the perception that Arizpe and other in Mexico worked so hard to overcome through the exhibitionary culture.

Foreigners came to Mexico in ever-greater numbers in the late nineteenth century. The expansion of railroad networks meant more reliable, cheaper, and safer means of transportation for eager travelers, especially Americans. These visitors—journalists, venture capitalists, photographers, and pleasure-seekers—presented an image of a "nation in a state of intense flux, constantly striving to live up to the Porfirian motto of 'order and progress' but also thwarted by a profound sense of backwardness."[42]

These visitors included writers like William English Carson, a British journalist who had earned a reputation as a controversial writer on social issues. He arrived in Veracruz for a four-month tour that did not make the work of Arizpe and other government boosters easy. Carson wrote that Mexico was a country of "picturesqueness, of romance, and of wonderfully strange contrasts." He noted the presence of technological advances (railways, telegraphs, telephones, automobiles, electric light and power) and conceded that the social milieu of the urban elite had been modernized, but he insisted that Indians with their burros, *cargadores* (carriers) with their burdens, and hidebound priests still populated the rest of the country. The old and new, he concluded, existed everywhere side by side. Carson found the transitions from the barbarically wild to the socially civilized jarringly sudden. He was not prepared for the nocturnal sight at the end of his railroad journey from Orizaba, in Veracruz, to the capital. After traversing "sandy wastes," maguey fields, and sprawling Indian villages, electric lights suddenly twinkled along the highways and an electric streetcar—full to capacity—flew by on its way to the city.[43]

Foreign travelers often encountered electric lighting in Mexico City with a disbelieving eye; they expected to find what they imagined was a premodern, even colonial place. Electric lights and trams did not fit in their imagined Mexico. In order to explain the presence of technology, their narratives often took a judgmental tone: the mere presence of electric lamps could not compensate for that backwardness and unhygienic habits of the lower classes dominating the city. The same logic became applied to the nation as a whole. Mexico City's ornaments of modernity did not eclipse what travelers judged to be the backwardness and material primitiveness of the countryside. *Mexican Financier* captured the contradiction when it described the capital as a man wearing a top hat, sandals, and a sarape. The top hat was representative of electric lights, paved streets, and railways, but it sat only upon the head. The rest of the body stood in stark contrast, clothed in sandals and a sarape.[44]

Travelers—and especially foreign travelers—seemed to see only a dualistic Mexico City in which modernity was a thin veneer over a fundamentally traditional society. Prolific American travel writer Nevin Otto Winter claimed that in no other city did ancient and modern customs present such an intimate contrast.[45] Men and women came into the city through streets illuminated by electricity but carried immense loads on their heads and backs instead of utilizing a wagon. Winter, like so many travelers to Mexico, compared the nation to the United States, the benchmark for progress and civilization in the Western Hemisphere. Winter, like others, came away with the impression that Mexicans appreciated mechanical devices and electrical inventions but preferred their traditional way of life. Winter considered them simple people, tied to their traditions, who believed that God intended to keep them happy.[46]

American author Fanny Chambers Gooch Iglehart, who lived in Mexico in the 1880s, promised her American audience a fairer account of the Mexican capital than those in the writings of those who briefly passed through the city. Yet even she described the capital as a scene of shocking contrasts. As bright rays from gas and electric lights shone across the wide streets in the heart of the city at night, she observed a large number of lower-class individuals sleeping along the Zócalo. They had only the sky as their roof, the stone pavement as a bed, and their rebozos or sarapes as cover. Despite living in Mexico for nearly seven years, Iglehart understood native people no better than Winter did. She found the indigenous people to be picturesque. Whether in their mountain homes tending to their animals, dressed to partake in a patron saint's day, or under the city's electric lights, their rags and ancient traditions clung to them and made them pathetic.[47] For Iglehart, electric lighting stood apart from native people, without granting them a veil of modernity by association. No amount of technology could overcome their limitations and render them modern. They were too strongly fixed by their class, their habits, and—in keeping with the thinking of the time—their race to enjoy improvement.

When foreigners looked around Mexico City, all they could see were contrasts. Familiar objects like a telephone or a steamroller in fifteenth-century streets became unusual sightings, according to Mary Elizabeth Blake and Margaret F. Sullivan in an otherwise "untranslatable country." Even electric lights gave an eerie look to Mexico City. They had always found electric lights unnatural, but they were "a thousand times more unnatural glittering above a people and a country as primitive as if the world were a thousand years younger."[48] Grace Owen Brown, an American traveling through the northern state of Sonora, found it preposterous to be writing by electric light in the middle of the countryside: the bulb shining next to her seemed out of place in the adobe cabin that she occupied. An oil lamp from biblical times would have been more appropriate for her surroundings.[49] Because of writers like these, audiences around the world came to see Mexico as outside the stream of time and hence incapable of attaining modernity.[50]

The Porfirian regime worked hard to overcome this stigma, but the portrayals by these foreign writers did serve some purpose. Being "backward" was a double-edged sword. On the one hand, the nation's pride suffered from its not being considered "modern" enough, but on the other, the accounts largely painted the country as a nation of untapped potential that required only foreign capital and expertise. Its exotic traditional portrayal lured tourists. As Jason Ruiz notes, the foreign writer's Mexico "allured because it represented a source of cheap labor and natural resources, a growth market, and a pleasure zone."[51] Mexican bureaucratic officials thus attempted to sell the nation's modernity—particularly at the controlled stages offered by international exhibitions, where manicured images delighted fairgoers—while sentimental accounts of foreigners sold potential.

Not everyone was swayed by the efforts to sell the country's potential. A series of sketches by W. J. Thompson in the New York–based *Electrical World* in 1887 painted Mexico as an untapped terrain with little potential. Although American investors remained curious about opportunities in Mexico, he argued, no one knew where those opportunities might be or what form they might take, Thompson argued. Knowledge of Mexicans' electrical needs and their purchasing capacity as consumers was even scarcer. While the series offered sketches of electrical prospects and possible remuneration in Mexico, it bluntly advised readers that there were "far larger chances of a profitable return in South, than in Central, America." After touring through Mexico, Thompson could not say whether the "outlook electrically is promising or the reverse," for it was hard to convey a picture of a country "so entirely different in all essential particulars from what we are used to in the United States." For him, Mexico was still the land of the Montezumas. Out of a population of roughly ten million, he claimed, "there are perhaps [eight million] . . . who practically do not produce or consume anything."[52] What could electricity do for a people so bereft of enterprise? Capitalinos' relationship with electricity during the late Porfiriato would prove Thompson wrong.

For technocrats to craft Mexico as a potential market for electrical applications required combating a long-held view of the country as stagnant and peopled by citizens averse to technology. Thompson reiterated a time-worn notion of Mexico as a land where people lived by a "do it tomorrow" ethos and time and labor counted for little or nothing. He called Mexico the "strangest, quaintest, and the most behind-the-age country." Unlike other foreigners who wrote that "it seemed all the time in Mexico as if it was the year 1640 and four o'clock in the afternoon," Thompson was inclined to place the country in a much earlier period. When he saw workers constructing a home or water porters lumbering down the streets, he could imagine that he was in the Holy Land in biblical times. He thus joined a cadre of foreigners who employed symbolic inversions that cast the Catholic country as ancient and primitive and its indigenous peoples as drunk, lazy, and dishonest in contrast to their own Protestant societies, which—by contrast—were implicitly seen as sober, industrious, and honest.[53]

Absent from this time-worm trope was the way thousands of Mexicans earned a living in the new electrified urban and rural landscapes in mines and factories that were as often Mexican owned and operated as foreign owned and operated, pursuing tangible or perceived opportunities. Despite his condemnation and his promises of higher returns elsewhere, Thompson believed that technology could take root and become profitable in Mexico, but only under American tutelage. He looked favorably upon the influence of American energy and enterprise but warned that "Mexicans themselves are not progressive. I am inclined to think that there will be considerable progress . . . whatever is done . . . must almost entirely [be done] by foreigners, and in spite of the opposition of the Mexicans." Hence Mexico required not just investment of foreign capital (which Thompson claimed, on little evidence, was already doing "much more in the way of improvements down that way than native" investments) but foreign expertise and, perhaps more importantly, the wherewithal to see the projects through. He noted that demand for electric lighting was concentrated in the cities and larger towns, not in factories, which generally ran day shifts only. He further cautioned his readers that the demand for electricity was concentrated around a demand for light. According to Thompson, Mexicans seemed incapable of seeing any broader use for the new power source: "I hardly think there is an electric motor in the Republic, and I do not imagine that there will yet be use down there for many."[54] Despite his ample warnings, Thompson's ill-informed series did not impede Americans' interest in Mexico. Indeed, by the time his series was published, all major American companies that produced electric material already had a foothold in the market.

While others, usually foreigners, represented Mexico as timeless and backward, Arizpe intended to sell Mexico's electrical future. It helped that the facts were on his side. Arizpe showed in *Estadística de las aplicaciones* that the number of electric street lighting installations in Mexico had quintupled since Best's compilation. Virtually every major city in the land had electricity, as shown in his tables of inauguration dates, the number of lights, annual service hours, and costs—useful data for those eager to enter the market. This was Arizpe's best chance to prove to the world that Mexico was ready for investment. He believed that development and progress would come through foreign investment, as envinced by the amount of capital invested in electrical installations for street lighting and private electric light that nationally reached 12,135,298 pesos.[55]

Despite its recent introduction, the nation's electric map had expanded rapidly. Arizpe, ever the optimist, claimed that important projects underway might improve his report's figures in the five months remaining in the century. The triumphal march to Mexico's progressive future was a question of not if but when. Arizpe recognized that the audience of *Estadística de las aplicaciones* was selective and that his writing was not elegant, but he hoped his facts would be received well among those, both local and foreign, who applauded and encouraged Mexico's advancement and progress.[56] The final goal of the information was not to be gathered, ordered, and forgotten in a government report but to find its way to an investor or else be exhibited on an international stage.

Arizpe's compilations evinced the technocrats' implicit double goal: to "make Mexico known in Europe [and] to make Europe known in Mexico."[57] The haste to collect, publish, and distribute evidence of the country's progress in the last years of the century to the scientific community emanated from the recognition that electricity, as a symbol of modernity and tool of modern life, would be well represented throughout the Parisian fairgrounds in 1900. The International Electrical Congress, held concurrently with the fair, would serve as an additional stage to broadcast Group V's work. As one of the most important technological gatherings at the dawn of the new century, the congress served as a platform for inventors to showcase innovations, propose new applications and theories, and debate them. For nations like Mexico, where manufacturing of electrical equipment was absent, glass cabinets offered the platform to showcase statistical compilations and historical accounts of technological development. Mexican officials thought that the attendees of the Congress were the perfect target audience for Arizpe's works.[58]

Measurement of the electrical industry's development and its promising future went beyond Arizpe's numerous tables and included two maps. Porfirian technocrats embraced cartography as vital to redefining the nation and capturing the attention of foreign investors. Surveyor maps, for instance, were seen as a paramount tool for breaking up public lands and communal landholdings and advertising them to potential buyers.[59] Mexican cartographers' efforts to define and describe the nation to their fellow citizens in the late nineteenth century were matched only by their efforts to craft images for international exhibits. For example, technocrats had ambitious plans for the maps that the nation would exhibit at the 1889 Parisian Exposition. They wanted displays that could document the nation's climate, urban life, and political and social order as well as the bureaucratic, financial, legal, and communication facilities that Mexico had established, which would guarantee the life and property of prospective immigrants and foreign capital. Embellishing the maps were illustrations that showcased the sheer number and quality of civic buildings and monuments that evinced the country's republican values and aesthetic achievements.[60]

The two maps that complemented Arizpe's work emphasized economic development and magnified investment opportunities. The first aimed to undermine the country's notorious image of inadequate transportation networks and widespread banditry that had been popularized between 1889 and 1891 by the famous writer and diplomat Manuel Payno in his serial saga *Los bandidos del Río Frío*.[61] The technological improvements visible in the 1900 map portrayed a well-connected territory. Besides the astronomical expansion of railroad lines that crisscrossed the national territory, this map showed the telegraph and telephone networks that made the territory into an imagined orderly and legible space. Maps like this permeated Porfirian visual culture of progress and efforts to portray Mexico at the turn of

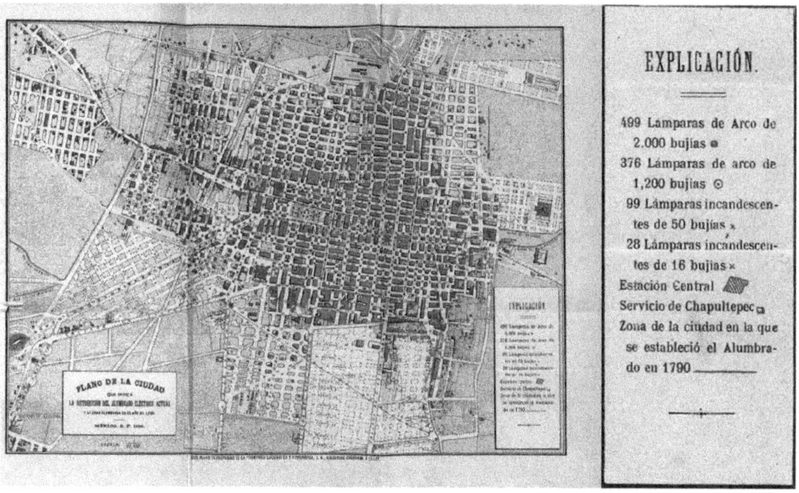

FIGURE 2.2. *1900 Electric Public Lighting Map. Arizpe,* El alumbrado público, *203.*

the twentieth century as technified space.[62] Mapping itself as a technological exercise was a tool to order, manipulate, and control nature.[63]

As a wizard of progress, Arizpe mastered what was intrinsic in the universal truths of progress, science, and industry: "form, style, and façade."[64] *El plano de la ciudad que indica la distribución del alumbrado eléctrico* (City plan showing the distribution of electric street lights) developed by the Compañía Litográfica y Tipográfica, complemented his 1900 statistical compilation. The expansion of the electric mantle over the city's lightscape stood as a material signifier of its modernity. To the observant eye, a well-defined and well-integrated territory—necessary attributes of a modern nation—as well as an inclusion of the capital's history of lighting technologies, evinced the steady and speedy steps that the nation and its capital had taken in the last decade. The capital's lightscapes no longer offered the palimpsest of various lighting technologies recorded by Best in 1889. The map offered an electrified urban space. The illuminated area constituted just under nine square miles, with a mixed street lighting system of arc lamps and incandescent lights.

The map (fig. 2.2) was at once a real and imagined representation of technology's expansion. It marked the area of the oil lighting system established in 1790 in red ink, underlining the long history of lighting the capital.[65] Arizpe also added more recent cosmopolitan accouterments, such as railroad stations to the northwest of the Zócalo and electric pumps on the opposite side, part of the latest effort to manage the city's historic floods.

The map revealed the recently inaugurated panoptic penitentiary and city slaughterhouse to the east—both celebrated modern additions to the city's infrastructures. To the west and south lay large swaths of undeveloped areas colored in green, but they would stay that way for only a few short years. By 1910 these areas would be neighborhoods for the well-to-do, part of an exodus of the wealthy from the city center.

Arizpe's accompanying comparative table of street lighting brought the city's efforts to illuminate its streets into focus. Although far from its Parisian or Berlin counterparts (with 11,028 arc lamps and 701,524 incandescent lights and 10,314 arc lamps and 229,858 incandescent lights, respectively), Mexico City still shone brighter in the night than cities like Brussels, which had 770 arc lamps and 47,371 incandescent lights despite housing two times Mexico City's population. Arizpe compared Mexico City exclusively to Europe for two reasons. First, Europe was still the benchmark of technological progress globally. Second, and more cynically, he hoped to draw the attention of Europeans and break the US monopoly in supplying machinery for the country's electric infrastructure.[66] Indeed, this was part of a larger economic policy of the Díaz government in which it craftily offered "some degree of preference" to Europeans investors to balance against the otherwise dominate American investments in the nation.[67]

A NEW LITURGY OF FESTIVE LIGHTING

The first dream of a centennial celebration of Mexico's independence began a full generation before the date. As early as 1880, Mexico's leaders understood the basic tenets of exhibitionary culture: the nation's most important celebration to date had to be big. Those most ambitious even dreamed of combining the event with a world's fair. Out of the efforts to make the centennial a global event came the drive to make Mexico City an ideal metropolis. Officials revised streets and avenues and embarked on a demolishing campaign associated with the "fight for nationalization laicization" (see chapter 1). In the place of what had existed before, "nationalist mythologizers" pushed beautification projects that included monuments memorializing liberal heroes.[68]

By the time of the centennial of independence in 1910, government officials, businesses, and ordinary citizens of the capital had developed a liturgy of festive electric lighting that became intrinsic to its patriotic ceremonial culture. Independence Day celebrations featured lavish amounts of electric lighting for many reasons: extending festivities well past dark; aesthetically enhancing public spaces; and signaling the country's progress.[69] Mexico City may not have become Berlin or Paris yet, but Díaz intended it to rank among the world's most advanced cities one day. This liturgy was not limited to civic ceremonies or official events. Capitalinos sought artificial lighting to coronate the country's patron saint, the Virgin of Guadalupe, and enhance private social gatherings.

As dazzling as the electrical spectacle was during the years leading up to the centennial, Mexico had a long history of adding flamboyant displays to public rituals. Oath celebrations during the colonial Bourbon period witnessed the presentation of ever higher, more numerous, and more ostentatious stages, arches, and monuments, opulently illuminated and bedecked with fireworks. The power of these displays was cogently articulated by an anonymous eighteenth-century author: "[T]he nighttime illumination forms the most eloquent Language with which one can persuade the People . . . the commoners of New Spain have such a passion for this type of nighttime Spectacle that it can form one of the tenets of their character."[70] Like the viceroy of New Spain before him, Díaz knew how to speak the language of the crowd, through light.

Given its value to enhance the spectacular quality of civic dramas, electric lighting became an essential physical and symbolic element in Porfirian public celebrations. Politicians at all levels of the regime recognized electricity's potential to enhance the theatricality of public civic rituals. Public ceremonies and processions were essential to the construction of the Porfirian commemorative culture. While the centennial itself was enormous both in importance and production and received a great deal of attention, it was far from the only opportunity for the government to speak—via lights—to the public. The openings of power stations, like the opening of railway and tram lines in the past, became ideal occasions to eulogize the virtues of material progress as well as the social order attained by the Porfirian administration.[71]

On November 12, 1904, Herbert Gibbs, associated with the London firm Antony Gibbs and Sons and now president of the board of directors of the CMGLE, welcomed President Díaz along with his cabinet members to the inaugural ceremony of the San Lázaro Power Station. The company's early efforts had been sustained. The new plant was the company's third power plant in the city and the replacement for the former gasworks at San Lázaro. In an eloquent speech, Gibbs declared how that new station represented the latest chapter of the Antony Gibbs and Sons confidence in the country's future. This unshakable faith, he explained, had arisen four decades earlier when the firm financed the nation's first railway line, the Veracruz. The firm had also worked on behalf of Mexico for the renegotiation of foreign debt and to introduce both gas and electric lighting in the capital.

Gibbs declared that if the CMGLE had been "outstripped by other companies which utilized better methods" (certainly referring to the CME), his company was resolved to carry out its operations "on a scale commensurate with the present extraordinary development of this country." Gibbs credited Díaz's leadership for the solid progress and substantial prosperity that he had witnessed in his travels throughout the republic. The government's stability and rectitude inspired "confidence . . . in foreigners," evidenced by "the eloquent fact that we have invested our capital in the plant." Gibbs then led President Díaz and distinguished guests, including Best and Arizpe, through a tour of the new power station. At the touch of a button, the president set a machine in motion, officially inaugurating the plant. After a sumptuous luncheon set up on the lawns, each visitor left with an album of views of the power station annotated with brief descriptions, bound in an elegant stamped cover.[72]

As in the power plant's inauguration a month before, electricity, progress, and Díaz's leadership were key themes for the festivities held to celebrate the new presidential term in December 1904. On the first night of the celebration, commercial houses and public buildings in the downtown area donned their ceremonial dress, creating a general feeling that one witness described as "heavenly electric lighting." A tricolor national flag made of incandescent lights on the Metropolitan Cathedral was visually marvelous, confirming Díaz's claim that he had put to rest older tensions between secular and religious powers. Capitalinos flocked to see this and the other displays.[73]

El Tiempo described the scene as a "torrent of light flooding the main avenues. . . . no one stayed at home, everyone came out to see the illuminations." The Zócalo dazzled attendees with what looked like "a succession of enchanted buildings."[74] The drugstore Julio Labadie y Compañía Sucesores outdid the rest, with rows after rows of incandescent lights framing the company's name, a painting of the president with trimmings, and the illuminated inscription "Pax" over the main entrance. The toast given by Alfred Chavero, a politician, archaeologist, and dramatist, placed Díaz in the pantheon of national heroes through an electric metaphor. The nation's history, he noted, could be compared to a teocalli, a Mesoamerican pyramid surmounted by a temple, typically expanded by building over prior ones. In Chavero's metaphor, the Aztec ruler Cuauhtémoc was the base, followed by Father Miguel Hidalgo y Costilla, who led the country's independence movement, and president Benito Juárez, representing liberty. Over all of them, Chavero continued, "you have built a temple to the Peace Goddess. Just as Benjamin Franklin stole the lightning from the clouds and Thomas Edison transformed it into light, Chavero praised Díaz, by saying, "[Y]ou have stolen the lightning of war and have converted it into the power force of peace."[75] At every turn of the celebrations, officials thanked Díaz for his leadership, with the implicit wish for his continued leadership going forward into Mexico's promising future.

Celebrations of presidents, religious figures, and national events brought out dazzling electrical productions, but the largest annual event was always Independence Day. By 1910 the celebration would outstrip the 1904 presidential festivities, but the first few electric displays were actually quite modest affairs. In 1889 the towers of the cathedral were lit up with a simple string of lights at the exact moment when the president proclaimed the Grito de Dolores (Cry of Dolores). The National Palace and the Ayuntamiento building, as well as the most important commercial houses in the business district, joined the cathedral under a luminous veil in 1893. Over the next seventeen years, improvements to incandescent lighting along with more accessible and cheaper spectacular lighting designs allowed more and more government officials, merchants, and private citizens to partake in festive lighting to show their modernity and patriotism.

The 1896 reenactment of the Grito de Dolores was especially notable for capitalinos. Díaz secured and installed the original Liberty Bell in the National Place in that year. This was the same bell that Father Miguel Hidalgo y Costilla rang in 1810 as he delivered his legendary call to rebel against the Spanish crown. Hundreds cheered in the Zócalo when President Díaz symbolically received the bell from General Sóstenes Rocha. Díaz eloquently declared that the bell was like a phonograph that preserved the voice of Father Hidalgo.[76] Through his friend Thomas S. Connery, Edison had sent Díaz the first phonograph to enter Mexico.[77] After hearing Edison's message, Díaz asked for a cylinder to record a response. He later visited Edison in New York, and they corresponded for decades. The president saw in his "good friend" Edison "a talented hero, a successful achiever who would later discipline the lighting snatched from heaven by Franklin."[78] Throughout his life, Díaz held Edison in the highest esteem, calling him one of the great benefactors of humanity, who created new sources of labor, welfare, and wealth through his work and intellect, symbolizing everything that the president wanted Mexico to become.

For the night of the 1896 Grito de Dolores, Díaz literally used the future to illuminate the past. In addition to likening the bell to a phonograph, he used his friend's incandescent lights to paint the bronze bell in patriotic green, white, and red lights, dramatically singling it out from the darkness. The patriotic symbol resembled a star when the lights were turned on, and a collective gasp escaped the crowd.[79] Despite the sublime incandescent lights, the much-awaited sound of the bell ringing was too weak to hear. Still, by all accounts, it was a rapturous moment. At 11:35 P.M., when Díaz stepped onto the presidential balcony overlooking the Zócalo, the multitude congregated below grew to such size that several individuals were reportedly crushed to death in the evening's commotion.[80]

Buildings flanking the Zócalo were bathed in light that night. Beginning with the rather modest attempt to illuminate the cathedral tower in 1889, each year the buildings along the Zócalo accumulated ever more elaborate lighting schemes. Not to be upstaged by the debut of the bell, not only was the cathedral lit up in 1896, but cascades of lights resembling colored fire poured down from its towers.[81] When the Liberty Bell's chime proved too weak to hear, the cathedral bells broke out: the square reverberated with effervescent cheers.

Electric lighting remained an expensive technology. Its presence at entertainment venues and private parties was a sign of distinction for ambitious elites. Some of the tension between those who had electricity and those who did not was so strong that it could be broached only by humor. A professor in one joke poses a question to his class: "What is the best insulator for electricity?" Without hesitation, a student replies, "Poverty."[82] Two people talk at a party in another joke. The first person says, "Gentlemen, have you ever seen such a miserable party? What annoying men, ugly girls, and such a scarcity of lighting, liquor, and pasta!" He declares, "I am leaving because this is ridiculous," asking the other person, "Would you join me?" The man sheepishly replies, "I would gladly do so, my dear friend, but I must stay; I am the owner of the house."[83] Electric lighting was more than a convenience. At the turn of the century, it was a marker measuring where individuals stood in Mexican society.

Mexico's exhibitionary culture was not monopolized by secular patriotic celebrations alone. A deeply Catholic country, Mexico spent time and energy revering religious themes and icons. For the 1895 coronation of the Virgin of Guadalupe, a group of devoted *guadalupanos* (followers of the Virgin of Guadalupe) wrote a letter to the Catholic daily *El Tiempo* to propose including all Mexicans in the celebration—from the millionaire who lights a magnificent mansion with bulbs to the tenant in the last corner of the *vecindad* (tenement house). The writers proposed that the directors of all Catholic colleges form a commission to organize the decorations of the city during the festivities. The commission would provide the means so that even impoverished individuals who wanted to place *farolitos* (little lanterns) at night by their door but did not have a cent to spare could do so. Funding for the commission would come from donations raised by the directors. Another subscriber to the paper made a second proposal with the same goal: Catholic college students could divide up the city by barrios and gather a list of individuals whose notorious poverty prevented them from at least hanging a lamp at each door or window of their home during the coronation festivities.[84]

The liturgy of light had a long history in Mexico. It went back to colonial oath celebrations, continued through Catholic festivals, and developed into the patriotic celebrations of the nascent state, encapsulating everything from modest lanterns to fireworks to electric lights. Like civic events, illuminations were a special component of religious celebrations. Governments, businesses, and individuals alike could participate and outshine one another. The *Álbum de la coronación* edited by the Catholic daily *El Tiempo*, for instance, detailed the extravagance of specific decorations for the festivities held for the coronation of the Virgin of Guadalupe in 1895. The most opulent decorations were found at the mansion of Doña Guadalupe Arango y Escandón de Escandón and her husband, Vicente Escandón, a well-known conservative capitalist who had bought a two-story house on the Plazuela de Guardiola on San Francisco Street next to the iconic Casa de los Azulejos.[85] The Escandóns converted the house into a modern palace with a three-arched portico, terrace, balustrades, and dog and lion sculptures. For the coronation, snow-white flowers forming the name "María" rested on the balconies that divided the colonnade. Three massive, electrically illuminated stars—two white ones and one in patriotic green, white, and red—shone from the front of the house. Artisans had wrapped green foliage bands around each of the columns that supported the entablature and the horizontal lintel and then placed electric lights over these bands, producing an enchanting effect, according to one report.

A house on Capuchinas Street had an extravagant design of giant electrically illuminated stars. Between these stars was an arch made of reeds, garlands, flowers, and elegant golden streamers. Above it all, in gently white light, the name "María" shone upon the entablature, illuminated by more than 350 lights. Witnesses found it magnificent and tasteful. The silent language of the decorations expressed the Catholic community's feelings loudly, but nonbelievers and liberals were conspicuously absent. Although almost the entire city donned its party dress, the *Álbum* noted, "Unfortunately, there are families in our society whose erroneous beliefs cast a dark shadow over us."[86]

By 1910 reports on the wonders of electric lighting had expanded from isolated articles to frequent publications that reached broad international audiences. Readers looked to journals like *Illuminating Engineer* to see what Mexico had in store. Like the audiences of international exhibits, they were eager to read about what could be done next with lights.[87] For months, coverage of the illuminations planned for September 1910 promised a city glowing in light. Page-long descriptions and maps of the planned lighting schemes complemented interviews with those responsible, helping to forge expectations of a potentially magical celebration. *El Imparcial* imagined that the regal lighting of the centennial would make the metropolis look like a fairy tale.[88]

When the festivities began, bright lights draped the city as never before. On September 1, a quarter of a million incandescent lights began shining, along with the ones already installed by private citizens and businesses, which brought the total to approximately a million lights. This staggering number of bulbs blazed within an area bounded on the north by Tacuba Street, on the east by the Zócalo, on the south by Cadena Street, and on the west by Humboldt Street. The other focal point for the civic festivities began at the Zócalo, running westward along Cinco de Mayo, San Francisco, and Avenida 16 de Septiembre toward Alameda Park via Avenida Juárez, a major artery connecting the city center to the Chapultepec Park and Castle. Along San Francisco/Plateros Avenue alone, seven arches colored the street a golden amber.[89] *Mexican Herald* described the city that month as an actual city of light crowning the festivities.[90]

Up close, lighted signs delivered a public lesson to the masses below.[91] Within the heart of the centennial city, the phantasmagorical Metropolitan Cathedral stood out. Its architectural beauty and imposing height overshadowed the National Palace and the City Hall on the other side of the plaza. During the month-long celebrations, ecclesiastical authorities joined the celebration, offering one of the city's most important Catholic symbols as background for the regime's otherwise liberal message. As one of the most important centers of worship in the country, the cathedral stood as a symbol of the Roman Catholic Church's power, both past and present. By embracing spectacular lighting, ecclesiastical authorities staked a place for it in the modern, cosmopolitan capital and even took pride in it. The Catholic newspaper *El Diario* dismissed the illumination of another of the city's ecclesiastical architectural jewels, the Basilica of Our Lady of Guadalupe, as being "in poor taste and ugly." Ecclesiastical authorities threatened a lawsuit against the company in charge of the installation, only to learn later that gusty winds had caused pennants to tangle up with the electric cables feeding the lights, made them go out. Given the accidental nature of the basilica's "dull appearance," ecclesiastical authorities dropped the suit and, with Christian humility, accepted their inferior display.[92]

In contrast, the Metropolitan Cathedral's lighting surpassed expectations. Frederick Starr judged it the most glorious sight on the night of September 15, 1910. The effect of the building against the night sky was "weirdly beautiful."[93] Starr's reaction was common; the Catholic daily *El País* noted that the prodigious spectacle dazzled the eyes of viewers. The sparkle and twinkle of lights looked as if a kind and aesthetic-minded genius had poured out the treasuries of Golconda, a region in southern India famous as a diamond trade center. In passionate tones, the paper compared the cathedral to Mount Vesuvius erupting in stars.[94]

The two towers that rose over the cathedral offered viewers a lesson in Mexico's history—as interpreted by the Porfiriato. The first tower served as a backdrop for signs reading "1810" and the word "Liberty." The second had signs emblazoned with the year "1910" and the word "Progress." Between the two towers passed one hundred years of Mexican history. In those one hundred years the liberty of independence had led to the progress of the Díaz presidency. Between the two towers, where a clock was ordinarily housed, organizers had stretched a lighted sign with the word "Peace." The language of illumination, so central to the Porfiriato, was simple and direct for the occasion: since independence, the nation had never enjoyed such peace or progress as it had under Díaz. His leadership would produce more marvels for the awed crowd.

Dwarfed by the cathedral, the National Palace nonetheless stood elegantly dressed in lights, ready to stage the main event of the night: the Cry of Dolores. Instead of serving as a canvas for a visual lecture, strings of lights framed each window and doorway on the palace. A sequence of lights in the shape of star-rosettes alternated along the structure's roofline. Eight thousand multicolored lights transformed the façade into a "splendor of lights" and highlighted the balcony, where the cry would be reenacted.[95]

Businesses also vied to participate in the capital's liturgy of light. The same logic that demanded competing nations to outdo each other in their displays demanded the same of competing businesses, on a smaller scale. Patriotic celebrations served as a stage to distinguish themselves from their competitors. San Francisco Street was particularly captivating. The strings of lights at each intersection resembled a luminous canopy. Starr, who had practically swooned at the spectacle of the cathedral, noted that shops and

FIGURE 2.3. *Frank Kahlo, "Mexlight Centennial Lighting." "Las iluminaciones."*

businesses on both sides of the streets competed in their presentations. Their designs, helped by the uniformity and lowness of the buildings, worked to concentrate the light cascading upon the street.[96]

Just as world's fairs used statistics as a barometer of progress, the city's newspapers attempted to establish a "who's who" in the business community based on candlepower. While the iconic department store El Palacio de Hierro led with 6,000 lights, La Esmeralda jewelry store came second with 5,000, followed by the 4,000 lights of the oil company Compañía de Petróleo El Águila. The cigarette company El Buen Tono came fourth, with an impressive 3,000 incandescent lights. As one of the leading promoters of spectacular lighting—and therefore under considerable pressure—the Mexican Light and Power Company was not going to be shown up. It fastened streamers to its three-story building and encased it in 6,000 lights (fig. 2.3). A luminous eagle crowned the building, standing out starkly against the darkness of the night sky, while luminous signs around it spelled out "Light and Power," "1810," "1910," and "Servicio a toda hora" (Service around the clock).[97]

Not only capitalinos could enjoy the magnificence of the Centenario. Picture postcards captured much of it and broadcast it around the nation and the world. Numerous American and Mexican firms had contributed to the "postcard craze" that was sweeping the United States and Europe by 1910. They were only too eager to sell images from Mexico City that September.[98] A handful of these postcards sharing the title "Las iluminaciones" documented the lighting designs of the National Palace, the Telegraph Office, the cathedral, and Mexican Light and Power's building, among others. As ephemeral witnesses of historical events, these postcards not only provide a reading of the past in their limited space but embody a sentiment, an impression, or an observation worth communicating. Even though they often traveled outside the city, they became part of the collective memory of the Centenario.[99]

"Las iluminaciones" was not just series of postcards in terms of exhibitionary culture. The author and photographer of the series, Frank Kahlo, was a German immigrant who began his artistic career taking pictures of the construction of the Boker Building at the turn of the century and quickly became a specialist in photographing buildings, factories, and machinery.[100] His timing was impeccable, for he opened his studio at the height of the Porfiriato, when the regime thirsted for images that documented signs of progress. As part of the many projects undertaken for the centennial celebrations, the Ministry of the Treasury, headed by José Yves Limantour, commissioned Kahlo to make a visual record of the constructions underway throughout the city. Kahlo responded with "Las iluminaciones," a confirmation of the role that electrical lighting played in the city's life. More importantly, this visual record could be exported around the world, which demonstrated how far Mexico had come. If modernity demanded pushing back against the darkness of the past, then the centennial celebration proved both literally and metaphorically how modern Mexico had become. Electricity was no longer just a novelty trotted out as a gimmick, as in the case of Axford and his glass hen. It was part of contemporary Mexico and, even more important, an element key to its future.

The electricscape was first planted with utilitarian lighting and germinated with the festive lighting of exhibitionary culture. From there it would grow ever more rapidly and widely, into homes, stores, workshops, and, finally, the very politics of Mexico.

PART II

| THREE |

Trapped under the Wheels of Modernity

A little after midnight in February 1902, Benjamín Gutiérrez, a driver for Mexico Electric Tramways Company (METC; a British-owned enterprise), spotted something on the rails and brought his tram to a sudden stop.[1] Accompanied by a few gendarmes with dim lanterns, Gutiérrez walked down the tracks. After a short distance, he came upon what appeared to be the remains of a human body: a set of human legs and, a few feet away, a torso with its ribs torn to pieces. A couple of meters farther along the track were scattered traces of intestines and finally, just beyond that, a severed head. Doubling back, Gutiérrez discovered a trail of remains that began near the Indianilla tramway warehouse and continued for about 200 yards. A police inspector ordered the remains collected and placed on a stretcher. The disfigured face made identification impossible, but small shreds of *manta* (white cotton fabric) adhered to the flesh and indicated that the victim belonged to the lower classes.[2] Details of what occurred proved as hard to assemble as the victim's identity. The victim appeared to have been struck in the wee hours of the night, then streetcars continued to run over the corpse, shredding it. At the Indianilla warehouse, the inspector examined tram cars servicing the Tacubaya line, where he found traces of blood and manta encrusted on the wheels of several cars.

The inauguration of electric traction in 1900 put Mexico City's transportation system on par with those of major North Atlantic cosmopolitan centers. Scenes such as the one Gutiérrez uncovered had become unremarkable a mere two years later.[3] While leaders and the mass of Mexican citizens celebrated modernity and believed that expansion of the electricscape could make their lives better, this sword had a second edge. Lighting at night, electric heat and power generation, and electric traction were welcome, certainly, but they all entailed costs. New technology required new ways of interacting with it. Some perished in the process, like the unnamed man above. As accidents took place on streets—most in broad daylight—they became public manifestations of the dangers of an electric metropolis.

The *tranvías eléctricos* (electric trams), locally called eléctricos, embodied both real and imagined notions of Porfirian modernity. Eléctricos made the capitalinos who rode on them feel as if they were as modern as any people in the world. Mexico City had as much to offer in this respect as London, Berlin, or Chicago. This feeling of having achieved modernity was seconded by city guides, literary works, and travel literature, which assured both Mexicans and a broader global readership—many of whom had read stories of Mexican backwardness—that Mexico City was no different than other cities in all important markers of modernity.[4]

Complementing the paved and lighted boulevards and avenues crisscrossed by telegraph, telephone, and power lines, eléctricos connected the city, from the crowded downtown to far-flung suburban areas and newly founded colonias. Providing a fast and comfortable means of transportation for a rapidly growing population, eléctricos allowed workers and merchants to reside in one municipality and work in another.[5] Eléctricos were central to the city's quotidian operation and the lives of capitalinos by 1909. *Electric Railway Journal* claimed that they entered "more intimately into the life of the citizens of the Aztec capital than probably any other of the improvements . . . introduced in the last 100 years." The journal was right; capitalinos depended on eléctricos for "every aspect of daily life": for entertainment, such as taking them to the bullfights; for spiritual goals, like participating in innumerable Catholic celebrations; for quotidian chores like shopping; and, ultimately, to take them to the cemetery.[6]

What is most important in the history of eléctricos in Mexico's capital is not necessarily the technological innovations that made them possible but the relationships between these electrified machines and capitalinos—that is, to use the language of cultural studies, their social reality.[7] Leading cultural and media studies scholars such as Roger Silverstone, Eric Hirsch, Metere Lie, and Knut Sørensen have reconceptualized the role of users in shaping technological objects. Their user-centered studies challenge the traditional distinction between production and consumption by reinstituting Karl Marx's claim that consumption *is* production. In this conceptualization, consumption is "not complete until users have defined the uses, meanings, and significance of the technology."[8] To establish user–technology relations, we must understand how machines became integrated into everyday life. In this reinterpretation of Marx, consumers are not passive agents and technology is not a universal standard.

We must see technological innovation through the eyes of those who witnessed it, including capitalinos riding the tram. For their generation, the idea of achieving a pace faster than a horse's trot was exceptional. Within a few years, however, what had been a day's journey became an early morning jog on this new technology. The tram brought them over distances that their parents would have found amazing. Even more remarkable, when the tram failed them (for example, someone was hit along the line), capitalinos felt that something essential—a right of transportation—had been taken away. Silverstone called this integration "domestication," which entails "a taming of the wild and a cultivation of the tame"; it is a process through which users and mediators transform a technology from "unfamiliar, exciting, and possible threatening things" into "familiar objects embedded in the culture of society and the practices and routines of everyday life."[9] Scholars of technology should conceptualize users not in terms of narrow user–machine interactions, a view that limits a priori the way individuals interact with a given technology but in terms of a broader set of social, cultural, and economic relations.

Of course, one generally overlooked aspect of those relations is the danger that technology poised. Like other thriving metropolises at the turn of the century, Mexico City became the stage of street accidents, resulting from the combination of increased street traffic, new technologies, and higher speeds in crowded urban spaces. As horrific as they could be (clearly, the accident that Gutiérrez stumbled across was gruesome), these accidents have much to tell us. The extensive documentation generated by accidents elucidates how eléctricos altered conceptualizations of time and space, body, self, and "other."[10] Once accidents are analyzed in terms of the victims' imagined race and class, their disposability becomes apparent. When streetcars struck a capitalino, riders were literally halted in their tracks (or the trams' tracks). Left with nowhere to go until the authorities resolved the accident, riders had ample opportunity to mull, debate, and discuss the merits of life in the city—particularly how they got around in it.[11]

Newspapers sensationalized accidents and officials tried to explain and devise solutions, but all such discussions touched upon public space, how it was configured, and how people were supposed to behave within it. There was a proper way to navigate streets: thus specific behaviors (walking, crossing streets, and social interaction in public spaces) needed policing. Disciplining the behaviors of users (drivers and passengers) and nonusers (pedestrians or bystanders) was, for the political, business, and social elite, paramount to a technified capital. Once tracks ran down a street, capitalinos had to act differently—even if that meant changing a lifetime of learned habits and cultural notions of how people lived together in the city—to accommodate the tram. To investors and those who ran the city, the tram was frankly more important. When people did not privilege the eléctricos and give them the right-of-way—that is, when they acted in an undisciplined way—they risked delays and shutdowns. In essence, they threatened the city's technological maturity. In official announcements, newspaper articles, books, everyday conversations, and a wealth of other forums, capitalinos negotiated how they should interact with the new technology. They had been forced to have conversations like this before, as when they discussed how much the suffering of arc lights was worth the light offered. As the electricscape grew and expanded, however, the stakes rose proportionally. At each level, these conversations dealt with relations among machines, bodies, and space.

I take to the streets in this chapter, transversing Mexico City aboard eléctricos to piece together how these machines transformed daily life. In particular, I search out the accidents that occurred all along the eléctricos tracks. Rather than offering a counternarrative of modernity—that is, questioning whether this new mode of transportation was worth the lives it took—capitalinos generally rationalized the violence that came with eléctricos. Modernity, for them, meant living with a growing degree of anxiety. Making eléctricos familiar took time. The domestication of this technology required a conversation that went beyond the hopes raised after its arrival, past the anxieties of accidents and the sensationalist coverage that fed them. It reflected the actual experiences of those who used the tram and risked life and limb while doing it. Celebrated both as a symbol of modernity and as a practical, necessary component of an industrialized Mexico, eléctricos were a double-edged sword. The way capitalinos came to handle that sword gingerly is a complicated story.

THE CONCERT OF MODERN CIVILIZATION

When Jacinto García arrived in Mexico City in 1909, he was mesmerized and wrote in lyrical tones of the capital's "beautiful and animated streets. As lively as Madrid, as urban as Vienna, and as populous as New York." A representative of the Argentine government, García was awed by all the accomplishments that business and government had achieved within the nation's cosmopolitan center, noting the city's "smooth cement sidewalks, impeccably paved streets and superb electric streetcars full of passengers." Along these same streets "hundreds of luxury and rental carriages, and thousands of automobiles form[ed] the concert of modern civilization and the backbone of human activity."[12] It was a hub of action and a wondrous spectacle for a businessman like García.

Like a musical score whose tempo is increasing, capitalinos could feel the pace of life in Mexico City picking up during the first decade of the twentieth century, yet the downtown streets' *prestissimo* tempo contrasted sharply with the largo tempo of "Old Mexico" still evident elsewhere. The shift between the two could be jarring. A 1909 *Electric Railway Journal* special report cautioned visitors to the city against snap judgments when they saw men leading donkeys straining under the burdens on their backs. Scenes like this fed into foreigners' narratives of Old Mexico and of having traveled back to medieval times, but in reality, the article cautioned, only the lower classes traveled like that. A closer inspection, the editors of the journal argued, would reveal that capitalinos rode some of the best trams in the world. Despite the nation's embrace of its past and traditions, Mexico City was a fully "modernized" ancient city.[13] These prophets of modernity saw no contradiction in the characterization.

Behind the acceleration in the pace of life were the eléctricos. Within a short period they ran along streets already crowded with horse-drawn coaches, mule-drawn trams (known as *tranvías de mulitas*), bicycles, and, in lesser number, automobiles (fig. 3.1). The city laid down its most important lines between January 1900 and the end of 1901. An astounding 70 percent of trams were electrical by 1908.[14] Yet the METC did not stop there. Besides converting existing animal-traction lines to electric lines, it continued to lay new, fully electric lines at an astonishing pace.[15] Although other companies managed horse-trams in the outskirts of the city, the METC ran only electrified lines.

Passenger traffic offers a telling picture of the competitive edge of eléctricos. In 1896 tranvías de mulitas carried over 21 million passengers across the city. A mere six years later—and only two years into full service—eléctricos carried twice that many. In another six years the new transportation was hauling over 70 million capitalinos. This dramatic increase was possible only because of both growing demand and supply for transportation. On the demand side, the metropolitan area experienced staggering growth throughout the Porfiriato: neighborhoods stretched out farther and farther. On the supply side, companies like the METC extended older lines and laid down new ones; improvements in electric traction allowed more frequent and faster service. Providers also found it easy to boost carrying capacity by connecting trailers to motorcars. Within a decade, traveling the tram went from a heart-pounding experience to a mundane part of life.

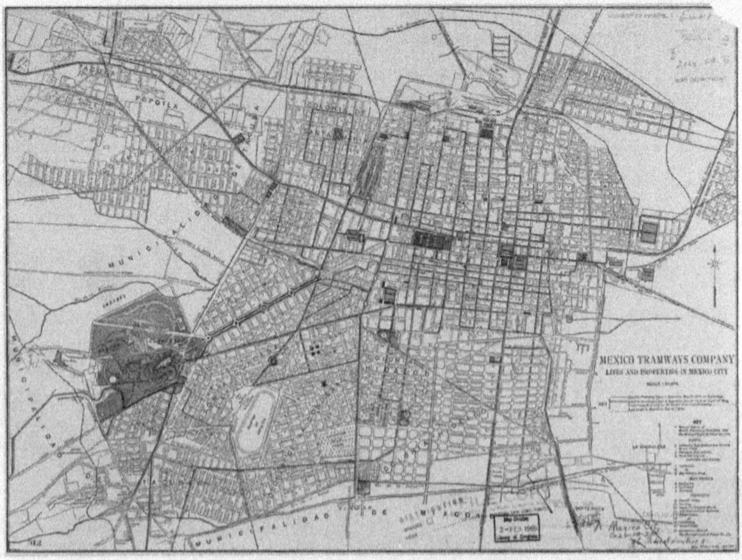

FIGURE 3.1. *Map showing the METC's extensive streetcar network and properties.*

Eléctricos offered not only faster and more frequent transport for daily activities like work and visiting friends but also several additional services. Tram companies devised other ways to get people on their cars. The METC offered a wide array of funeral services, with prices ranging from 3.75 to 140 pesos, which bought the grieving party both cars and decorations. For instance, 8,000 funeral processions traveled by tram every year between 1900 and 1906. During the same period, hauling general freight brought in twice as much income as funeral cars. After observing funeral corteges heading to the Dolores Cemetery, Grace Owen Brown noted that the endless processions made the avenue a true Via Dolorosa.[16] A black platform car capped with a cross carrying the coffin on a stand led the cortege, while enclosed cars carried mourners. Although the very rich still used private hearses, tramway funerals were common among the middle and upper classes but varied by degrees in terms of elegance and expense. A more expensive funeral procession had special cars and could be several blocks long.[17] While the wealthy could buy more, a clause in the METC's concession guaranteed access to its funeral service regardless of social standing: the corpses of paupers were carried free of charge. Class distinctions extended beyond the living.

While they made only a fraction of the money that funeral cars did, special cars rented out to prominent families for "trolley parties," as parties on trams became known elsewhere, were practically an advertisement for eléctricos. For one trolley party that took place on a summer night in 1901, families from San Ángel boarded a double-decker car flooded with patriotic green, red, and white electric lights.[18] Departing from the Zócalo, the car carried a musical ensemble that played throughout the entire journey, enlivening the ride. When the tram circled back to San Ángel, an orchestra awaited the passengers, who proceeded to dance the night away.

Yet for all their spectacle and the way they revolutionized lives in the capital, trams also reinforced and even exacerbated existing differences. As in the earlier horse-tram service, first-class and second-class carriages were distinguished by color: orange for first class and green for second class.[19] Mary Barton, an American who lived in Mexico during the early twentieth century, initially criticized her "haughty" friends for letting second-class cars pass while waiting for a first-class car. After accidentally boarding one, she came to understand her friends' reluctance. "The car was very crowded with dirty people," she wrote, "and the floor . . . indescribable."[20]

THE URBAN IMAGINARY: LEARNING TO "STEP ASIDE!"

El Universal captured the scene on the day electric trams hit the capital's streets:

> The bell went *tan-tan*! The cable rustled, the corner filled with people, and the eléctrico appeared. The gendarme removed the lantern placed between the rails; a carrier with bundle in arms overtook the tram; the shopkeeper from the store and cantina "El Puerto de Santander" jumped over the counter, at risk of having the merchandise wiped out, and opened his maw in astonishment; the *rapa-barbas* [barber] across the street left a parishioner half-shaved and walked out with his *matona* [pistol] in hand—everyone imagined he would revenge the victims of del Valle [railroad wreck]; the housekeeper took the opportunity to meet the eléctrico and to hang out with a cousin leaning against a post—fighting off the vapors of six *shomas* [maguey stalks] of pulque . . . [together they formed a] crowd of rebozos [shawls], *tápalos* [scarves with bright patterns], blouses, berets, palm leaf sombreros, hair rollers, and mops of hair.[21]

By all accounts, excitement rippled up and down Mexico City's streets as trams moved along, augmented by the sounds of bells, rustling cables, and the crowd's shouts. The occasion brought the gendarme, carrier, shopkeeper, barber, housekeeper, and even drunk relatives together shoulder to shoulder to witness the tram's first run.

Undoubtedly, it was a moment that would stay with them all. Yet, as the *El Universal* article confirms, life continued:

> Half a minute later, the tram takes off after a *tan-tan*! The odor of burned beans from the housekeeper's house reaches the corner; the barber's parishioner shouts from the door making a comical gesture; clients have left "El Puerto de Santander" without paying, and the gendarme's cape is missing! But in the end, all of them now know what electric trams are.

Those who could spare the time spent the morning going from corner to corner, awaiting the car's return, pausing only to lunch on enchiladas and pulque. With arch sarcasm, the paper noted, "That's how we progress!" Curiously absent from the crowds congregating on the corners that day were top hats, enormous hats with feathers, and parasols—items associated with the elite classes. Rather, by singling out trades, foods consumed, and clothing items, *El Universal* associated the disorder caused by the tram's arrival with the lower classes. The paper criticized the crowd; instead of "naturally" heading back to work, they remained unproductively—comically, even—on the streets with their "maws" opened.[22]

For newspapers like *El Universal*, the wide-mouthed amazement among the crowds that morning was proof of a conviction widely shared: the new transportation would put the city's unsophisticated masses at risk. Indeed, *El Universal* in particular predicted that "[t]he press will soon publish the thousands of accidents that will take place." These accidents, the paper predicted, would mostly involve the poorer and less educated classes, those least acquainted with technological advances. They would pay the costs of progress, as *El Universal* noted, with "many *pellejos* [lives]." Rather than question the safety of the new technology, the newspapers underlined the importance of personal freedom. Editorials advised those who valued their lives to carry out "trials with electricity": those who did not want to become victims of civilization could plainly "step aside!"[23] In a few years, with so many trams racing across the city, it would have been reasonable to counter: step where?

Eléctricos not only transformed the city's physical environment and altered its riders' conceptualizations of time and distance but also had profound repercussions on the way people thought about and navigated space. Urban scholars refer to this as the "urban imaginary": how "physical change is reflected and manifested in collective behavior, thought, fears, and desires that in turn transform the city spatially and culturally." Few urban transformations have had the impact that eléctricos did in Latin American cities. People thought of themselves differently after the arrival of eléctricos: now they could take jobs farther from their homes, go out shopping on a whim, or be part of crowd witnessing a spectacle much more easily than before. However, eléctricos also brought chaos and danger. Both intimate and public, the cars offered encounters with strangers and violence and injury for those not paying close enough attention—or just unlucky.[24] The tension between what eléctricos offered and their riskiness emerged frequently in reporting about accidents.

UNDERSTANDING TRAGEDY: DATA AND METAPHOR

Like other nations during the same period, the Porfirian state prided itself on being rational, modern, and above all scientific. *Científicos* (referring to positivist-oriented technocrats and close advisors to the president) earned a prominent place in the government and pushed for "measurable scientific methods."[25] Technocrats understood and deployed statistics as a technology of ruling. Recently established professions such as medicine, psychology, and criminology likewise relied on statistics to validate their practices as scientific and therefore above reproach.[26] For these newly minted professionals, data collection and analysis would be the key to reveal the origin of social problems—such as suicide and crime—and determine their solution.[27] These groups sought information on every aspect of daily life. The tragedies of eléctricos accidents were no different, for they understood and deployed them as a sign of modernity.[28]

Eléctricos fell under the authority of the Secretaría de Comunicaciones y Obras Públicas (SCOP; Ministry of Communications and Public Works). This ministry provided the data regarding the technology to científicos. The ministry compiled two data sets on injuries and fatalities in the years just after the arrival of the tram. The first data set, focusing on the years between 1900 and 1903, showed annual injuries from eléctricos accidents at just over 200. In the second set of years, 1904–1906, injuries grew to over 300 annually.[29] Even in 1900, the first recorded year, eléctricos were responsible for about 80 percent (227 out of 286) of railway accidents registered in the capital.[30] There is a considerable difference between the deaths reported by these sources, however. While the SCOP reported 29 fatalities for the first year—which had more than doubled to 68 by 1906—the federal courts that compiled the data claimed that almost half of the victims in 1900 perished. According to court data, eléctricos were responsible for over 100 deaths in the first year of service.

Official data had two shortcomings. In calculating the number of deaths, courts included only those who died at the scene of accidents, leaving unaccounted those who later perished from injuries. These numbers had to be reported by family and friends of the victim, but failure to report was a significant problem: unless they were so immobilized they could not move, many accident victims who had suffered otherwise severe injuries walked away. For instance, a tram full of passengers was headed back into the city from outlying barrios in 1903 when the driver saw in the distance a man on horseback traveling too close to the railway.[31] He blew his whistle, but the frightened horse panicked, crossed the tracks, and was hit by the tram, causing it to derail. While the tram remained on the railway with no lights on, a special car from San Ángel headed directly toward it, unaware of its presence. The collision was catastrophic. The second train ran headlong into the first, completely wrapping around it. Surprisingly, according to official records, only two injured passengers arrived at the hospital, but this number was misleading: most passengers chose not to reveal their injuries to authorities. They would have suffered needlessly by the time authorities actually got them to the hospital for treatment. Instead, they chose to leave with family and friends, who took them home to see to their wounds immediately.

In the fictionalized accident in Gamboa's novel *La llaga* (The wound), a coach driven by the novel's main character, Eulalio, collides with a car and an eléctrico.[32] Eulalio lies unconscious on the pavement, next to one of the coach's horses, until authorities come to take him to the hospital. At this moment, Nieves, his lover, makes her way through the crowd to him and yells, "No, not the hospital! The wounded man has a home; he has me!" In order to spare him the hospital, she lies and says that Eulalio is her brother. He is so badly injured that she cannot bear having him die alone in the hospital. Though this account is fiction, there were other real-life Eulalios, missing in the official statistics.

Despite these shortcomings, government data on injuries and deaths due to eléctricos accidents provide real insights. In their reports, officials recorded the victim's relationship to the tram: passenger, employee, or unrelated individual. Passengers suffered more injuries than employees from 1900 to 1906, but the real victims of accidents were bystanders or pedestrians. With 70 percent of all injuries, bystanders were—again, according to government statistics—the most likely to be hurt in a tram accident.

Over time, however, this number dropped, with bystanders constituting 60 percent of all injuries by 1903. Even more shocking is the number of bystanders who died from injuries. Already high from 1900 to 1902, averaging four deaths out of every five injuries, the percentage of bystanders killed increased in 1903 to 85 percent of all those who died in tram accidents in 1903 and rose to 91.2 percent the following year before leveling at just over 80 percent.

Those driving or riding in the eléctricos, ironically, were not the most at risk. It was often those who were simply near eléctricos who suffered. In practical terms, that meant that nearly all capitalinos were potential victims. With so many people around the tramlines, injuries and fatalities were inevitable (tables 3.1, 3.2). Table 3.2 compares tram injuries and fatalities with those of the three major railroad lines for 1904–1906. Notably, the numbers for the eléctricos closely resemble those for the Mexican Central Railroad, which had over 3 million miles of track compared to a mere 149 miles for trams. However, most of the Mexican Central lines were in open, sparsely populated areas, while the METC ran through the densely populated capital.

Physicians had a front-row seat when it came to appraising personal injuries. Some thought the official records were far too conservative, hiding the carnage taking place daily. Doctor Manuel S. Soriano, a member of the capital's scientific community and a prolific writer (who discussed everything from prostitution to potable water to the country's history of medicine), served as editor of the National Academy of Medicine's *Gaceta Médica*. Soriano declared that every day "capitalinos see accidents, derailments, *machucados* [crushed bodies], and disabled individuals . . . in the hundreds."[33] Tram service across the city was only a few months old, yet doubts about the government's accuracy had already arisen. The public could not be fooled about the scope of the problem: eléctricos were dangerous.

Five years later, Domingo Orvañanos argued that the annual number of accidents surpassed 600, a number much higher than the number reported by the SCOP. Orvañanos was also a member of the medical elite. He served on the Superior Board of Health of Mexico and had been elected president of the American Public Health Association (APHA, with members from the United States, Mexico, and Canada) by 1906. He published extensively on sanitation and public health and established himself as an expert on the relationship between tuberculosis and the climate. He also fervently believed in the efficacy of statistics, promoting the creation of the Section of Vital Statistics at the APHA. Called upon to compare the effect of eléctri-

TABLE 3.1. INJURIES IN RAILWAY ACCIDENTS

Year	*Railway company*	*Km in operation*	*Total passengers in all lines*	*Injured passengers*	*Injured company and govern- ment employees*	*Injured persons unrelated to service and operation*	*Injured indi- viduals*	*Ratio of injured passengers to total number of passengers*
1904	Mexico Electric Tramways	241.006	42,602,194	63	55	196	314	1:676,225
	Interoceánico	1,062,164	1,141,655	3	84	26	113	1:380,551
	Nacional	2,448,252	1,325,847	14	106	55	175	1:94,703
	Central	5,055,422	8,941,998	60	174	86	320	1:49,033
1905	Mexico Electric Tramways	241.54	47,757,440	88	48	218	354	1:542,698
	Interoceánico	1,020,693	1,481,743	25	63	24	112	1:59,269
	Nacional	2,527,344	1,885,069	31	95	45	171	1:60,808
	Central	5,056,028	3,371,281	70	198	92	360	1:48,161
1906	Mexico Electric Tramways	249.986	54,562,725	123	40	203	366	1:443,599
	Interoceánico	1,188,664	1,483,629	6	67	22	95	1:247,271
	Nacional	2,527,344	1,782,802	20	6	34	155	1:89,140
	Central	5,134,728	3,806,536	38	20	79	364	1:100,172

TABLE 3.2. FATALITIES AND INJURIES IN TRAM AND RAILWAY ACCIDENTS

		Deaths					*Injuries*			
Year	*Railway company*	*Passengers*	*Company and government employees*	*Persons unrelated to service and operation*	*Total number of fatalities*	*Ratio of fatalities to total number of passengers*	*Personal injuries*	*Ratio of personal injuries to total number of passengers*	*Ratio of personal injuries of passengers to total number of passengers*	*Ratio of number of accidents in general to kilometers in operation*
1904	Mexico Electric Tramways	2	3	52	57	1:21,301,097	371	1:114,831	1:655,418	1:0.650
	Interoceánico	–	20	13	33	–	146	1:7,819	1:380,551	1:7,276
	Nacional	2	14	18	34	1:662,293	209	1:6,343	1:82,864	1:11,714
	Central	13	34	48	95	1:226307	415	1:226,307	1:40,301	1:12,182
1905	Mexico Electric Tramways	5	6	46	57	1:9,551,488	411	1:116,198	1:513,520	1:0.588
	Interoceánico	2	17	18	37	1:740,871	149	1:9,810	1:54,879	1:6,850
	Nacional	3	22	26	51	1:628,356	222	1:8,491	1:55,443	1:11,384
	Central	14	33	57	104	1:240,805	464	1:7,266	1:40,134	1:10,897
1906	Mexico Electric Tramways	6	7	55	68	1:9,093,787	434	1:125,720	1:422,966	1:5,076
	Interoceánico	–	15	9	24	–	119	1:12,467	1:247,271	1:9,987
	Nacional	2	17	29	48	1:891,401	203	1:8,782	1:81,036	1:12,449
	Central	2	34	118	154	1:1,903,268	518	1:7,348	1:95,163	1:9,912

cos upon public health and safety, Orvañanos concluded, "Neither tuberculosis nor yellow fever caused the mortality rate [of eléctricos]."[34] The writer Benito Muñoz-Serrano, under the pen name "Khit," echoed Orvañanos's assessment in his poem "Barley on the Tail of the Dead Donkey!":

> I don't know, but there's no epidemic
> nor could I think of any infection
> that takes the lives that
> trams of this city take.
>
> Without counting the one-armed ones, the disabled,
> the one-legged ones we now see,
> and who owe their missing oars [extremities],
> to their terrible encounter with a tram.[35]

Physicians, called to care for the injured bodies of accident victims, also contributed to knowledge about the consequences of these crushing accidents. Dr. Tobias Nuñez addressed crush injuries from trams in a 1906 article.[36] Until recently, he declared, animal-drawn vehicles had rarely caused crush injuries.[37] When they did, a wheel generally passed over feet, causing bruises, contusions, or fractures, but only exceptional incidents resulted in immediate death. Severe crushing injuries appeared more frequently with industrial machinery, particularly *nixtamal* (boiled maize) grinders, which typically affected the upper extremities and could destroy everything from fingers(s) to an entire arm. The *machucamientos* (crushing accidents) produced by trams were more varied and severe. Common injuries ranged from the loss of one or several toes to amputation of legs and even, ultimately, death.

Nuñez was further troubled by the quality of medical attention that such victims received. He explored aspects of diagnosis and treatment that he believed other doctors had missed, including overreliance on amputation, lack of attention to internal injuries, and disregard of a traumatic shock state, which could make amputation more dangerous. Nuñez believed that the nature of the injuries called for a modification of traditional protocols for severely injured extremities: victims with such injuries should be in a separate category of trauma victims.

The language of medical experts contributed much to the social conversation about the dangers of eléctricos, but its vernacular was too esoteric and abstract for broad popular audiences. When writers wanted not just to relay the facts of an accident but to spark an empathetic note of dread and disgust, they typically relied upon images of food. One favored metaphor, perhaps unsurprisingly, was ground meat. When *El Popular* tried to describe the body of Carlos Alcocer, who died after three eléctricos ran over him—ironically, on the inauguration day of a *salvavidas* (a type of fender attached to the trams' front to lessen or prevent bodily injury). The paper could only say Alcocer had taken on the indescribable aspect of *albóndigas* (meatballs).[38] Albóndiga soup, a traditional Mexican meatball dish, called for ground meat rolled into small balls, often mixed with minced onion and spices: similarly, clothing, dirt, and the debris on the street were brutally mixed with flesh and organs when trams pressed bodies down and over the rails.[39]

A few days after Alcocer's death, Felipe Estrada also tested the salvavidas. According to *El Popular*, "although it was almost certain that the tram was going to make picadillo out of him, the safety device worked with such precision that, in less than a second, it tore his leg to pieces."[40] Picadillo, a beef hash or stew made with minced meat, potatoes, and fresh tomato sauce, had the same central ingredient as albóndigas: ground meat. Just as the cow was unrecognizable in the ground beef, the victims of particularly gruesome accidents were transformed into a wholly new substance. Indeed, observers even bestowed a new sobriquet upon the remains: *machucado.*

Other food metaphors were used to express shock at the frequency and severity of bodily injury. Because burst bowels were a common sight in the aftermath of accidents, guacamole became a favorite trope for such injuries. Guacamole at the turn of the century differed from our modern green appetizer. An 1895 dictionary of *mejicanismos* (Mexicanisms) defined guacamole as a stew of *aguacate* (avocado) and mole. Guacamole in this period had the reddish-brown color of mole, with the chunky consistency of smashed avocado. *El Popular* called on the common image of guacamole to describe the products of "disemboweling trams," wondering who would tire first: individuals being made into guacamole or the eléctricos that made human guacamole?[41]

In starting a conversation about the impact—and consequence—of eléctricos, *El Popular* in particular was prone to dismiss statistical parlance in favor of food metaphors that resonated more effectively among its audience. When estimating the number of such violent accidents in 1903, it stated "If the machucados' torn flesh were stacked up, it would be as high as the Peñon" (a rocky hill northeast of the city). Perhaps concerned that the image did not strike the right horrific chord in readers, the paper added another metaphor: "a pot of seafood salad and guts the size of the Zócalo." Instead of dry numbers, readers had a mental map populated with recognizable guideposts to visualize the extent of the trams' carnage. In another effort to calculate this carnage, *El Popular* compared the endless butchery on the streets "to Christians killed as rams in slaughterhouses."[42] References to slaughtering echoed the earlier sensationalist stories about the Peralvillo Slaughterhouse, which, after its catastrophic 1897 opening, was marked with unsanitary conditions, inefficient meat processing, and a dangerous work environment. The slaughterhouse closed abruptly after dailies announced that human blood had been "baptizing" its machinery.[43]

THE LANGUAGE OF DISCIPLINE

Mexicans celebrated electric traction as a demonstration of the nation's modernity, offering faster, more reliable, and more comfortable transit. To become faster, more reliable, and more comfortable, however, it demanded much from the people and infrastructure, including scheduling of regular stops; new traffic regulations; skilled conductors; and, most important, passengers who displayed "proper" behavior in and out of the cars and pedestrians who knew how to navigate the city's transformed streets. All these things had to come together to make the new system work, which often did not happen, as demonstrated by numerous accidents. Rather than look at systemic failures, writers and officials focused on particular figures: *imprudentes* (passengers or pedestrians who were run down because of their own risky behavior). As captured daily in gruesome descriptions, arbitrary and abrupt tram accidents threatened Mexico City's claim to be the Paris of the Americas, But the press and the courts shifted the blame from the system and mechanism of eléctricos to human error.

The contemporaneous construction of imprudentes and *motoristas* (tram drivers) drove the conversation away from the machine itself and toward irresponsible pedestrians and passengers, whose foolhardy *imprudencias* (reckless actions) caused accidents. Everyone knew of the dangers of machinery in places like mines and factories, but it was jarring to think that the same kind of trauma could happen in front of people's own homes. Stories about eléctricos were vague on details but provided clear protagonists and villains. These narratives largely centered on the human element of the tragedy and not the endemic risk in the tram system. Newspapers focused first and foremost on those who were run down and the tram drivers. Rarely did they have much to say about the machines themselves.

Across the political spectrum, from conservative Catholic newspapers to popular liberal papers and even the penny press, journalists mostly refused to stoke the flames of a Luddite-style denunciation of machinery but actively steered their readers' attention away from technical aspects and toward the actions (often mistakes) of individuals, primarily of the lower class. These journalists exploited brutal scenes of mangled and dead bodies and broken machines in the wake of the eléctricos. In large part this is because the papers knew that explicit depictions of injuries, blood, death, and suffering sold well. Shocking material was an essential component in the model of mass-circulation journalism that had developed over the previous decade. The new model moved from "the short-print-run, politically inclined press with specialized editorials and reflections directed at a politicized public" toward a press proclaiming political impartiality.[44] Professionalized reporters in this industrial press advocated impartial journalistic standards yet wrote literary and news articles designed to attract new readers and advertisers.[45] Then as now, people found it hard to turn away from a gruesome scene. But there was another reason for the sensationalism: the Díaz regime wanted people to blame an unfortunate moron or callous driver rather than a system of transportation that required enormous sums of foreign investment.

Díaz, his officials, and the wealthy investors he favored preferred a narrative of modernity that did not question its benevolence or inevitability. They wanted people to ignore systematic critiques and instead focus on the faults and failures of identifiable individuals. Fortunately for them, as the historical record shows, this was just the kind of story that people wanted to read. Sensational accident coverage concurrently documented and amplified such daily experiences as both thrilling and extraordinary.[46] This was a way to sell papers but also shaped how people thought about everyday life, the ills that came with modernization (industrial and railway accidents and perceived increased criminality), and rapid urbanization.[47] It made disaster both everyday and sensational.

Although the potential for accidents was widely known, accident reports fixated on tragic failures to reform personal habits. Responsibility for preventing accidents lay with citizens: commentators dismissed the predicted problems as nontechnical and outside the METC's ability to change. According to sympathetic reporters, the company had gone to great lengths in the months before electrification of its lines to ensure that "everything is under control, that design and operating procedures have been rigorously checked, and maintained by careful inspection. . . . everyone and everything is following carefully defined rules of mechanical, electronic and human performance, in which uncertainties are peripheral."[48]

Newspapers were far more eager to educate the public on the technical aspects of electric traction. *El Imparcial*, for instance, ran front-page articles detailing the machinery that powered and controlled the tram and its complexity, often accompanied by illustrations.[49] The paper's admiration for the new system was immense. It was concerned only about the unsophisticated masses in the streets: eléctricos would "contribute to many misfortunes if traditional ill-judged actions, such as trying to outrun cars, that characterized the public were to continue."[50]

Capitalinos were not ignorant: they simply had to acclimate to the new speed of city life. They shaped their responses to the introduction of eléctricos based upon their experience with previous transportation systems. Mule-drawn trams had begun to circulate in 1856 from the city center to Tacubaya. For the next forty-four years, residents became accustomed to the soundscape of "rolling cars, the heaving of wagons, the resonant trot of horses and mules, and the hoarse and strident sound from the cornet of mule-drawn trams."[51] Jasper Moses, president of the Christian Institute in the northern city of Monterrey, in 1907 lamented the disappearance of the mule-drawn cars that had once graced the streets.[52] He fondly recalled the drivers' sympathetic character: they would wait for passengers to walk (not run), sometimes a block, or more, so they would not have to wait for the next tram—whose arrival time was unpredictable. These drivers assisted ladies, carried their market baskets, and bid them farewell with a sweeping bow and wave of their cap. Moses was not simply putting on rosy-tinted glass as he recalled the past, for many of these practices had been part of tramway regulations since 1877. Whether or not these regulations were religiously followed is debatable, but courtesies such as those that Moses described were officially dropped when the regulations were amended for eléctricos.

Formal stops for mule-drawn trams had remained unfixed: passengers could flag a ride when they came upon one. To the dismay of many, eléctricos introduced regular stops, altering the relations between passenger and public transportation and interrupting what Anthony Giddens calls "the implicit assumptions that individuals share about how to 'get along' in their daily interactions with each other."[53] After electrification, the equality between passenger and driver was lost: capitalinos continuously complained that motoristas fully stopped only at their own will.[54]

Not everyone accepted this change. Conflicts between passengers and drivers over where to stop for pickups or drop-offs were common. A gentleman who requested a stop at the corner of Cinco de Mayo, hoping to cut short his walk in the rain, was told that the dead-end street had no stop. He would have to wait for the next stop on Plateros Street.[55] Word of this encounter reached *El Imparcial,* which noted that it was nonsense for drivers to argue that corners of dead-end streets were not stops, forcing passengers to walk an additional block in the rain. Sometimes the consequences were more dire. On August 8, 1902, a man aboard the first-class car of a Tacuba line eléctrico asked the driver to make an unscheduled stop.[56] When the driver refused, the passenger tried to get off but ended up under the wheels of the second-class car. To everyone's surprise, he walked away with only minor scratches and ruined clothing.[57]

Cómico fictionalized these encounters in "The Eternal Dispute":

—Hey driver, stop!
—Not until the next corner.
. . .
[Minutes later, at the station, the passenger approaches a supervisor.]
—Listen, you should energetically reprimand the driver; he made me rush across the street. We are sick and tired of bossy drivers; here everyone wants to be lord and master. . . .
—Yes, sir.
—Well, no, sir! Why disrespect the public? We have enough with . . .
—That's the order. They only stop at corners.
—*¡Qué orden ni qué canastos!* [An order, gosh!] Do you think I reside at the corner like gendarmes?
—Who knows!
—Hey, little gentleman, careful with the jokes, I'm not willing to tolerate them. Anyway, . . . you are a herd of savages; it seems incredible that after two years of electric traction we are still served by mules . . .
—Thanks, you flatter me.[58]

Cómico by then had an expert pen for accident coverage. Departing from its counterparts that catered to middle and upper-class audiences, the periodical had unswervingly characterized the locomotive "as a symbol of death and disorder . . . [and its] stories and poems often emphasized the violence, confusion, and complexity that resulted from technological change."[59] In covering tram accidents, however, the daily joined with its readers in lamenting the loss of the drivers' previous deferential customs.

Fixed stops enhanced the driver's power to the detriment of the power of class-conscious passengers. First-class passengers and those without the wealth to afford private coaches demanded the same deference that they had long enjoyed before the new regimented system. With elétricos, even middle-class riders recoiled at the "offending behavior" of arrogant and insubordinate motoristas. The regulations of animal-drawn trams reflected a society where generally lower-class drivers would defer and provide courteous treatment to their often middle-class and upper-class passengers. Thus, when eléctricos arrived, most passengers boarded them with the same assumptions and expectations that they had when boarding earlier public transportation. They were "acting neither blindly nor strategically, but simply going about their daily business, routinely reproducing the familiar."[60] With the arrival of eléctricos, the social world seemed to change with startling speed. It would take longer for the passengers' expectations to shift and become modern, in the language of the time.

In order to facilitate the speed of the new tram system, people would also have to "modernize" other aspects of their behavior. Traffic superintendent A. J. MacDonald noted that one of the significant impediments to the eléctricos traffic was the extreme politeness of capitalinos.[61] Before exiting, a Mexican "felt obliged to shake hands with everyone he knew in the car." This took even longer whenever there were "any particular friends present, [as] he stopped to embrace them." MacDonald, who began his career as a conductor in the Montreal street-railway system, moved to curtail this ritual by instructing drivers "not to wait for these civilities, and thus the custom was broken up."[62] The weekly *Ki-ki-ri-kí* seconded the superintendent's opinion, noting that "we [Mexicans] get lost in *excesiva galantería* [excessive politeness]" and agreed with the idea of restricting it "*just a little bit*."[63] No clear record exists as to whether the superintendent's instructions entirely eliminated these so-called ill-timed demonstrations of friendship. All the evidence, however, points to the increased stress that ordinary people would have felt as they navigated their obligations to their friends and neighbors and demands for efficiency of the tram system.

Efficiency was a byword of modernity, and an efficient transportation system was one where passengers boarded and disembarked quickly. Yet not all could do so, especially women, children, and elderly passengers. For decades, the 1877 Federal District Urban Railways Regulations had governed the circulation of mule trams for decades with article 9, obliging "drivers always to help the elderly, disabled, unaccompanied ladies, and children to board and exit the cars."[64] This rule suggests a social expectation that these individuals merited particular attention, but whether it was obeyed is a different matter. Within two months after inauguration of the eléctricos, *El País* claimed that they "inspire a real terror, especially among women, because, given the short stops, they could not board/exit *con comodidad* [comfortably]."[65] Long, full skirts and dresses made it challenging to get on and off quickly. María Peña had only put one foot on the running board in late

1900 before the driver took off, leaving her dangling from the vehicle.[66] When she fell, her dress got tangled in the car's side ironwork. Dragged for five meters, she walked away with only several minor injuries and her dress shredded to pieces. Peña's experience was common, and many passengers risked injury as they stepped on or off the running board. She was lucky, however: in many cases passengers ended under the wheels of the second-class trailer cars.

In contrast with the tugging of a mule, the trams ran smoothly and with such apparent absence of effort or expenditure of energy that pedestrians imagined they were not moving fast. Only when they tried to step onto the running board did they realize their mistake. Artillery sergeant J. Reyes Ortiz was one of those who made this miscalculation. He attempted to board a second-class trailer car after the tram started but fell backward and struck his head against a big stone on the side of the track. He died on the spot. *Mexican Herald* claimed that the sergeant, like many others, had miscalculated the tram's speed: "they tried to board the moving train only to discover that they had made a blunder." By then, as in the case of Reyes Ortiz, it was too late.[67]

The METC's new time regime demanded that passengers comply with the expectations of the company and not the other way around. People would have to change their behavior before the company would change its machinery, infrastructure, or policy. Indeed, the consequences for not adapting could be dire. The METC went so far as to include illustrations of the correct ways to enter and exit cars in *Guía de pasajeros a bordo de los tranvías* (Guide for passengers on board the trams), its 1913 publication on routes and timetables.

For the next twenty years, Mexican readers could routinely open their newspapers to hair-raising accounts of death and mutilation. Literally and figuratively, the crushed body of the victim lay at the center of the stories. Generally of working-class background, these unfortunate people either brought the injury upon themselves or else were the victims of the actions of a tram driver—himself a working-class figure. Two lessons were implicit in the stories. First, the upper classes were too smart and sophisticated to fall victim. Second, only dull or immoral people became victims. Readers knew to which of the two groups they themselves belonged.

Social critics, journalists, wordsmiths, and lithographers approached eléctricos accidents with a vast repertoire of textual and visual tropes. In their influential re-creations, they shaped the urban imaginary: the streets became stages for bloody dramas between trams and people, though only people of a certain class. Even as they drew on existing tropes, they simultaneously developed a specialized new vernacular. These writers and artists sought to capture the tragic, intimate details when the machinery of the tram impacted a human body and invented an ever-expanding list of descriptors to do so. They sought honest, modern journalistic depictions and prose that would draw their readers in again and again. In these stories, an initial impact could be a crash, trampling, or being knocked down. Bodies then endured being thrown from the scene of the accident or dragged down the tracks. The trauma sustained could range from fractures to crushing and grinding (machucamiento) to complete dismemberment.

Papers like *El Imparcial* blared out headlines like "Mother and Son Ground by an Electric," followed by a story in which María Encarnación and her son, Cosme Villeda, were reduced to a "uniform mass" after being struck by an eléctrico.[68] Their fate was exceptional, but the lethal injuries that the papers described entailed decapitation, bursting entrails, shattering and jumbling of entire bodies, and flesh and blood embedded in the wheels and undercarriage. The newspapers were singularly inventive in highlighting the tram's power source. Snapped overhead wires electrified everything they touched: humans and animals were *chamuscados* (burned), *achicharrados* (scorched), *carbonizados* (charred), *electrocutados* (electrocuted), and *fulminados* (struck down).

Capitalinos routinely read headlines like "Painful Scene: Elderly Man with No Feet," "Bloodbath by an Eléctrico: A Woman and a Baby Crushed," "Horrible Spectacle: Two Eléctricos Ran Over a Woman's Body," or "Hecatomb: Derailment and Alarm."[69] Through sheer repetition, people became inured to the shock of death and injury, particularly as coverage focused on the victims' failings. As accidents were reported daily, more generic phrases popped into circulation, such as "Another Victim of the Eléctricos," "Always the Eléctricos," and "Another Decapitated Individual in Chapultepec: Which Tram Killed Him?"[70] Muñoz-Serrano turned the news of the day into verse in his column "Actualidades" (News) and frequently articulated a counternarrative through sarcasm and ridicule. Naturally, tram accidents repeatedly entered his column. In "It Is Well Known!" he facetiously told his readership not to speak of tram accidents: "because that is to talk about the sea, and its recounting would be a never-ending story."[71] To calculate the number of accidents was a futile exercise, he argued, for the numbers never stopped growing.

Writers like Muñoz-Serrano attributed the suddenness and randomness of tram accidents in part to the consequence of uncontrolled machinery. Speaking on traffic circulation in 1901, the newspaper *Cómico* noted that "pedestrians go on sidewalks; wagons, automobiles, and beasts of bur-

FIGURE 3.2.
"Today's Victims."
Cómico, *April 7, 1901. Hemeroteca Nacional Digital de México.*

den on one side; and tranvías go everywhere."[72] Derailed trams amplified the perilousness of life in the capital. After a catastrophic derailment, Muñoz-Serrano exaggerated for effect the grim thoughts of passengers upon boarding a tram. As the verse "To What Extent?" noted, "A passenger knows well / that it is very likely that / his trip will end at the graveyard." The verse magnified the possibility of death: "[W]e'll have to travel with the holy sacraments!"[73] In Roman Catholicism, of course, the last rites are meant to purify the dying individual's soul to enter the kingdom of heaven. No Catholic would want to die without them. Thus, in this satire, boarding a tram meant risking one's life.

Perhaps even more than words, illustrations captured the chaotic scenes of mangled iron and bodies, the crushing injuries, and the horror of spectators. Two illustrations from *Cómico* showed the dangers posed by eléctricos in graphic detail. In the first, a bird's-eye view of an accident published in a section titled "Today's Victims" exposed the damage done by a snapped trolley wire and heavy iron wheels (fig. 3.2). Various body parts lie scattered around the car in a cloud of smoke, likely from the burning of the mangled flesh. The accompanying text comes from a fictional "son of Tajimaroa," a predominantly indigenous town in the state of Michoacán. The narrator lists the numerous calamities that he endured during Holy Week in the capital, beginning on a frightful Monday when "walking at mid-street, over ten trams were about to run me over, one after another." The follow-

FIGURE 3.3.
"On the Eléctricos." Cómico, *August 11, 1901. Hemeroteca Nacional Digital de México.*

ing day, he recalled "falling off an eléctrico and losing his sombrero. [I] was told a gentleman picked [it] up, but I haven't find out his name or where he lives."[74] Life in the city was full of dangers for provincials like him.

The next image provides a close-up of an accident in which a victim is pinned under a tramcar as one of its massive iron wheels is about to sever his head (fig. 3.3).[75] In the bottom left corner, the *Cómico* illustrator drew himself into the picture, rendering the scene autobiographical. He, like the tram's driver and another passenger, appears captivated at the sight. In the caption, the draftsman asks the victim to "wait up" so he can capture the details. Quartered bodies and shocked bystanders paralyzed in horror became stock choices for the visual language of crushing accidents because capitalinos could relate to the feeling of helpless passivity in the face of these accidents.

Like the *El Universal* article describing the crowds that first greeted eléctricos, reports of accidents also featured the spectators' reactions to the tragedies. Reports tried to capture for readers the facial expressions of those in the crowd as they witnessed a body contorting, smelled scorched flesh, or heard the victim's lamentations. The voyeuristic nature of the reporting allowed readers to become like members of the crowd at the scene, although papers like *El Imparcial* often used the phrase *gente de pueblo* (lower class) to describe the crowd, thus associating it with the lower end of the social order. Federico Gamboa fictionalized an account of a wagon caught between a tram and an automobile in the novel *La llaga*. Gendarmes arrived as the initial crowd formed, followed by street rogues looking to cause trouble. Behind them came a mass of pedestrians and gawkers, who talked endlessly of the unfortunate event but never tired of staring at it.[76] As horrible and even unimaginable as these accidents were, capitalinos clearly enjoyed them or at least felt bound to imagine themselves in them.[77]

According to the sensationalist journalism of the period, danger lurked overhead in the electric wires, underground in the buried cables: oncoming in every direction at high speeds. A 1902 article titled "Horrible Spectacle!" reported how a tram driver, Félix García, put the car he was driving back in motion after the Portacoeli stop as an unidentified man tried to cross the rails.[78] He first tried to warn off the man, ringing the tram's bell and blowing its whistle, before slowing down by closing the speed controller handle and engaging the *contracorriente* (countercurrent braking system). When García's car struck the man in his chest, he fell backward: his body lay "horridly wrapped around the rails." He stayed pinned against the tracks for forty-five minutes, only to find that the jacks sent to lift the tram were too small when they finally arrived. Desperate for a rescue, officials decided to back up the tram. In the meantime, the inevitable crowd congregated. According to one estimate, two thousand people gathered on Portacoeli and surrounding streets. The gendarmes had to push back against the crowd to clear the area where the man was lying.[79] At one point, the "immense *muchedumbre* [rabble] expressed their indignation and horror by shouting and throwing rocks" at the police.[80] Gendarmes responded with swinging clubs. The crowd's anger conveyed that bystanders could easily imagine themselves in the place of the victim.

IMPRUDENTES

Reports of scenes like the one at the Portacoeli stop were exceptionally rare; much more common was top-down coverage that encouraged people to accept accidents as inevitable or else as the fault of the "public." *El País* published a report, "Imprudent Acts," three months after the inauguration of the eléctricos, declaring that the public's own recklessness had caused the large number of tragedies and advising the public to "appeal to good sense."[81] *El Imparcial* later ran an editorial blaming victims' distraction or rashness for at least 90 percent of accidents.[82] This echoed official findings. Earlier that year, the Ayuntamiento had issued a list of preventive actions in response to the increasing number of accidents. Officers of the gendarmerie traveled on the platforms of eléctricos for two months to observe the sources of accidents. They reported that the imprudencias of the victims and drivers who disputed the eléctricos' right of way caused 90 percent of the tragedies.[83]

When commentators wrote about the failings of the public, they did not mean all capitalinos. They had in mind the lower classes, whom they saw as unsophisticated, too traditional, or even dumb, in the face of technological change. The expansion of Mexico City's electric mantle accelerated before the turn of the century despite the public's fear of electricity.[84] Even as late as the turn of the century, electricity was a mostly unknown, invisible, and generally frightening force for all but a select few at home and abroad.[85] In the summer of 1897 *El Popular* reported that Rufino Mora, *un hombre del pueblo* (a lower-class man), picked up a snapped power line on 4th Street of Real de San José and paid with his life.[86]

La Patria covered the death of Manuel Monsalvo in a similar tone in an article titled "The Boldness of Ignorance." The paper claimed that Monsalvo was a hardheaded man who did not know what electricity was and ignored what he heard about its effects. Friends had supposedly warned him about electricity: "the thing that runs through the lines kills people." Unable to see anything on the line, Monsalvo made a bet. When he grabbed the line, the "indignant current tossed his dead body aside."[87] The lesson in all these reports was that only the foolish or foolhardy had anything to fear. Naturally, readers of the papers imagined that they were neither and therefore had nothing to worry about.

Capital dailies blamed only the reckless behavior of the injured for any misfortunes. In 1900, for instance, a snapped telephone cable at the Calle Real de Santa Ana fell over the streetcar power lines and was left dangling above the ground.[88] Gendarmes were unable to cordon off the area to prevent pedestrians from approaching, and six individuals received severe burns for their reckless actions. *El Imparcial* vividly detailed how one lower-class man stepped on the crossed cable: the terrible shock brought him to the ground, howling. At the same scene, two women suffered grave burns before being rescued from certain death. The paper chastised the gendarmes for being unable to prevent such rash acts—"whether due to inadvertent danger or ignorance"—but in the end it was the victims who endangered themselves.[89]

Despite efforts to portray electrical accidents as solely a consequence of the ignorance of the lower classes, the reality was that accidents befell capitalinos across all class lines. A month before the tragedy at Calle Real de Santa Ana, an American electrician (twenty-four years old) had been working on a CMGLE post across the National Theater when he slipped and fell over numerous power lines:

> The cries; the contortions of the unfortunate individual burning alive; the sparking toasting flesh gave off a strong odor; a dense column of smoke from his clothing igniting. Together all of these horrible circumstances attracted thousands of spectators. Their faces expressed the fear and natural anxiety of the[ir] inability to aid the victim. Horror and desolation increased as time went by.[90]

The man writhed and contorted in a burning spectacle for over twenty minutes before CMGLE employees arrived. Delays meant not only suffering for the victim—and a greater likelihood of death—but time for crowds to gather, rumors to spread, and reporters to begin asking questions and documenting the scene for an even bigger audience. Once CMGLE employees arrived, they climbed ladders, cut lines, and disentangled the American electrician. In a gesture of humanity, the nearby Opera Hotel provided a mattress for the electrician's lifeless body. Like other papers, *El Popular* chastised the police, accusing them of being ill-trained and calling for more efficient and prompter aid after electrical accidents. Even after four years, however, not much had improved. In 1904 physician Domingo Orvañanos complained that electricians still took one to two hours to isolate snapped wires—plenty of time for multiple accidents to occur.[91] Given the increasing likelihood of electrical accidents, he emphasized the importance of sending doctors and other medical practitioners to precincts to train gendarmes on rescue procedures. The first step was to isolate the body if it was still in contact with the electrified line. In those cases, police should cover their hands with some thick wool, making sure the wire did not touch the face. He also recommended that gendarmes in particular areas carry rubber gloves and wooden-handled shears to cut wires. Finally, he demanded that all gendarmes be trained in artificial respiration. This way, aid would be immediate and suitable without having to wait for the electrician's arrival.

No evidence exists to suggest that gendarmes underwent such recommended training or received equipment to aid victims. It was simpler to blame the failure of an individual and move on. In the eyes of the Díaz presidency, the social needs of the people took second place to the demands of a modern commercial metropolis. For decades—and even centuries—the capital had been renowned for its street life and its liveliness in even the narrowest of spaces. But by the first decade of the twentieth century, as one historian claims, "sidewalks and streets were only to fulfill a role as channels for the movement of vehicles and pedestrians, while social life was to be relocated to parks, gardens, or inside houses, to the domestic sphere."[92]

Pedestrians competed with other pedestrians, horses, wagons, and a growing number of automobiles for scarce space on city streets in the first decade of the new century. This problem was only exacerbated by eléctricos. Competition became especially fierce in the city center. After a parade on San Francisco Street, Mercedes Godoy, the daughter of a Mexican diplomat, pointed out that only three cars could fit on the narrow street. With a few notable exceptions, this was true of most of the streets in the commercial and business center.[93] The limited space at intersections was particularly challenging for the METC double-deck trams, which had to use reverse curves in some places to round corners.[94] Ethel Tweedie, a prolific English travel writer, vividly captured the issue of the narrowness of some streets in the "strange performance" she witnessed: "four men, each holding the end of a pole, trot along with a heavy wardrobe suspended therefrom . . . the traffic considerably congested by tram-cars, the width of a wardrobe passing along does not tend to make matters less difficult."[95] In this battle for space, the authorities granted ample room to eléctricos, but "a city that calculated everything," a critic noted, "apparently forgot where passers-by would go."[96]

The eléctricos exacerbated pedestrians' peril. In one satirical piece titled "A Gift to Provincials: Rules on How to Behave in the Capital," a writer offered useful advice on how to avoid dangers in the city. Visitors were always to walk in the center of the street; otherwise, they would run the risk of being knocked down by the carriages, trams, and other "objects of destruction."[97] Other satires pointed out that the visitors' situation did not improve on sidewalks. A letter by a fictional provincial character named Pantaleón shows his envy of his friend Antón for not having to suffer Holy Week in the metropolis—a fantasy of wonders only in the imagination of provincials who never experience it.[98] Pantaleón finds himself stuck at mid-street despite having arrived a few days before, unable to step to any side to make his way to a hotel. He has learned that large trenches, which dwarf the ditches he knew for irrigating wheat back home, are for sewage. The previous day he saw workers finish a sidewalk, only to find to his astonishment that it was torn up the following morning because they forgot to lay electrical cable. While the sidewalk was being repaired, a new crew of laborers with shovels and crowbars that arrived to lay telephone lines was in turn interrupted later by those sent to lay sewage pipes. Clearly, the city was under constant construction: the focus of their work was the sidewalks, where pedestrians were supposed to move safely.

Streets were also sites of commerce. Authorities frequently reminded people that stopping in the middle of the street or gathering to obstruct the circulation of vehicles and animals was prohibited, but they faced an uphill battle.[99] One problem was that many people earned their living on the streets by selling food, drinks, clothing, and second-hand items as well as engaging in waste scavenging, prostitution, lending services, advertising, and public entertainment. As the historian Mario Barbosa shows, these activities not only went unregistered in official labor statistics. When officials said anything about them, they dismissed them as nonproductive and condemned those who earned a living on the streets as "an amorphous, frightening mass of insignificant customs."[100]

Since the late nineteenth century, municipal reforms had targeted street vending in particular and "moved, removed, and otherwise marginalized [many vendors] from a legal means of making a living."[101] These vendors' best chance for a sale was heavy foot traffic. They customarily set up their booths in crowded areas of the city, ideally on sidewalks but also on streets close to railways. Female street vendors from mostly indigenous or peasant backgrounds—commonly referred to as "Marías"—sold food in makeshift tents or *petates* (palm straw bed mats) that municipal authorities blasted for "invading" sidewalks and roads and slowing or blocking traffic.[102] According to Susie Porter, Marías countered their critics (and their frequently sexist undertones) by emphasizing their right "to inhabit the public sphere and defied being characterized as immoral due to public activity."[103]

Working-class men had nearly as much trouble as the Marías. *Cargadores* (haulers) also experienced infringements on their ability to work. Considered an obstacle to smooth and fast circulation, cargadores drew a great deal of municipal wrath. Authorities tried a variety of measures to push them off the streets, particularly in the crowded areas of the central district. The battle between the city and the cargadores, like the battle with the Marías, was at best a draw. By 1904 the Federal District government allowed cargadores along the Plazuela de la Concepción as long as they left space for pedestrian traffic.[104]

In this modern world of trams whizzing by, even the mere action of walking down the street, which capitalinos had done all their lives, drew attention and criticism from social commentators and authorities. With complete sincerity, a 1903 *El Popular* article titled "Racial Vices: When Will We Learn to Walk?" claimed that there were "certain peculiar customs . . . true vices of education, among which was the harmful, condemned, and frequently hazardous and residents' lack of skills to walk the cultured metropolis, frequently with serious consequences." While the physical aspects of the city had been rationalized, the author lamented, those who walked within it remained artless and immoral. Honest, if simple, citizens walked open-mouthed, looking at everything but their path. Others, like the *mujeres del pueblo* (lower-class women), "carrying the indispensable *bulto* [sack] containing mostly things harmful to other people's clothing: baskets with all sorts of food or drinks. They march in God's hands, not giving thought toward others, and at the slightest distraction . . . ram into pedestrians, striking them with baskets."[105]

Distracted individuals abounded: pedestrians had to watch out for street readers immersed in their books, lovers embracing right in the middle of the sidewalk, and the unthinking masses when stepping onto the street. Such constant ineptitude was the source of calamitous accidents. The only solution was for the paternalistic hand of the regime to step in. According to the article in *El Popular*, "we are *un pueblo niño* [a young people] needing all sorts of basic training and, like children, will thank imposed healthy customs that save us from vicissitudes and difficulties."[106] The author ignored the way the paternalistic authorities had contributed to the problem by prioritizing how the limited space in the city could be used and by whom. Instead, the problem lay in some intrinsic premodern characteristic of capitalinos. According to this logic, the greatest obstacle to modernity was the Mexican people.

The author of the 1903 article shared a conviction with many scholars of the period that distraction in an urban environment was a real and growing problem worthy of study and policy reform.[107] He found a sympathetic ear in Guillermo de Landa y Escandón, governor of the Federal District, who issued measures in 1905 to regulate pedestrians.[108] People were "not to stand idle" on Plateros, San Francisco, Independencia, Coliseo, Refugio, and Tlapaleros Streets, "interrupting the traffic and bothering the families and señoritas that circulate through those crowded avenues." The police were also under no circumstance to allow "*lagartijos* [dandies] and bums to stop on sidewalks, nor any individuals who stopped before shopping windows; everyone was to keep on moving."[109] Violators were to be taken to the district's precinct. Not everyone was impressed. The weekly *Gaceta de Policía* mocked the effort, criticizing it for ordering residents "to circulate, walk, move . . . to break our traditional custom to stand still."[110]

How could anyone take the effort seriously when it made the very common practice of standing into a crime?

Behind all the laments, criticisms, and reforms lay one simple truth: eléctricos had priority over animal-drawn cars, coaches, carts, and, most importantly, pedestrians on Mexico City's streets.[111] Article 15 of the city's 1900 traffic regulations stipulated that the eléctrico had the right-of-way when the bell rang, so the sound of the bell became a powerful signal. Whenever a crowd of people or a tangle of traffic blocked the way of an eléctrico, a police officer was to step in. Everyone—vehicles, pedestrians, even passengers—was obliged to obey him, and the tram would get through. Motoristas and gendarmes were empowered to detain, fine, or arrest individuals who ignored the eléctrico's bell and violated the its right-of-way.

TRAVELING AMONG THE UNSOPHISTICATED MASSES

According to the press, imprudentes were members of the lower classes, indigenous, or provincial, and the lines were often blurred. For example, shortly after the eléctricos arrived in the capital, *El Imparcial* reported that in many cases victims had attempted to outrun the eléctricos or refused to get out of their way despite the drivers' frantic whistling and warnings. The paper identified these careless or indifferent people who got struck down as members of the lower classes. The article echoed the contemporary elitist view that the victims of tram accidents were individuals who found themselves in "a deep state of mental engrossment and digression, so uniquely found in the country's indigenous and poor classes." Sometimes, the report added, the problem lay in people's poor spatial choices, such as forming crowds right on the railway lines or "mothers in barrios allowing children to play over the rails, to say nothing of the drunks who fell under the cars' wheels." It even censured blind and disabled individuals whose "frightened guide dogs flee away, leaving their owners under the wheels."[112]

The report saved its most vehement condemnation for lower-class men who sought to prove their valor by dodging cars, hesitating until the last minute to avoid a tram or otherwise zigzagging among coaches and eléctricos to capture people's attention. All these negative habits needed reform and correction, *El Imparcial* declared, for only then would the city "succeed in reducing the accidents and many other *males* [wrongs]."[113]

If a picture is worth a thousand words, engravers had a great deal to say about crushed victims. Their illustrations depicted victims as ignorant or unaware of the force and speed of eléctricos. They often appeared distracted, simple, and naive, without fear or respect for the machinery they were confronting and thus ultimately responsible for their own fate. El Buen Tono, a leading cigarette manufacturer, contributed to this visualization. With its strong record of developing advertising novelties, the company was at the forefront in making a spectacle of city life and creating an urban mass culture.[114] The company established its own lithograph shop—an achievement that few competitors could replicate—to print packs, labels, wrappers, posters, flyers, and *historietas* (cartoon strips), beginning in 1904. Juan Urrutia joined the company in 1899 as an artist and lithographer and worked his way up to second-in-command of the lithograph department, where he led the team responsible for the design of the historietas. A staunch urbanite himself, Urrutia naturally ridiculed recently arrived peasants. His early historietas focused on love stories, tram accidents, provincials in the city, and health. However, he also questioned technological innovations and scientific discoveries.[115]

One historieta follows the adventures of rancher Timoteo López in the capital (fig. 3.4). The famous equestrian statue of Charles IV of Spain in the first panel squarely establishes the setting as López traverses the city's streets in boorish, open-mouthed amazement. At the bullring, Don Tancredo leaves a deep impression on the newcomer. Tancredo was a real-life bullfighter who stunned audiences by donning a white robe and climbing on top of a small white platform in the center of the arena. There he stood like a statue, with his arms crossed over his breast. At the blare of trumpets, a heavy wooden door opened. As one impressed witness described the scene, "a huge black fighting bull rushes into the arena, bellowing furiously. Its bloodshot eyes look around for a victim, its long, pointed horns dig into the ground . . . [and] stamping fore hoofs raise clouds of sand and dust."[116] Excited crowds roared with anticipation. Once the bull perceived Tancredo, motionless, spectators imagined the unfortunate bullfighter "tossed to the skies, spitted, crushed, transformed to a bleeding lump of flesh," but the matador did not flinch. He remained calm, his eyes fixed on the bull as it rushed toward him. Unexpectedly, the charging animal would slow down, "standing still before [him], as though hypnotized, [it] noses him repeatedly, and looks restlessly about . . . until . . . it turns its back to the man and glares around after other enemies." The audience's thundering applause would finally break the spell.[117]

After seeing Don Tancredo, López sets out to replicate his performance on the city's streets. He enters a typical street, where he finds threats familiar to capitalinos: a rabid dog, a rickety *calandria* (carriage), and a

FIGURE 3.4. *El Buen Tono "historieta."* El Mundo Ilustrado, *July 24, 1904. Hemeroteca Nacional Digital de México.*

furious cyclist. López sets his sights on an approaching eléctrico. Once on the tracks, feeling scared, he lights a Primores cigarette to gain courage. To the bystanders' astonishment, the approaching eléctrico, which has left a trail of destruction, comes to an abrupt stop. The caption announces a prohibition preventing drivers from running down smokers of Primores. Satisfied that he is the equal of Don Tancredo, López sets out for his village with a box of Primores fastened to a mule.[118] Any sophisticated reader of the historieta would have laughed at the bucolic figure of López and his dumb luck. Those, like López, who sought to test their unrefined manliness against technology were thus irrational, backward, and unfit for urban life.[119] By contrast, rational, sophisticated, and urbane Mexicans, the lesson implied, knew that they had to make way for machinery.

Illustrators enjoyed imparting this lesson through characters like López. In another historieta, the Gutiérrez de Tajimaroa family arrives for a visit to the capital.[120] Decked in full archetypical attire, these provincials flee at the sight of a tram. Don Trini, along with his family, runs away as if the tram were the devil himself. However, when they come across a steamroller, they stand in its path, intrigued. Having never seen one—and being slow-witted—they stay there until the machinery runs them over, leaving them flat as sheets of papers. The driver promptly fills a bellows with smoke from one El Buen Tono's Canela Pura cigarettes and blows it into the crushed bodies. Happily, everyone comes back to life. The driver offers them additional cigarettes to forget their *susto* (fright).

In their ridicule of provincials, historietas exemplified the values and characteristics of ideal urban dwellers: prudence, alertness, attention, and bodily discipline.[121] These historietas were ultimately a demonstration and celebration of urbanity. One way to capture that urbanity and avoid being seen as a rural bumpkin was to consume the right products. Cigarettes worked to prepare or "regenerate" individuals for modern living in these advertisements. López smokes to gain valor, while members of the Gutiérrez de Tajimaroa family benefit from their regenerative value: the consumption of cigarettes saves and transforms them. Publicity campaigns had reinforced a popular association of progress and the machine-rolled cigarettes, which became the Porfirian symbol of modernity par excellence. Conversely, one way for the many rural migrants coming into the city during the period to demonstrate that they were wise to city life—had become real capitalinos—was by smoking the right kind of cigarette.[122]

The claim that victims were responsible for the majority of accidents did not go unchallenged. Muñoz-Serrano's verse "Run Them Over!" sarcastically remarked, "It is outrageous that not a day goes by without a citizen being cut in pieces by an eléctrico. *El Imparcial* sees this as the natural consequence of the uncultivated state of the masses, which lack [even] a 'pinch' of good habits, since no known risk intimidates them, or no car's wheel fails to mangle them." Sarcastically embracing the elitist notion that victims were responsible for their own accidents, he concludes, "it seems that it is a thing of 'pleasure' to go one-armed, to be frightened, to arrive home two hours late without someone saying that you have been a coward for letting stops go by in fear of exiting [the eléctrico]."[123]

These horrific "accidents" encapsulated many events and could reflect a variety of human motivations. The vast majority of injuries and deaths by eléctricos were accidental, but bodies sometimes came into contact with the machinery deliberately. Some people tried to end their lives by jumping in front of approaching trains, while others incorporated eléctricos into their love dramas.[124] Gendarme Francisco Váldez escorted Enriqueta Vásquez, Miguel Vidal (her lover), and the driver Joaquín Sánchez to the police station on February 5, 1901.[125] A pedestrian had notified Váldez that a couple was quarreling on Tepozán Street. When the gendarme arrived, he found the sixteen-year-old Vásquez under the wheels of the Peralvillo line eléctrico. She later declared that in the heat of an argument she had threatened to jump in front of the approaching eléctrico. Vidal managed to rescue her from the first tram that crossed the street, but she was then struck and thrown under a second tram heading in the opposite direction. Incidents like this were so common that lovers' quarrels ending in a tram accident became a common trope in poetry and fiction.[126]

It is impossible to determine how many other crushing accidents veiled dark secrets. Certain railway lines became known for leaving victims so mangled that they were impossible to identify. The darkness of the night may have helped to cover traces of criminal activity, but one woman tried to use eléctricos for her own purpose in broad daylight. Manuel Montalvo and a group of workmates were standing at the corner of Comonfort and Pensamiento Streets when they noticed a decently dressed young woman carrying a package under her arm.[127] A few minutes later, as a car of the Don Toribio and Santiago line approached at full speed, the woman knelt close to the railway and dropped the package over the rails. The prompt reaction of the motorista prevented the car from running it over. Montalvo hurried over to inspect the package only to discover a newborn wrapped in old clothing. The woman was arrested as she tried to escape. As tragic and dark as stories like this could be, they were only a small part of the conversation about how people were to behave and comport themselves around eléctricos and the unforeseen uses capitolinos made of them.

"LOS MATARISTAS"

The only other group who routinely shouldered blame for eléctricos accidents were the drivers (motoristas). The public chastised drivers and accused them of many faults: traveling at excessive speeds, not making full stops, failing to ring the bell when approaching intersections, inaction when an accident was foreseeable, and drunk driving. The conversation about drivers never reached the scale and popularity of the rhetoric concerning lower-class pedestrians and passengers, but it was nonetheless an active one. In newspapers, magazines, songs, poems, illustrations, and other outlets, motoristas became scapegoats, convenient human representatives of abstract modernity itself.

The arrival of eléctricos coincided with a growth in scientific inquiry into social condition and human behavior. Researchers took a keen interest in the impact of machinery upon both society and individuals.[128] Doctor Manuel Flores, a founding member of the Society for Psychological Studies and director of the Escuela Nacional Preparatoria (National Preparatory School), from 1901 to 1911, was particularly concerned with how the human body and brain reacted to speed. In an article titled "The Driver's Psychology: Influence of Speed over Character," Flores contended that the habits and tendencies of those who drove vehicles that traveled at high speeds revealed their mysterious and insolent character. He posited that this could be a major cause of accidents.[129] Flores argued that motoristas exhibited an altered mental state, a so-called psychology of the motorman. He alleged that the habit of speed, devouring space, inevitably had an adverse effect on the drivers' mental state, leaving them impatient, bad-tempered, and brutal. The faster the vehicle, the more exacerbated these traits became.

This was observable every time a race occurred: "The brute sees nothing, feels nothing, loves nothing. . . . Intoxicated with speed and space, he could run over his own mother's body!" Jockeys, cyclists, chauffeurs, and motorists, according to Flores, showed these characteristics to an alarming degree. Borrowing from the biological theories of the Italian criminologist Cesare Lombroso, Flores invited his readers to study the motoristas' physiognomies, claiming that they would find "frowning foreheads and never a smile; bristly beards and never trimmed mustaches; fierce eyes and never a pleasing gaze; blasphemies on their lips and never madrigals; growls in their throats and never sighs in their chests." A motorista standing in the glass cage of a powerful machine looked "more like a beast than a man . . . aveng[ing] his condition as a driver by obliterating pedestrians."[130]

Flores wrote for a small and select audience, but *El Imparcial* had a much wider circulation. The newspaper also put motoristas under constant scrutiny. This did not mean that the paper intended to shift attention from accident victims to motoristas: it continued attacking both simultaneously, as they supposedly shared many of the characteristics that made them a threat to life in the modern city. Both came from the lower classes and behaved similarly around eléctricos: they both failed to understand how to interact appropriately with the technology.[131] Early on, the daily newspaper hoped that electric traction would offer punctuality and speed, despite being dangerous, but it never thought that eléctricos would claim more victims than yellow fever, Mexico's great scourge. After three years, the paper could no longer claim that reckless drunks or careless individuals were responsible: it was clear that the conductor and the motorista, "in their savage desire to harm, knocked down peaceful travelers by taking off before time."[132] Echoing Flores's assessment, the article claimed that drivers, overwhelmed by some vertigo, went through town in a frenzied race, even in the riskiest areas of the city.

With accident victims and motoristas made into scapegoats, the conversation about technology in the city safely steered away from the capitalists that ran the tramlines. Casting motoristas as villains was one manifestation of a broader phenomenon occurring in Mexico at the time. Since the late nineteenth century criminologists and police forces had popularized the notion that they could describe the nature of criminals and make sense of their motivation. In actuality, they largely turned broad stereotypes of the lower classes into markers of criminality. According to the nascent science of criminality, visual clues signaled the motoristas' primitivism and innate criminality.[133] Illustrators drew from Lombroso's theory of "atavism," "which linked criminality with the physical traits of 'primitive' races." Lombrosian criminal anthropologists concluded that a combination of anomalies such as "small skulls, sloping foreheads, broad noses, and jutting jaws . . . [were] invariably associated with atavism, or biological regression to a less evolved racial type."[134] Media profiles repeatedly bestowed one of more of these visual markers on motoristas. The period's most famous engraver, José Guadalupe Posada, caricatured motoristas in several satirical verses and illustrations known as *calaveras* (skulls), which mocked lower-class professions. A profile of a driver with an oversized kepi-hat exaggerates the driver's small skull, with practically a compendium of Lombroso's facial characteristics: broad nose, prominent ears, teeth protruding over the lips, and scanty facial hair.

El Imparcial built on these new conventions of crime journalism and research, offering them as evidence of drivers' brutality and criminality.

FIGURE 3.5. *"The uniform they wear, how they should dress."* El Diario, *January 20, 1908.*

Drivers' testimonials—as recounted by reporters and letters to the editor—purportedly revealed that motoristas consciously struck down people with no fear because of their impunity. In one shocking story three young men overheard three motoristas laughing together over the hair-raising details of their gravest accidents.[135] One of the drivers told a story of a woman crossing the rails as his car approached. When the heel of her shoe got stuck between the rails, he accelerated to scare her, ringing the bell repeatedly. The poor woman made the sign of the cross, called upon all of the heavenly court, and even asked the tram to stop. When the driver tried to stop, it was too late. The woman lost her foot.

Because motoristas did not hold themselves accountable, some people called for making their infamous records visible in other ways. The author of one letter to the editor claimed that the motoristas' uniform was more feared than those of employees of the Gayosso Agency, the city's leading funeral house.[136] Just as soldiers announced their bravery with medals and decorations, this writer suggested that motoristas should announce their mistakes and cruelty by wearing badges for each accident that resulted in an amputation. If a leg was lost, a leg-shaped pin would adorn their right lapel. Decorations would continue until their coats were covered with these badges of shame: pedestrians would then know what to expect. Calls were made to dress drivers in a manner that faithfully represented their trade. *El Diario* juxtaposed two caricatured motoristas in an illustration (fig. 3.5). The tall, smiling motorista on the left wears his distinctive real-life uniform, including its kepi-hat with a plaque and number representing years of service.[137] The shorter motorista on the right wears what the paper offered as a more faithful representation of his profession: his jacket is similar, but skulls and crossbones replace the plaque on his hat and two more skulls and crossbones adorn his lapel.

Other outlets took up *El Imparcial*'s tarring of motoristas as villains to varying degrees. Some took to publishing popular songs and humorous verses that colorfully described their malevolent exploits. The 1906 song "Los motoristas" described them as cunning and violent, readier than anyone else to inflict harm. It sarcastically noted the drivers' promptness in getting to their stop—but only after running over two or three pedestrians on the way. The masses had nicknamed them *mataristas*, a play on the words *motorista* and *matar* (to kill):

> They are called mataristas
> among the vulgar people
> and they don't like them the least
> for their sweet abruptness
>
> We ought to tell the truth
> it is not they who are responsible,
> but the pedestrians who don't sniff
> where the motorcar is coming from.

The song sarcastically excuses the "poor mataristas" for running down individuals and blames pedestrians for "jumping between rails to see if the train or motorista could do a miracle."[138]

Humorous verses played with the sense of submission that pedestrians and passengers felt in the presence of motoristas. "The Terrible Ones" advised riders to assume a subservient position to save their necks and address the motorists thus when ready to exit: "Young man of pale hat and canary-colored uniform, in the name of your mother who had the pleasure of carrying you in her womb, please stop the convoy, suspend your speed of light, and, if you allow me, I'll get off at the corner."[139] Muñoz-Serrano echoed this sense of powerless. In an untitled verse he declared that motoristas were "the referees of so many citizens, sovereign authorities inside their cars." Those who dared speak up received "a reply with a sea of atrocities, worse if the offended party was a woman because she heard what she had not [heard] or hoped to in her lifetime." Unsurprisingly, motoristas repeatedly receive a suitable death in Posada's popular calaveras.

> Agustín the motorista,
> *Tín, tín, tín, tín*, in a hurry
> Opened the controller to full speed
> without stopping, not even for
> the best-dressed lady.
> That's why today the Grim Reaper
> took him by the feet,
> increasing his speed.
> Requiescat in pace, Amen.[140]

The Drunken Matarista

Many of the accusations leveled against motoristas were as unfair as those leveled against pedestrians and passengers, but one especially damning charge had some substance to it. In the aftermath of many accidents authorities discovered that motoristas had been intoxicated. The issue received special attention after a municipal government study identified alcohol as a leading factor in accidents. A 1904 decree tried to solve the problem by prohibiting METC employees from entering establishments that served alcohol during their shifts, arguing that driver ineptitude, which repeatedly played a role in accidents, increased with alcohol consumption.[141] The Ayuntamiento had invested motoristas with the duties and powers of police officers in order to uphold the Railroad Regulations inside cars since 1894.[142] Therefore, drafters of the new decree reasoned, the Reglamento de Bebidas Embriagantes (Regulation for Intoxicating Beverages) that prohibited uniformed police agents from entering cantinas and other establishments that served alcohol also applied to motoristas.[143] The decree also prohibited establishments that sold alcohol from serving uniformed employees. Motoristas who violated the decree could face a fine of three to ten pesos (which would be deducted from their salary) or one to ten days in jail, and the transgression would enter their record. Those who served them (or the establishment's owners if it proved impossible to determine who poured the drink) faced a fine of five to fifty pesos or five to twenty days in jail.

Efforts to police the motoristas' alcohol consumption were part and parcel of a more extensive municipal campaign to curb vice in the lower classes. Municipal authorities, doctors, and social critics considered consumption of alcohol a grave urban social problem, as evidenced by the disproportionate number of alcohol-serving establishments. In 1902 the city registered 924 pulquerías, 650 cantinas, and 285 inns and taverns: a total of 1,859 businesses officially sold alcohol. In comparison, Buenos Aires, with a population three times larger than the Mexican capital, had only 1,097 legal establishments. Critics described Mexico City as the "City of Pulquerías" and its inhabitants as "a quarrelsome pueblo," "a permanently soaked sponge" of pulque and moonshine.[144]

Authorities classed drunkenness as a lower-class vice and focused on policing pulque consumption in those neighborhoods. To curtail crime and disorders linked to alcohol, city officials limited "the operating hours of cantinas and pulquerías, prohibiting gambling and music on their premises, and banning the sale of alcohol during festivities." Curbing drunkenness thus became yet another "rationale for further official control of people's movements"—or at least the movements of certain groups.[145]

The presence of cantinas and pulquerías near railway lines heightened the possibility of accidents caused by alcohol. This was evident in the area of Mixcoac, which had three pulquerías in a row on Molinos Street—mere steps away from an eléctricos line. The last establishment sat on the corner of Molinos and Avenida México and had two windows that served passing customers but was less than a meter away from the tracks of the San Ángel line. The patrons were at risk of being "ground by drivers who at that spot had already sacrificed dogs and pigs to the Electric Civilization."[146]

With many businesses so close and in competition with each other, motoristas easily found ways around the 1904 decree. The simplest solution was to pour drinks inside the eléctricos instead of serving liquor at the bar. The pulquerías were only too happy to oblige; as one observer noted, liquor was taken to the motoristas "by one of those thousand accomplices unconscious of the menace the drivers' intoxication represented." The official decree could accomplish little, papers like *El Imparcial* concluded, for drivers were fundamentally representatives of the lower classes: they were simply dressed-up ordinary, unsophisticated, boorish workers. "Without the uniform, the motorista returned to his natural environment."[147]

Drivers recognized that alcohol was a problem for some individuals in the trade—a response to the stressful working conditions they faced daily—and some sought to police the problem internally. This is implied in a letter that a handful of METC motoristas, conductors, and other employees sent to the city council in 1912. The letter protested the removal of Juaquina Corral, a poor woman who for years had a booth on the south side of the Zócalo that provided an important service to company employees. The clean water she sold, METC employees explained, deterred workers from drinking dirty water or, more pressingly, bottles of pulque, to satisfy their thirst.[148] Her service, they noted, was particularly important during the hot season, which had just started. When authorities prohibited Corral from setting up her stand, they both unfairly prevented her from securing an income for herself and her elderly mother and denied motoristas a reliable way to quench their thirst. Instead, they would either have to drink dirty water or carry a bottle of pulque. Unless the Ayuntamiento was willing to risk intoxicated employees, the signatories asked that Corral be allowed to return.

The Immoral Matarista

Few of these discussions and criticisms of motoristas bothered to explore the pressure that the drivers were under. They had a schedule to keep and machinery to look after. Failure in either could spell fines or losing their jobs. According to confessions from drivers—substantiated by complaints from passengers—company rules to meet predetermined schedules meant that drivers often skipped stops or could not make full stops or decelerate when crossing intersections or switching tracks. They were motivated, above all, by an effort to *emparejarse* (make up lost time). Some asserted that this was a choice and that the motoristas had the "option" to uphold the METC's regimented demands, but few actually bothered to ask the drivers if this was the case.

One exceptional editorial weighed the drivers' obligations as employees against their duties to their fellow citizens.[149] On the one hand, the exigencies of time schedules encouraged their "habit" of driving in the downtown area at the same speed as they would in thinly populated areas. On the other hand, they knowingly violated regulations and drove at full speed as soon as they could. The simple reason was that this was in their own best interest. They privileged their own paychecks over human life, as the slightest delay in their schedule translated into fines or suspensions.

Commentators labeled the motoristas' decision to adhere to schedules at the public's expense immoral. *El Imparcial* questioned whether the company based the schedules on electric power generated at the Indianilla or the speed at which engines could run without considering the city's unique circumstances. It was absurd to base the circulation of trains in the capital on the schedules and speeds of American trams, which journeyed through wider boulevards.[150] Mexico City's eléctricos journeyed through narrow streets invaded by a multitude of individuals who, the papers claimed, did not know how to walk and spilled over from sidewalks until they occupied the whole street. The trams had to navigate around individuals riding burros, who, in their drunkenness, planted themselves in the middle of the railway. Their strict schedule corresponded to a commendable desire to provide regular and fast service, not to saving money, which would be punishable. Regardless, a new detailed study was needed. Based on their findings, prudent and consciously refined schedules were feasible. If the METC's schedules forced drivers to violate traffic regulations, they were obliged to obey these regulations before company orders. Drivers had two simple choices: protest against those orders or, if scrupulously honest, leave the job.

As matters stood in the first decade of the twentieth century, drivers often did not bear criminal responsibility for any harm done to people, but they bore financial responsibility to the METC's bottom line and faced various fines for a wide spectrum of issues. Even before they earned a wage, motoristas had to purchase uniforms from La Bella Jardinera for 21.50 pesos and make a 5.10-peso deposit for possible minor breakages of the company's equipment. The combined sum was nearly a month's pay for drivers, who received at most 1.50 to 1.80 pesos per day in 1900. The METC calculated wages by the hour, which consequently led workers "to exceed their capacity to work to earn more, resulting in an increased number of accidents due to carelessness and tiredness."[151] Motoristas organized a strike for the Day of the Dead in 1902, the busiest day of the year.[152] Among the complaints was that they had to pay five pesos for the badge on their kepi-hats. Although the strike did not occur, charges and suspensions remained, along with the motoristas' resentment of them. Over a decade later, the 1914 bylaws of the tramway workers' union specifically called for the elimination of burdensome and arbitrary fees for damages in cars.[153]

Despite the accusations of indifference and carelessness, motoristas enjoyed little legal protection and did bear responsibility for accidents. *El Universal* alleged that motoristas—caught between reduced salaries or even potentially losing their jobs and risking accidents—chose the latter, a selfish decision. No matter what pressures they were under, the paper wrote, motoristas were still subject to the law, which did not authorize anyone to take a life simply to earn a few more cents. To excuse them from responsibility, warned the article, would be like excusing those who stole to satisfy their needs. It would destroy the principle of justice and the rule of law. "We believe that a claim to fines would be sufficient to make the author of an accident responsible not for a crime of negligence but for a premeditated crime."[154]

Unfortunately for the victims, the law did not meet all their needs. Take the case of Rodrigo Flores. Benito Muñoz's eléctrico ran over him in 1910, causing several injuries and eventually the amputation of his right arm. The courts found Muñoz responsible for the injuries and sentenced him to twenty-four days in jail.[155] A few months after the accident, Flores was back in the courts, requesting (and obtaining) the right to litigate as a poor person, which exempted him from the stamp tax, in a civil suit against METC. The loss of an arm made his work as a plasterer for the Palacio de Justicia del Ramo Penal very difficult, and he demanded the highest compensation that the law afforded him.[156] Article 323 of the 1871 Penal Code granted the injured individual not only monetary damages for injury and losses but also additional compensation stipulated by the judge (based on social status, sex, and the body part left disabled, crippled, or deformed). Flores argued that owners of "carriages" were responsible for the accidents of their dependents that took place during their service and demanded 1,500 pesos from the METC plus trial expenses. The METC denied Flores's right to compensation, citing a precedent that it could not be held responsible for the actions of its employees. No record of the case's result has been located.

Miguel Sobrino filed a similar lawsuit in 1906 for damages against the METC after an eléctrico crashed into his wagon.[157] The detained motorista declared that the accident happened because the motorcar's breaks were in poor condition and malfunctioning, but he drove it anyway for fear of being dismissed. In this case, the METC's lawyers argued that motoristas were indeed under the company's authority, but the METC would need to have been in a position to prevent said actions to be held liable for them. That had not been the case, so the company was under no obligation to compensate. The courts agreed. Sobrino, or at least his legal counsel, should have predicted the court's ruling based an earlier case involving Eduardo Cumming, who demanded 150 pesos for the repairs to his Fiat after a collision with an eléctrico.[158] In this instance, the courts again found that the plaintiff had failed to produce evidence that anyone but the motorista was in a position to prevent the accident, so he alone was guilty. The courts demanded a high standard of proof that the company had a role in accidents. The burden that this placed on plaintiffs meant that compensation would not be forthcoming, no matter what the law said. The law worked to insulate the owners from the horrors of accidents and the public's exasperation. Instead, working-class drivers alone would shoulder the legal, financial, and social burden.

Motoristas bore other expenses too, including the costs of property damage to their cars incurred while traversing an increasingly crowded city. On March 16, 1907, while driving through Revillagigedo Street, the trolley pole of motorcar 318 hit and broke one of the Mexlight's streetlights.[159] Fearing arrest, the driver, like many others involved in accidents, abandoned the car. Conductor 264 stayed behind, only to be arrested and confined to the sixth district precinct. His brother scrambled to gather the 17.50 pesos for his release.

The conductor probably did not realize it, but he was an unfortunate pawn in a larger battle between the METC and Mexlight. As the established player in the field of electrical service—and having previously installed its own electrical infrastructure—Mexlight took a dim view of the METC and its demands to have its own infrastructure accommodated in the crowded city. The tensions had grown into threats of lawsuits by 1904, over the costs of having to move Mexlight lampposts so METC's overhead trolley lines would not keep breaking the globes. That year, Mexlight directly challenged the METC's reluctance to cover the costs to relocate lamps: "in the United States, as well as in Europe, it is usual for a new electrical company entering the field with its installations to pay the expenses of all changes necessary made to prior electric installations." METC's general manager coolly responded, "If you think it is a good business policy to bring a suit against us, I have no objection."[160]

The problem between the two companies extended beyond the establishment of tramlines. As the driver of motorcar 318 knew, eléctricos routinely damaged streetlights as they made their way around the city, but it would not be the METC that bore responsibility. Just as the METC and its managers argued that they had no liability for personal accidents caused by their trams, they also argued they had no liability for property damages. If Mexlight wanted to pursue compensation, it would have to squeeze it from METC's employees like conductor 264. Mexlight saw the futility in this. Its operating manager, using the occasion of the broken globe on Revillagigedo Street, appealed to the METC, asking, "[I]s this not a question between the companies concerned and should we not refund the 17.50 pesos to conductor 264?" No evidence exists as to whether any money found its way back to the conductor or whether he faced a legal action. By the summer of that year, however, Mexlight apparently recognized the ineffectuality of either holding motoristas liable for damages or awaiting the METC's change of heart. The Mexlight manager wrote to his counterpart at the METC. In the case of damage done to any of their property by the trams,

"I have again issued instructions to all the corresponding Departments that no arrests should be made in cases of this kind, and I trust that hereafter no such inconvenience will be caused to any of your employees."[161]

The Ignorant Matarista

When not being blasted as callous and immoral, the motorista was abused for being ignorant. By 1902, even before portrayals of the drivers' bloodlust became popular, newspapers were already writing about the poor quality of workers on the eléctricos. The *Mexican Herald* praised the work of general manager Charles Clegg, highlighting how difficult his work was because of a lack of skilled, diligent drivers. The paper claimed that "many were stupid, rude, and entirely useless. However, employees resembled the rest of the population." *El Imparcial* persistently critiqued the METC for hiring men unable to secure a job elsewhere, who, pressured by need, accepted a job with a miserable salary. The daily joined other critical voices warning that the METC would attract only poor workers without changes to the motoristas' salary.

The debate over the motoristas' skills gained traction in early 1903 when it became known that 500 accidents with at least some physical harm had taken place in the preceding two months. *El Imparcial* argued that the METC had taken on employees who lacked the necessary skills to drive the trams to fill the demands of the rapidly expanding tram system. Consequently, uneducated workers could not drive the eléctricos as they had driven earlier horse cars, for the machine was far more complicated.[162]

The French daily *Le Courrier du Mexique* proposed a wholly reasonable solution: if ignorance was at fault, motoristas needed special technical instruction and should pass a test before working in the streets. *El Imparcial* reproduced the proposal, arguing that, even if these measures would not eradicate accidents, they "could at least *reduce* daily accidents." The daily considered the idea easy in theory but less so in practice and undoubtedly unrealistic at the time. Electric traction was too recent in Mexico for its service to be perfectly organized. Anywhere else, the measures would take some time, "but in our conditions, that period was to be considerably longer." The city jumped without transition from "that awful, rudimentary, and primitive horse tram service to electric traction just like most advanced cities of the world, adopting the latest procedures without having the least notion of what an eléctrico was."[163] Although motoristas had to coordinate their hands, legs, and knees to drive a tram—often with counterintuitive motions—no detailed evidence on their training has surfaced.

Qualified personnel were available in the city, but not at such a low wage. *El Imparcial* criticized the METC's plan to offer a salary that would appeal only to those without an education—or common sense, it added—and then send them out onto the city's streets, expecting they would become polite, genteel motoristas in charge of a large, complicated electric tram. Anywhere in the world, not only in Mexico, "it would be impossible to provide a service of average quality with miserable salaries." The company was in a position to offer better salaries, for profits had been enormous under Clegg, with a 12 percent return on investment in the first few years alone. Management needed to recognize, concluded the daily, "that [their enterprise] was not a business of usury; [they needed] to reconcile . . . [their] interests with those of the public." They had a right to profit, but they had to earn it by attracting more customers and serving them better, not merely by cutting costs and grinding down wages till employees would run people over rather than risk a fine.[164]

Commentators ridiculed the METC for hiring low-skill workers and then attempting to pass them off as respectable. Donning a uniform, these critics implied, did not suddenly imbue drivers with skill and make them up to the task of handling expensive machinery on city streets. In one poignant *Cómico* illustration, the METC made sure to take off the motoristas' peasant clothes and sombreros but left on the sullen frown so commonly associated with indigenous people and the rural poor.[165] Despite the METC's efforts the driver had trashed and gotten mud splattered over the company jacket, kepi-hat, and pants within a week. No matter how they were attired, the drivers were still part of the dirty, poor underclass in this critic's eyes. The satire ended with the drivers once again back in their traditional peanut-colored garb, with no shoes and a palm sombrero.

Elite views, parroted in the dailies, engaged in mental gymnastics to avoid confronting the reality of eléctricos in the capital. They dismissed or shot down reasonable suggestions of training and certification for drivers, arguing that such efforts would have no effect. In a limited sense, they were right. More training for drivers might not have led to fewer accidents, because the crowded conditions and demands placed upon motoristas by the METC would remain. Instead, elite capitalinos lamented the abstract, innate lag that Mexico faced while trying to catch up with the world. The machinery of modernity had arrived in the nation, but the Mexican people were still—somehow—not up to the task. That argument served to protect Clegg and the METC's investors but not capitalinos trying to get around their city.

With so much bad publicity, pressure from the company, and bad pay, the position of motoristas suffered a high rate of turnovers. Many employees stayed only a few weeks, others only seasonally. As long as the METC refused to increase salaries, many held that people with technical backgrounds would never agree to the work. Instead, those who accepted the jobs were almost entirely illiterate. Without any investment in training, the company could never expect to make a skillful electrician or a trained mechanic out of a common laborer in a few days. Likewise, even unskilled workers would not take on a long traineeship that paid only a few cents a day. Clegg, always with an eye on his bottom line, appears to have reasoned that it was more profitable for the company to have high turnover and less-trained drivers even if that meant risking injury and death to passengers and people along the streets. As long as the courts insulated them from paying compensation to victims and having to take responsibility for their choices, this was not an unreasonable decision.

Motoristas were not stupid, of course, but lack of training may have been the fairest of the accusations against them. In report after report of accidents, witnesses swore that the eléctrico sped up rather than slowed down before a collision. Some could not help but conclude that the driver must have wanted the accident to happen. Yet the simpler truth is that these drivers, many of whom were new to the job, were just repeating the same mistake. When driving a tram, a motorista resembled nothing so much as a one-man band, playing several instruments by using his feet, hands, and mouth.[166] At all times, the motorista had his hands on two separate handles, which simultaneously controlled motion, direction, and braking. With his feet, he operated a bell or whistle. He also had to swivel his head every few seconds to check on the pole that connected the car to the electric circuit overhead. In an emergency, this already complicated arrangement became positively chaotic. The motorista had to operate the bell or whistle with a foot or knee; his left hand had to reach for a lever to reduce the car's speed, while his right sought out a separate lever to destroy the car's acquired speed. On top of all this, in order to stop the car the driver had to rotate two sets of handles in opposite directions, a difficult, non-instinctual task.

In a rare note of sympathy for the plight of motoristas, one paper explained, "When the dread of an immediate catastrophe overcomes one, or the surprise of an unexpected event is experienced, any complicated mechanism is a defective one. Only with years of practice, which motoristas lack, could one give those movements the unique automatism and instantaneity capable of preventing a disaster."[167]

Most people had no idea of what a motorista had to do, so they alternated between or fused two conclusions. Either motoristas had sufficient training, in which case the accidents they caused were the result of indifference to human life and suffering, or eléctricos were a black-box of technology, a complicated fusion of machinery too esoteric to be understood by all but the most highly trained experts. After all, other areas of life like medicine and governance were making similar claims in this period. Modernity, in the popular imagination, was the realm of the educated and the expert, and eléctricos were part of that. The common motorista was neither, in their view. Eventually, the two threads of conversation combined: drivers became ignorant, indifferent, and incompetent. No one ever checked to see if any of this was true. The SCOP, the government agency in charge of regulating the tram system, simply took the METC at its word that it had secured enough operators and provided them with training. The SCOP never demanded that the training be documented. No evidence exists that it ever occurred.

CONCLUSION: BLAMING EVERYTHING BUT THE SYSTEM

Accident reports in the first decade of the twentieth century reveal much more than the citizenry's experience with eléctricos. They open a window into how accidents served as a channel to express broader anxieties of the period. From the first announcements of the plans to electrify the tramways, newspapers expected accidents and explained that they were the price to be paid for progress. Faster travel was too vital, too valuable for modern life to be reduced—even for the sake of lives—so it went hand in hand with accidents. It was up to citizens to decide between "modernizing" their habits and becoming victims of eléctricos. However, not all capitalinos would be sacrificed for the sake of modernity. Only lower-class and indigenous men, women, and children—those that officials, writers, and illustrators accused of holding onto the habits of yesteryear—were really in any danger. In their writings, drawings, and announcements, members of the elite made it clear: while it was unfortunate to see these people injured and maimed in accidents, it was not a real tragedy. The attitudes and actions of these lower-class groups were an obstacle to progress. In this respect, accident reports reveal a powerful fear swirling around Mexico's efforts to be modern. Although the exhibition culture encouraged Mexico to display its march toward modernity to itself and to the world, what if this progress was not real, substantial, or sustainable? That is, aspects of the electricscape may have made Mexico look modern, but it might be only a veneer, covering up what many judged as a backward, tradition-bound society. The nation could import and apply technology, but would it ever reach the rarefied status of a truly modern society? This anxiety was powerful enough that officials and commentators had already started looking for a reason to explain why this was possible. They found it in the lower classes.

The middle and upper classes believed that the poor, indigenous, and working classes were the antitheses of themselves: immoral, ignorant, and drunk, not to mention unsophisticated and superstitious. In blasting the lower classes for their failures, they made clear that they were not only their opposite but also the only ones in Mexico ready for modernity. If progress was stalled, it was not because of them. Whether it was the motorista who was incapable of commanding modern machinery or pedestrians who did not know how to comport themselves around it, whatever problems arose from eléctricos had nothing to do with the machines, government and company policies, or the wealthy who demanded profits from it all. The problems lay squarely with the unsophisticated masses. It was a very convenient narrative for the METC, for it essentially silenced any criticism of its policies. After all, so the reasoning went, only the ignorant and unsophisticated were in any danger, so only these classes would have any cause to voice concern. In the end, why would anyone with any real influence want to be associated with those groups?

| FOUR |

Ladrones de Luz

A SCRIPTED ELECTRICSCAPE, 1901–1918

In early December 1901 an eclectic "inspecting team" stormed into the Ambos Rumbos Hotel in downtown Mexico City, demanding entry to room 35.[1] The team—made up of Oliverio López, an electrical expert witness for the municipal government; Gabriel Ortiz and Guillermo Feuss, legal representative and installation supervisor for the Compañía Mexicana de Electricidad, SA (CME; Mexican Electric Works, Ltd.); and a police inspector—accused the hotel of stealing electricity and sought to determine the source of the theft. The hotel manager objected that only the hotel owner, Francisco Torres, could give consent to such a search, but he was absent. When the team brushed aside this news, the manager explained that the room was occupied and the guest had left with the key. The team was not dissuaded and secured permission from the adjacent room's occupant to enter through the balcony. As they made their entry, someone abruptly exited from room 35, trying to make his way to the roof terrace. The police inspector apprehended the suspect, who was none other than Torres, the hotel owner. With the owner in hand, the team members entered the room and found what they were looking for. Two cable ends lay severed on the balcony, strategically hidden from the street by a mattress and sarape resting against the balcony railing.[2] Confident that an electrical connection had been deliberately cut, López and Feuss joined the cable ends to those found inside the room: the hotel instantly came to life. Fifty-seven lights illuminated the hotel's stairs, office, lobby, hallways, and thirty-five guest rooms.

Historians of daily life have emphasized the importance of looking at the subtle aspects of ordinary matters to appreciate the rules that govern behavior.[3] Like the cases of dozens of other Mexico City residents suspected of power theft, the arrest of Torres and the resulting trial offer a fruitful avenue to document the nuances of the capital's electrification and "reconstruct the texture of crime as experienced in everyday life by those who formed the majority of offenders and victims."[4] No matter how legal experts interpreted the law, testimony from the trials reveals that ordinary people held their own notion of what was fair, reflecting the kind of lives they wanted for themselves. The accused certainly framed statements in terms that lessened their responsibility, but the judicial records are still valuable in revealing how they drafted narratives.[5] What did people consider a viable explanation—or excuse—to secure electricity for themselves?

The exhibitionary culture that took root in Mexico in the 1880s had become exceptionally effective by the turn of the century. The demand for electricity and all it could offer reached up and down the social ladder of the nation. By this time, people saw that electricity could offer more than just large-scale public goods like street lighting or mass transportation: it could come into their homes and businesses. Yet not everyone had access to electric service, could afford it, or saw a need for it. When capitalinos stole electricity, their reasons were varied.

Here I discuss these cases of theft to confront our understanding of technological change in the Mexican capital and how, on the everyday level, capitalinos grappled with rapid urbanization, expanding industrialization, and technification of spaces. For many capitalinos, securing even a little electricity meant a few bulbs in the home or a new, more efficient piece of machinery. Trivial though those things may seem now, they provided a vital connection to improvement of both their own lives and the nation. Power theft reveals the subtle choices and aspirations of ordinary people that accompanied electrification. By grounding electrification in everyday life, capitalinos emerge as agents of technological change who understood the importance of electricity in transforming their lives. The responses of capitalinos charged for power theft displayed "dispersed, tactical . . . makeshift creativity" and constitute an "antidiscipline" that a top-down approach misses.[6]

Therefore I offer another set of narratives swirling around electrification in this chapter. Unlike previous conversations, these were more like an active argument. Like the capitalinos who stole from them, the power companies had their own interests and corresponding demands. They had little interest in the give-and-take of a conversation with the public. With the complete support of the government, they scripted how much power people could have, when they could have it, how they would use it, and, most important, how much it would cost. Electrification amounted to large investments of capital entering into people's private spaces. Electrical companies, with the backing of foreign investors and wealthy Mexicans, would bring power to people's homes, but subscribers within were expected to honor their contracts. Judicial records, however, demonstrate that users, authorized and unauthorized, resisted this scripted consumption and frequently subverted it altogether. They would not bow to these demands, especially when they considered them unfair—or, at least, when a better deal was available. What constituted theft in the eyes of the law seems to have received communal sanction from ordinary citizens. For many capitalinos, large electric companies' expectations of how they should engage with electricity were less a final script than a first draft, subject to revision.

Technology has been an important field in recent Latin American history. Scholars have privileged the study of technologies in their sociocultural context, "factoring in local communities of expertise, hybrid knowledge, and domestic technical capacities." Historically, "the adaptation of foreign technologies and the invention of new technological devices are embedded in social, economic and epistemic dynamics in a given society, while at the same time constrained by the relations between societies."[7] Latin American historians have expanded their scholarship on crime and punishment at the same time, but the two historiographical currents have yet to cross-fertilize.[8] The city's industrialization and urbanization created social tensions. Urban designers' orderly plans for the proper use of the city—much like power companies' scripted consumption—clashed with residents' quotidian demands.[9] If one important idea has emerged from these studies, it is that the law always allowed marginal groups to contest, elude, and even use the law for their benefit (or at least reluctantly accepted this). Yet, though cities were prime sites for the introduction of technologies, we know little of the juncture of technology and crime.

I attempt to correct that lack in this chapter. The study of power theft contributes to the historiography of crime and modernity by introducing technology to the equation.[10] This particular type of theft allows us to explore how notions of policing, private property, space, honor, and even decency influenced the ways people secured and used electricity. Capitalinos could flip seamlessly between the elitist, scripted, "proper" use of electricity and the ad-libbed, "improper" use that fit their needs in specific circumstances. In these cases, the categorical difference between criminals and innocent people breaks down and is replaced by a spectrum of those who could secure electricity reliably and affordably and those who could not.

Using evidence from both the National Archives and Mexico City newspapers, I explore sixty-three cases brought against individuals indicted for power theft and tried before the Tribunal Superior de Justicia del Distrito Federal (TSJDF; Federal District Higher Court). These records document the frustration, indignation, and resentment that the electric companies and the police felt toward capitalinos who could, with such ease, violate their script and switch seamlessly between being paying customers and illegal pilferers (often, as shown below, both simultaneously). For them, the capital was nothing less than a city of suspects. Yet these legal records provide more than the accusations leveled by prosecutors, company officials, and eyewitness experts. They also contain the voices of defendants, their attorneys, and their friends and family. These records raise powerful questions that push us past the celebrated images of electrification and the exhibitionary culture it powered and into experiences of those who lived through it: Who would get electricity? Could some have preferential access? Was it ever fair to steal it? Was stealing electricity even a crime? Mexico forged answers in these trials.

POLICING SCRIPTED CONSUMPTION

The Porfirian regime saw technology as a tool to transform Mexico into a modern nation and therefore instituted legal reforms in the 1880s and 1890s that "strengthened private property rights and foreign investors' power vis-à-vis the state."[11] The sanctity of private property would be the bedrock of a capitalist Mexican society and an assurance to all-important foreign investors. Theft of electricity—of property, in the mind of the regime—transgressed the foundation of the "modern" Mexico that Díaz envisioned. He made sure that private companies knew that they had the full support of the state to curtail this violation.

As enthusiastic partners, the companies actively policed consumption, with entire departments devoted to detecting illegal use. Investigation of fraud fell upon inspectors who set out to police power lines, carry out unannounced inspections, identify and verify clandestine connections, set up surveillance, and detain alleged *ladrones de luz* (power thieves) with the aid of gendarmes. Uncovering theft had become the exclusive task of Mexlight's Inspectors Department by 1907. The department coordinated thirty-five inspectors, who walked the streets day and night with their eyes fixed on the lines that crisscrossed the city. In addition, undercover inspectors and informants aided their work.[12] Besides surveying overhead power lines, inspectors carried out unannounced inspections of commercial and residential houses, checking for illegal hookups into power supply lines, unauthorized reconnections, or meter tampering. It was rare for public and private agencies in Mexico City to coordinate their activities efficiently in the early twentieth century, but not when it came to power thieves. It is a measure of how important electricity was to the government's vision of modernity that catching this kind of larcenist could happen with startling speed.

In the case of the Ambos Rumbos Hotel, suspicions of theft emerged when company employees noticed that the electric meters' readings were higher than expected, given the clients in the area. The CME legal representative, Gabriel Ortiz, approached the Ministerio Público (Public Prosecutor's Office) to report power theft on Vergara Street and asked the authorities to intervene to locate the fraudulent connection. The very next day company inspectors fingered the Ambos Rumbos Hotel. Ortiz demanded an immediate *vista de ojos* (visual inspection): if the theft was proven, charges must be pressed.

Mexlight came to dominate the capital's electricscape with the inauguration of the Necaxa hydroelectric plant. After months of advertising its 40,000 horsepower at dirt-cheap prices (slashing rates by 50 percent), Mexlight flooded the market with cheap power, driving independent companies out of business.[13] The consolidation of the electric power market in the city by 1905 was a double-edged sword: more people had power, but authority over that power fell to fewer companies. The Mexlight monopoly simplified matters for the leaders of Mexico City, but employees of smaller competitors were now available to lend their services to those who wanted to bypass the monopoly. It would be a cat-and-mouse game for the next decade.

Of all their responsibilities, the easiest task for inspectors was uncovering illegal hookups to power lines. These illicit connections were often stunningly rudimentary, consisting of nothing more than a wire running from a home or business to the nearest central power line. Inspectors could see the telltale sign of stolen power in the crude splice that joined the two. If the owner of the suspect property could not immediately show the inspector a contract, Mexlight would arrive within hours to take the unauthorized line down. Inspectors were diligent and efficient, but Mexico City was already large and constantly growing. Illegal connections could go undetected for days, months, and even years. Pedro González, the owner of an ordinary tailor shop on Corpus Cristi Street, tapped into a nearby power line for six months without being detected.[14] Rita Arias had been a Mexlight client until late October of 1906, when service was interrupted for nonpayment.[15] For two weeks thereafter, according to *El Imparcial*, she slung a *gancho de hierro* (metal hook) over the nearby uninsulated main power line nightly and drew a steady current into her home. Inspectors discovered the maneuver one morning when she forgot to take the hook down.

Arias's method of securing electricity was especially crude, but any unauthorized tapping of power lines was inherently dangerous. Stealing electricity risked fire, in addition to the obvious possibility of death. Fear of conflagrations intensified in the late nineteenth century as the establishment and expansion of factories and workshops within city limits brought hazardous materials, flammable substances, and machinery prone to sparking and explosion.[16] Like kindling, these dangerous elements sat in factories and workshops that were often exploiting unauthorized connections with rigged wiring and faulty circuits. Exposed wiring touched ceilings, tapestries, and bed canopies at the Ambos Mundos Hotel, but luckily nothing ignited. Other thieves were not so fortunate. A hair salon in the area of Santa Ana nearly went up in flames in 1901. The owner used bell wire (for low-current, low-voltage applications) to light a mere five incandescent bulbs, by tapping into a power line servicing a nearby drugstore. Probably through ignorance, the salon owner had chosen a wire categorically unsuited to the job: it overloaded, igniting and partially burning the roof before the fire was put out.[17]

The rudimentary methods and lack of care associated with unauthorized reconnections kept inspectors in a constant state of alert. Night was their greatest ally. Inspectors walked the city in the darkness, looking for bright electric lights where there should not have been any. Especially at night, it was not only workshops or penny-pinching business owners who stole power, as Inspector Gabriel Abrego discovered one night in 1912. Abrego appears to have been exceptionally diligent in his work—his name surfaces again and again in the early years of Mexico City's electrification. During one of his nocturnal rounds, he noticed electric light coming from the home of Miguel Palacios, whose service had been interrupted for nonpayment a month before.[18] When Palacios finally confessed to his crime at the police station, Abrego also found out his motive: he wanted to host a dance party for friends later that night.

Not every case was so easy and straightforward, and sometimes the inspectors did not win. The same year that Abrego ruined Palacios's party, Inspector Manuel Rodríguez Miramón stopped at the home of Federico Jah, a German citizen living on Patrotismo Street.[19] During his rounds Rodríguez Miramón noticed what he suspected was an illegal installation, so he rang at the door and was greeted by a maid. Under city ordinances, the maid had no legal obligation to let him enter, but Mexican custom demanded that guests be invited into the *zaguán* (formal, public portion of the house). The maid ushered him in but balked when he demanded to inspect the house. She said that the house had no service and would not allow him to go any farther. Rodríguez Miramón, however, had seen enough. The presence of exposed wires in the hallway all but confessed to Jah's theft. Though the maid was right that electricity was not flowing through them at that moment, Rodríguez Miramón could tell that the wires could readily be connected and disconnected from local power lines. He went back after dark, accompanied by another employee and three gendarmes, but this time the maid denied him entrance. Rodríguez Miramón and his party could only watch from the street as electric lights were turned off and candles lit.

In a similar turn of events, Inspector Alfonso Tovar walked by Enriqueta Ruiz's house at around 8 P.M. and noticed an incandescent light in one of the rooms.[20] When he requested permission to inspect the installation, the maid denied that the house had any electricity and refused him entrance. Tovar watched from the street as the electric light dimmed, replaced by a flickering candle flame.

The inspectors' work was not easy. Property owners—or their proxies, like maids—could deny them entry, as they did in the cases of Ambos Rumbos Hotel and the Jah and Ruiz residences. Inspectors had to be equal parts attentive, creative, and tenacious and were often all three in their stakeouts. They conducted surveillance to confirm their suspicions and document the extent of a clandestine pilfering. In 1913, for instance, Carlos Paliza, one of Mexlight's most astute and active inspectors, received instructions to investigate the suspiciously low consumption levels reported at the Monte Carlo and Salón México theaters. After staking out the establishments, Paliza caught Ignacio Nápoles, an electric company employee, in flagrante manipulating the electric meter at the Monte Carlo.[21] Paliza was relatively lucky; some company inspectors had to stake out locations for days. Manuel Muñoz, identified only as a technical informant in the court records, watched the urban estate of Jesús Arias Sánchez for three days to confirm that he lit his home with illegally obtained power.[22] When he returned to his supervisor, Gabriel Abrego, he provided not only photographs of a series of electric lights but also a detailed map of the property.[23]

Besides catching thieves brazenly lighting their properties at night, the simplest and most effective way to uncover electricity thefts was to go into homes and find the wires and tampered electric meters. It was an open question, however, whether police and company agents could do this. The answer came in 1905 when Isidoro Díaz, a storeowner, brought a 12,000-peso lawsuit for slander, *allanamiento de morada* (breaking and entering), and property damage against Ernesto Lobo, an inspector of the Compañía de las Fuerzas Hidro-Eléctricas de San Ildefonso (Hydroelectric Power Company of San Ildefonso, soon to be absorbed by Mexlight). Lobo had approached the sixth police precinct, demanding Díaz's arrest for the fraud occurring at his Dos Estaciones store. He claimed that Díaz was using a *brincador* (jumper cable) or wire that bypassed the meter and prevented it from recording consumption. Worse, according to Lobo, he was illegally feeding power to two homes with no contract. A judge agreed and imprisoned Díaz for eight days. Upon release, however, an enraged Díaz brought his lawsuit against Lobo.

Lobo based his defense on three arguments. First, the seventh clause in the electric lighting contract authorized company employees to carry out installations and inspect the electric meters at any time. Second, eyewitnesses who testified that Lobo had carried out the inspection were lying, for it was actually his underling, Carlos Olaguibel, who entered Diaz's home. This was a merely academic point, as Lobo pointed out, because Díaz had already conceded that company employees had the right to inspect. Third, according to Lobo, article 637 of the Penal Code stipulated that breaking and entering entailed entering a house, dwelling, or bedroom used as *habitación* (living space). A convenience store "is not used as a living space. It was instead an establishment open to the public." Thus, even entering the store would be no crime: anyone could "penetrate" a commercial establishment. It took two years, but in 1907 the Federal District Higher Court confirmed the lower court's verdict that Lobo had committed no crime. The court unanimously agreed that no breaking and entering had occurred because company employees had the right to inspect installations based on the seventh clause of the contract. The consent to "penetrate" homes or establishments was granted at the moment when the contract was signed.[24]

It was a victory for Mexlight, but not as complete as it might seem. Although subscribers were contractually obliged to allow an inspector's entry, they could be delay access to their property, demand conditions for entrance, or deny it altogether if the home or business owner was absent. More frustratingly, courts determined that contract clauses did not apply to nonsubscribers, so homes or businesses siphoning power from legitimate subscribers were under no obligation to allow inspectors to enter.[25]

Despite Mexlight's established legal right to penetrate certain private spaces, capitalinos pushed back. They not only questioned whether these intrusions were fair but argued that they violated gender, class, and moral understandings of space. Capitalinos went beyond the question of legal rights and asked a more profound cultural question: Was it decent for the inspectors to command entrance into private spaces? Mexicans attached moral values to their spaces: as the historian Sonya Lipsett-Rivera notes, residential homes "were a vital part of the way Mexicans defined themselves as moral beings." She identifies the main door as "the most potent signifier of morality and proper behavior." An open door during the day signified that the home was respectable and had nothing to hide, while that same door had to be closed at night because no respectable family carried on business so late.[26]

The threshold itself stood as a liminal space (associated with but not entirely part of the house), a physical demarcation between the private and the public. When a property owner was confronted by an inspector, it was only proper—indeed, moral—to allow the inspector into these spaces, but the moral calculus flipped when inspectors (with the force of the law behind them) demanded to go farther. Inspectors and gendarmes awaited permission to conduct inspections at the threshold, as in the case of Enriqueta Ruiz. In larger homes, they were addressed from high balconies or made to wait at the zaguán, an element of traditional Spanish architecture. In practice, these spaces served a gatekeeping function. If theft was indeed taking place beyond the zaguán, the thieves' indignation that inspectors might dare to violate their privacy became a convenient pretext to deny access.

Gender played a considerable role both in thefts and in the justification for sending inspectors away. In ten of the sixty-three cases located, inspectors accused women. Although only one was identified as a widow, many of these women were likely heads of households in charge of securing lighting sources for their homes. Like many societies in the early twentieth century, Mexico imagined that the home was a safe, ordered refuge run by virtuous women. It stood in stark contrast to the disorder—and even violence—of the streets. The capital historically had a high incidence of female-headed households. From the late colonial period to the mid-nineteenth century, they accounted for up to a third of the city's households. Female-headed households remained common during the Porfiriato and even expanded through the revolution, when migration to the capital by women far outpaced that of men.[27]

Indeed, the disruptive nature of war deepened internal migration and the overall vulnerability of women in the capital. When inspectors demanded entrance into one of these homes, women felt entitled to refuse, given that the cultural consequences of giving permission could be high. Women's refusal to allow inspections is understandable within a gendered honor system centered around the home. Women's *honra* (honor conferred by virtue) was intimately linked to sexuality: their reputation was bound to "their chastity as maidens and then their loyalty as wives and discrete respectability as widows." A woman who did something that could be seen as unvirtuous risked tarnishing not only her reputation but the reputation of the men in her life. Women were, theoretically, under their control, so a stain on women was "an attack not only on male honor but also on the very masculinity of their husbands, brothers, and fathers."[28] Women's reputations had far greater implications for their whole family than men's reputations. Women fought hard to guard their reputations, going so far as to legally and physically confront those who endangered them.[29] Respectable women did not allow unknown men into their homes. When an inspector demanded entrance into a woman's home, it was inescapably a test: would she allow her virtue to be questioned by the community?

Women alone in a house felt empowered to deny access—in violation of the law—based upon a notion of virtue. The penetration of private spaces by company employees, particularly when accompanied by gendarmes, could be—and was—interpreted as an insult to the residents' honor. The maid at the Jah house denied entry, citing the absence of the man of the house. The realm of its patriarch, the house was not to be penetrated by other men in his absence. No capitalino batted an eye at denying access on those grounds alone.[30] Some inspectors and gendarmes who sought to inspect the installation of city councilman Ángel Montaño's house in 1918 were asked by his wife to wait in the zaguán while she sent for her husband.[31] Tellingly, the inspectors obliged.

Not every capitalino had a servant to stall an inspector, let alone a gated wall or a balcony from which to converse. Many were too poor to rent or own a *casa* (house) and resided in either a *vivienda* (shared space within a building, ranking below houses) or in *cuartos* (rooms, the humblest form of lodging and one of the most common). For these capitalinos, as Lipsett-Rivera explains, "The private was often very public."[32] The absence of a stark separation between public and private spaces meant that people's actions were under greater scrutiny. Having so many people living in such close proximity could be a blessing or a curse for those accused of theft. On the one hand, while poorer people lacked servants to conceal questionable behavior, they might have caretakers or friendly neighbors who carried out the same role. On the other, many more eyes were watching for opportunities to settle old scores.

The detection of electricity theft thus required increased vigilance. Inspectors Léopoldo Fernández and José Saldaña carried out a nighttime inspection of the house of Manuel Castro, where they suspected a jumper cable was in use.[33] They arrived at Castro's house at 9 P.M. along with a gendarme. They had warned him that if lights went off when they introduced themselves, that would be evidence that a jumper cable was in use. A woman came out to the balcony but went back inside after learning their purpose. Before she came to the door, the lights in the house went off and immediately turned back on. The inspectors confronted her and demanded that she hand over the illicit cable or explain the interruption. She adamantly refused their requests. The inspectors and the police officer left: the woman was not the patriarch of the house, so they would take a great risk by pursuing the issue.

Cognizant of these moral connotations, Mexlight attempted to frame these inspections not as policing but as a safety measure. Not long after the higher court ruling, in late 1907, Charles H. Cahan, Mexlight's manager, claimed that a considerable number of commercial and residential houses were at risk of fire.[34] Business owners placed themselves in harm's way, Cahan declared, by carrying out installations themselves or hiring individuals whose credentials could not be verified.[35] Inspectors estimated that 3,000 of the nearly 30,000 Mexlight clients had dangerous installations, which, given the city's fire risk, put whole neighborhoods in danger. The company had sent letters to warn customers and explain that corrections could mean considerable savings, but only 300 clients had responded by the time Cahan made his plea. He concluded that people failed to understand that inspections were for their own benefit. Customers might well have understood his offer but nonetheless saw opening their homes and businesses to inspection through a different lens.

Cahan was right that faulty circuits were dangerous. If capitalinos would not voluntarily allow company workers to correct them, Mexlight would take advantage of its newfound legal right and send inspectors for unannounced visits. Clients, in turn, went to greater lengths to camouflage the cables. An electrical goods store at San Felipe Neri and Isabel la Católica had two engines connected to its meter, which were used to test the merchandise sold there.[36] It was an impressive arrangement, so complex that for two years it hid the fact that the primary job of the two engines was to mask an illegal connection between the store and a nearby hotel to power its sign.

The inspectors were as diverse as the city itself and reflected the migration of Mexicans from the country to the metropolis. While the energetic inspector Paliza claimed to be from Mexico City, Gabriel Aranda (thirty-nine years old) was a widower from the central state of Guanajuato. Gabriel Abrego (forty-two) was married and hailed from the city of Chilpancingo in the state of Guerrero. They also reflected the rise of foreign immigration under the Díaz regime. Inspector Javier de la Vega (forty-four) was born in Santander, Spain. The supervisor of inspectors, Juan Emilio Boesch (thirty-one), hailed from Switzerland.

Like many capitalinos, the inspectors not only came from everywhere but lived all over the city—an important fact, given that inspectors were most effective in their own neighborhoods. *El Imparcial* narrated the apprehension of Enrique Lara, who owned a diner along Callejón del Garrote and had contracted for lighting service but stopped a few months later.[37] He secretly reestablished the connection, which an inspector believed had been stealing power for about six months, at a value of over 200 pesos. The report never named the inspector. Perhaps coincidentally, the inspector general, Carlos Paliza, resided next door to Lara on Callejón del Garrote at building no. 2, apartment no. 5.

Inspectors were normal people who held real power over their neighbors and fellow citizens by threatening fines and jail time. As the case against Antonio Vázquez Suárez demonstrates, inspectors could be as susceptible to abuse as anyone else. A rumor reached the offices of Mexlight in 1912 that someone in the El Chorito area had been making and installing "combinations," possibly jumper cables, to cheat meters. Tasked with locating these installations, inspectors Ignacio Magaña and Manuel Rodríguez Miramón later wrote that they had learned that Vázquez Suárez made and sold cables at a rate of 75 cents each and found one at his house. Magaña immediately left to find two gendarmes to bear witness, while Rodríguez Miramón stayed behind to guard the evidence.[38] After an inspection by court personnel, police arrested Vázquez Suárez, who maintained that he knew absolutely nothing about lighting, let alone electricity. The brincador found at his house, he added, must have been planted by one of the inspectors before calling the gendarmes. He suspected Rodríguez Miramón, claiming that he had quarreled with him in the past.[39]

For his part, Rodríguez Miramón testified that he had never met the accused. Although he could not provide the court with a list of names and addresses of those who bought jumper cables, he maintained that Vázquez Suárez sold them in the area. The accused denied the allegations, claiming that vengeance motivated Rodríguez Miramón—an accusation made credible when two character witnesses testified that the inspector and the accused had known each other for years and been rivals for the same woman.

An accusation of theft was no minor matter. With it came searches, possible trials, and, frequently, jail time. The spectacle of raids made questionable private behavior public knowledge. Neighbors and friends who witnessed these scenes constituted an "insider audience" who held unique "power" over others, for they "had a collective memory for a person's reputation and honor."[40] Questions of trustworthiness and righteousness had direct social and financial consequences in a world without credit scores or other objective means to establish risk, as the previously discussed case of Isidoro Díaz shows. Díaz brought a lawsuit against the Compañía de las Fuerzas Hidro-Eléctricas de San Ildefonso for 12,000 pesos in 1906. He had been accused of electrical theft (a charge he denied), which led to the temporary closing of his grocery store. In calculating his demands, he asked for 11,000 pesos in compensation for "moral angst."[41] As uncomfortable as his brief stint in jail had been (the physical distress he referenced), the real suffering would come to his future business. Any blemish on his moral standing in the community would mean higher costs to do business.

The myriad scenarios emerging from these inspections reveal that when suspicious inspectors—who believed in the justness of their work—knocked on doors and demanded entry, they were not simply trying to "penetrate" homogeneous, interchangeable "structures." Rather, they approached spaces embedded with meaning for those who carried out their lives within them. It mattered what that space was and who occupied it. An open commercial shop may have been "public" enough to warrant easy access with little public pushback, but capitalinos bestowed a degree of sanctity and privacy upon even a busy apartment vestibule—and certainly a gated casa. Likewise, who answered the inspector's knocks mattered, for they were not all mere "clients." A poor man living in a room might not have the moral prerogative of a widow or the financial heft of a wealthy businessman, but even he had honor and standing to protect and had to weigh carefully whom he allowed into his space. Inspectors were likewise humans who had their honor to protect and morals to uphold and could, like anyone else, transgress both.

LADRONES DE LUZ, *DIABLITOS*, AND THE SEMIMECHANIZED WORKSHOP

A plague of robbers—all members of the lower classes—had descended upon the metropolis—at least, that was the perception of social commentators at the turn of the twentieth century.[42] This conversation about theft paralleled the rhetoric of pedestrians, motoristas, and eléctricos: modernity would be thwarted only by the unwholesome and fundamentally backward actions of the unsophisticated, poor masses. In his diagnosis of criminality, jurist Miguel Macedo insisted on the urgent need to install "a levee against the criminal wave menacing to bury our pueblo."[43] In order to help his fellow capitalinos, Macedo classified people via an easily identifiable marker: their clothing. According to his classification, "individuals wearing plain shirts were the lowest class, followed by those who wore jackets and, at the top of the scale, the group that wore frock coats." In Macedo's remarkably simple world, all people had to do to avoid crime was to steer clear of those in plain shirts and gravitate to the frock coated-classes. The problem was that not all criminals could be so simply categorized. Unlike the common perception of *rateros* (thieves) hailing from the lower classes, most power thieves sported jackets, and a few even wore frock coats.

Data on the extent of power theft in the capital are sparse. For once, capitalinos had found a crime that did not incite spasms of dread and anxiety. No study of power thieves examined them as a single criminal class. Then as now, little is known about them. Impressions and unsupported assertions have had to stand in for facts. Reflecting on the magnitude of power theft in the city in 1907, for instance, manager Cahan claimed electrical thefts so widespread that ladrones de luz could overflow the Belem jail. Therefore the company charged only those against whom it had substantial evidence. Cahan and Mexlight reached individual settlements with all others, who carried out petty theft or defrauded the company without malice. To give some measure of the problem, Cahan explained that company inspectors had allegedly discovered more than 300 theft cases in the preceding months, but only a handful were significant enough for criminal charges. The manager's admittedly self-serving assertions indicated that anyone charged by the company must have done something egregiously wrong.

By the 1910s, the number of owners of commercial establishments brought to court closely approached the numbers of residential owners. As the problem grew in the commercial and manufacturing sectors, Mexlight shifted from Cahan's more moderate stance. His successor, Haro Harrsem, took an aggressive position: if the company brought more criminal charges, it could curtail the abuses of those acting in bad faith.[44] Despite the company's surveillance efforts and reforms to the 1871 Penal Code, cases continued to proliferate. Without disclosing its methodology, *El Demócrata* argued that as much as 80 to 90 percent of energy subscribers in 1924 used "diablitos" (little devils), illegal devices used to divert electricity from power lines.[45]

The electrification of factory work increased industrial output while reducing both the cost and time of production. It also altered the city's physical nature by enabling the expansion of industrial life in the city center instead of in scattered hamlets around the Federal District.[46] The growing footprint of electrical consumption in the city had significant consequences for the nature of work in factories and workshops generally, but nowhere was electrification more transformative than in semimechanized workshops.[47] Electrical power brought new machinery, which led to a new workshop organization in which unskilled workers undermined or altogether eliminated the need for traditional artisans.[48] Small and medium-sized workshops remained in operation, but many crafts came to be dominated by a few companies organized around large, semimechanized workshops that, in turn, placed pressure on smaller enterprises and traditional artisans.

Factory owners such as those from San Antonio Abad textile factories studied the introduction of electricity. Electric power and the introduction of electric machinery notably transformed the traditional textile industry by drastically reducing the number of hand weavers and applying pressure upon smaller businesses that lacked the means to acquire the new machinery.[49] As smaller companies folded, in their place came corporations and more massive factories. Along with new machinery, personnel had to be trained or else recruited to run and service it.

Many factories engaged in only a partial introduction of electrical machinery, particularly when it was financially justified, as in the case of San Antonio Abad. Located within city limits, far from water sources that might have provided hydraulic energy, the San Antonio Abad cotton textile factories had relied on steam power since their inception, but forward-thinking management had always run the company. In 1883 a steam engine fed a Brush dynamo that, in turn, powered forty-seven arc lamps (1200 CP) and lit up the factory.[50] Thirty-five lights illuminated the weaving room, while the rest lit the machinery room, machine shop, boiler room, garden, entrance, and other strategic sites. The weaving room alone was a massive structure—almost the size of a football field—and housed 337 looms. With artificial lighting the work lasted an extra five hours every day, beginning at 5 A.M. instead of 6 A.M. and continuing to 9:30 P.M. instead of 5:30 P.M. The cost of this lighting system averaged 35 pesos daily, which included firewood, carbon, oiling, and personnel, while the cost of the 150 gaslights that preceded bulbs was nearly 55 pesos a day. Electric light also provided better illumination and did not raise the inside temperature.[51]

Electric lighting did not lead to the electrification of the factory's machinery. Coal-fueled, steam-generating boilers produced the horsepower necessary to feed the factory's machines until 1906. The factory counted itself in the minority of enterprises that enjoyed a reliable and affordable coal supply line from the Río Frío forests, which allowed the factory to delay the introduction of electric power by twenty years. The engineering professor Daniel Palacios toured the factory with students in 1903 and concluded that the delay was justified: the cost of steam horsepower stood at three cents per hour, while the electric horsepower reached five cents and over six if the operator's wages entered into the equation.[52] With its longer workdays, lower costs, and machines that did not require skilled workers to operate, the San Antonio Abad factory was in an enviable position. Its competitors realized that they too would need electricity if they were to have any hope of competing, so they sought to secure a power source—even if that meant stealing it.

The power thieves did not constitute a monolithic group; if they shared anything, it was an inability to be easily caricatured. However, this group did have an economic hierarchy. On the top were factories, hotels, and theaters. As was the case in San Antonio Abad, factories employed large amounts of electricity to power countless types of machinery for extended periods. Electric companies offered several types of contracts for these establishments. Customers could request light, heat, a power service, or a combination of all three. The resulting contract would guarantee either a predetermined horsepower or a given number of electric lights, depending upon the services requested. Of course, clients—even large, well-established businesses—did not necessarily honor contracts. Rodolfo Schuzandubel, the CME manager, accused owners of the La Michoacana tobacco factory of theft in 1903.[53] During a visual inspection, Schuzandubel found that although the factory's lighting contract prohibited electric energy being used for anything other than lighting, workers had illegally connected to a power line to run machinery. This violation, Schuzandubel charged, constituted theft.[54] A more subtle form of business theft involved tampering with electric meters. Nápoles, the electrician caught in flagrante at the Monte Carlo theater, admitted that it was a common practice for theaters in the city—and other businesses—to reverse the meter's dials. Given their energy needs, it is not surprising that other large commercial establishments also engaged in substantial fraud.

After factories and theaters, the next largest category of power thieves was mid-sized businesses, often sole proprietorships. A plethora of shops (grocery stores, dairies, diners, drugstores, hair salons, and tailor, shoe, and mechanic shops) are represented in court cases. From these accounts it is clear that, as restrictions on entrepreneurship lessened in the second half of the Porfiriato, more people began business ventures, competition intensified, and entrepreneurs did all they could to keep an edge on rivals. While the numbers of middling business owners grew, so did the power of monopolies: business owners collectively lost political, social, and economic influence.[55] Sole proprietors existed on razor-thin profit margins and scrutinized every expense. In many cases, establishing safe and professional installations, signing contracts, and paying for energy consumption fell on the wrong side of profitability.

Finally, at the bottom of the hierarchy of ladrones de luz were homeowners who siphoned off electricity to improve their quality of life. They most commonly used power to light domestic spaces at night but could find other uses too in a world of expanding electrical possibilities. Vázquez Suárez, accused of making and selling brincadores, mounted a telling defense. He claimed that he had no need for such a device because he took only enough power to light his home for half an hour every night and invited police into his house to look for a jumper cable. He directed authorities to pursue those who had a heater, appliances in their bathroom and kitchen, or an iron for their clothes.

One last, shadowy group must be considered part of the ladrones de luz: those who eked out a living making the illegal connections and mechanisms to tamper with electric lines and meters. Adept individuals such as mechanics, electricians, and current or former electric company employees sold or bartered their skills in handling electric lines. These local professionals joined with a cadre of homegrown inventors to facilitate the expansion of electrical use—just not in the script that electrical companies encouraged. Sometimes their efforts were as simple as splicing or throwing cables over a line. Even when Mexlight disconnected both legal and illegal service, capitalinos could simply reconnect the wires or arrange for someone to do so. However, some individuals channeled notable amounts of ingenuity and skill into devices to steal electricity.

The most sophisticated case emerged in 1911 with a man named Palmer (his first name is never given) and his device to tamper with meters. According to *Nueva Era*, Palmer, a former Mexlight employee, had invented a miniature transformer known as the "Palmer transformer" among electricians, dubbed a diablito by "dishonest clients."[56] Once installed, it moved the meter's dials backward, lowering the recorded amount of electricity consumed. Thieves had previously employed hairpins or magnets for tampering, but those tricks were easy to spot because meters typically broke down. Inspectors Julio Manuel Muñoz and Felipe E. Paniagua allegedly discovered a Palmer diablito at a hardware store. When they arrived to carry out an inspection, the owner's nephew, who did not understand who they were, naively explained to them how the apparatus worked. The inspectors found one in the attic. When the uncle arrived, he tried to force himself past them and violently remove the diablito, but Muñoz overpowered him and secured the device. The newspapers made no mention of whether the company filed charges against the uncle. From the perspective of the broader electricscape, he, in a small way, was "promoting" the adoption of electricity as a source of light and power but outside the script of the power company. The power company did not appreciate his efforts, for an electrified Mexico following their script would primarily benefit them.

How many diablitos, brincadores, and simple illegal splices were there in Mexico City? The inspectors and the electric companies never knew, and neither do we. Some made best estimates, others made wild guesses, but, tellingly, no one dismissed the problem as insignificant. What we can say is that electric companies considered the problem large enough to justify sending squads of inspectors to stop it. Capitalinos, on the verge of a technologically transformative moment, were not going to accept that this life-altering resource would be rationed out only to those who could pay. They believed that they had the right to a modern, comfortable life, like anyone else, and would defy the inspectors to secure it. The companies, in turn, would go to further lengths and called on the judiciary to safeguard their right to profit.

Theft brought forth a growing number of individuals who tried to police the quotidian use of electricity to little avail. Company inspectors, lawyers, and jurists tried to demarcate what constituted an unlawful use of electricity and the price to pay for those who engaged in it. Courts became arenas in which daily dramas revealed how users went off script and how they understood and defined electricity in its context of use.

WHEN IS STEALING NOT A CRIME?

Power theft challenged the Mexican judiciary system, jurists, and lawyers. For over three decades before the Ambos Rumbos Hotel raid, article 368 of the 1871 Penal Code provided the statutory definition of *robo* (theft) as "the appropriation of *una cosa mueble* [a movable thing] without right or the consent of its legitimate owner."[57] This article served as the basis to prosecute theft, so defense lawyers naturally probed its limits. Under mounting challenges and criticism, jurists raised calls for reform. As an editorial in *El Imparcial* explained, although the Penal Code had been "an efficient, adaptable, and deliberated body of rules when issued, it no longer met the progress of modern life."[58] Civilization had created a diversity of crimes unknown until recently, so the code needed reform. At the center of these calls was the problem of electrical theft.

Lawyers challenged the prosecution of electrical theft under article 368 on two intertwined grounds: the fact that no penal law proscribed power theft and the nature of electricity. Jacinto Pallares, a prominent jurist of the Porfiriato and lead attorney in the defense of Francisco Torres, the owner of the Ambos Rumbos Hotel, claimed that his client had violated no penal law, although his actions were reprehensible.[59] The prosecutor's application of article 368 to his client's case, he insisted, violated constitutional article 14, which barred interpretative judgment of penal laws and demanded that laws be applied accurately. For an act to be considered felonious and punishable, the penal law had to include it in literal, exact, precise, and unequivocal terms. As defined by article 368 (and 370, which defined the precise moment when an action turned into theft), theft constituted the appropriation of someone else's movable thing that the thieves literally held in their hands. Given the legal precept, Pallares asked whether it could indeed be possible for a thief to hold electricity in his hands or for a thief "to handle a current that shocks, often knocking down the careless or stupid individual."[60]

Judge Ismael Elizondo, who heard the case against Torres, found the articles applicable, based on his judgment that electricity was a movable thing.[61] In legal parlance, he claimed, *cosas* (things) and *bienes* (properties) were synonymous and constituted all that made up a person's estate. As Elizondo explained, "whether called a fluid, current, energy, or what have you, no one could deny that thanks to the century's prodigious progress and man's activities, man had imprisoned it [electricity] between his hands, using it as a lucrative force."[62] It was no longer a matter of debate whether electric current was a thing because it had been commercialized and was thus susceptible to appropriation. It was indisputable that electrical current could be transmitted and transported from one place to another, regardless of the medium, and therefore it had to be deemed movable.

Pallares, reading the law literally, challenged Elizondo's definition. Pallares contended that a thing needed to be subject to being physically taken for there to be theft. Indeed, in order to be considered movable, an item had to be susceptible to being handled. In other words, it had to have a physical body that could be "isolated, dispersed, individualized, and independent of other bodies."[63] Legally, there was no theft unless the item passed from its legitimate owner to the individual who committed the crime. Furthermore, the essence of theft consisted of physically seizing an item and claiming it: only then did it violate the rights of the legal owner. The simple act of physical seizure did not constitute theft, which also required taking the item away from the owner, into the possession of the thief. Only physical seizure could meet the fundamental description of theft as defined in articles 368 and 370.

Prominent jurist Demetrio Sodi stepped into the fray and supported Pallares's stance, arguing that two cardinal elements had to be present to constitute theft. First, the object had to be seized from the owner—the substance of the crime—without being handed over voluntarily or by error or fraud. Second, the object taken by the thief had to be movable. Sodi found it inconceivable to think that someone could steal electric current by the mere act of installing a connection from the wires of the electric company. He reasoned that the electric current remained in the "full and absolute possession" of the company that generated it, which could at any time suspend, reduce, withdraw, or cut service. Basically, Sodi reasoned, the essence of theft—that is, physical dispossession—was impossible, given the nature of electricity.[64]

Torres's case brought forth difficult questions over the regulation of the new energy. Despite Pallares's specific challenge to Elizondo's interpretation, an appellate court upheld Torres's sentence and denied his *amparo* (writ of protection).[65] Judge José Saavedra argued that Torres had committed theft, even if the legal definition did not apply to electrical theft.[66] It was evident to him that the authors of the Penal Code had never imagined that electric power could exist, so they could not have legislated what did not exist. He held that lawmakers had simply not yet had time to legislate on the recent advancements. As if in response to his ruling, calls for reform intensified in the case's aftermath. Still, there were holdouts: traditionalists who held that power theft was indeed included and penalized in the 1871 Code. Judge Enrique Peña y Aguayo was one of them, but even he considered it convenient to add a particular article to avoid "twisted interpretations" and doubts.[67]

A consensus crystallized as the reform project developed. Prosecutors and jurists called for moderate changes to the articles that until then had regulated theft. The public prosecutor, Francisco Cortes, claimed that a simple change to article 368 was enough. He proposed adding the phrase "theft is committed by one who appropriates someone else's corporeal or *incorporeal* thing, without right and the consent of the lawful owner."[68] Prosecutor Salvador Ferrer suggested adding a new statement to article 381 that punished the wrongful use of electric current with eighteen months in prison.[69] Judge Emilio Tellez stressed the need either to include a new article or to broaden article 368 to define theft in such a way that courts could try such thefts merely as civil infractions.[70] Public defender José del Castillo, who believed that incidents of theft were reaching alarming levels, recommended streamlining article 368 by suppressing the word *mueble* and called for harsher punishment.[71] If he had his way, those with sentences of more than a year would go to penitentiary colonies and would take up residency away from urban centers under police surveillance upon release.

Though they were on different sides of Torres's case, Elizondo and Pallares agreed on the need for reform. Having argued for the applicability of current articles, Elizondo recognized that the definition of theft needed to include the pillaging of all currents and fluids, incorporeal and intangible, to avoid controversies and prevent the crime—which he feared would grow only more frequent in the absence of clear laws.[72] Only then, he argued, would the interests of companies and individuals be guaranteed. Pallares agreed but thought the solution could be arrived at more easily. He pointed out a single independent article regulated water, the most movable and easiest thing to transport. Something similar could be developed for electricity.

Everyone involved with the Torres case agreed that he was no saint and that he took electricity for which he did not pay. The question was whether he had actually broken a law. His lawyers made a shockingly effective case on this point, though Torres personally lost. Prosecutor, defendant, judge, and third-party jurists all agreed—to one degree or another—that the law was unclear: given the demand for electricity, capitalinos would likely try to find any loophole through which they could siphon power. Something had to be done, or no investors, domestic or foreign, would dare invest in Mexico. However, as the electric companies' own inspectors perhaps could have warned them, crafting reforms would be easier than enforcing them.

COURT DRAMAS

Act I: La Velocitan

In 1906, five years after the juridical chaos of the Torres case, La Velocitan leather tannery joined the infamous list of enterprises accused of power theft.[73] Located in the area of Santa Crucita, less than two miles from the Zócalo, La Velocitan was no small enterprise. It was the largest tannery in the country, with a capacity to produce 300 leather soles daily and, according to the Ministry of Development, had a mere dozen competitors.[74] It was also a power thief: in order to defraud Mexlight, the company had developed ingenious plans and specialized employees.

The tannery owed its name and industrial capacity to the Velocitan process of quick tannage, which promised to save time, trouble, and money.[75] La Velocitan Fábrica de Cueros, established as a public limited company in 1898, mirrored the city's manufacturing companies. As was the case with much of the commerce, real estate, and large-scale industry in the capital, a tight-knit group of mostly foreign-born investors controlled the company.[76] Spanish and French shareholders based in Brussels had acquired rights to exploit the Durio brothers' patent on the Velocitan process. In addition to La Velocitan in Mexico City, they established the Velocitan Compañía Anónima de Curtición Ultra-Rápida in Barcelona. Both companies specialized in the manufacturing of belting, soles, and harnesses.[77]

Internationally, the increased market for leather coincided with the new quick tanning process and the introduction of new energy sources. Industry commentators could not hide their glee at the changes. "The modern tanner," claimed American tanning expert Charles Thomas Davis, "cannot stand still. . . . Electricity is driving out steam power . . . to stand still is to be passed by more enterprising men and completely distanced in the commercial race."[78] Armed with technology and financial muscle, the Velocitan entered the Mexican market at a time when large quantities of bovine leather were needed to make bands for new industrial machinery. Its founders, clearly sharing Davis's thoughts on the "commercial race," wanted the company to enter with an enormous lead over its rivals.[79] The tannery's preferred method involved soaking skins in a revolving drum filled with a liquid that accelerated the handling of the unhaired hide, a modest modification that paid handsome dividends.[80]

Inspector General Paliza brought a complaint against La Velocitan to the public prosecutor's office in March 1906.[81] His suspicions of theft emerged after a simple examination of the business's accounts; the estimated power needed to run the factory's machinery was much higher than the meter readings, evidence that theft had been happening for some time. He had no way of being sure a crime had taken place or how it was carried out, but he estimated loses of no less than 20,000 pesos. An inspection by the police superintendent of the sixth district precinct confirmed his suspicions and resulted in the arrest of manager Juan Caffarel, technical director Eugenio Heinriche, and mechanic Reyes Granados.

During his first interrogation, Granados (twenty-four years old) admitted that he had overseen the factory's electric engines for years and explained how the fraud was carried out. Although three electric engines of varying horsepower supplied 1,260 volts, a fourth engine supplying 3,000 volts—which would have demanded a newer, higher-capacity voltage meter—had replaced an older 1,260-volt engine whose output passed through the existing meter. The police submitted to the court a diagram showing how the company was able to install the more powerful engine without registering more electrical use. The meter for the new engine required a two-phase connection to work properly, but mechanic Granados had disconnected an essential cable running to the meter (represented in red in the police diagram), thereby preventing it from registering energy consumption without impeding the engines' operation.

With the crime clearly established, officials next set out to assign blame, but that process would involve less clarity and far less fairness. The first thing the courts did was to commission the Oficina Electro-Técnica Mexicana, an agency tasked with verifying electrical installations, to estimate the amount for damages. It arrived at a sum of 43,455 pesos, which was paid to Mexlight by La Velocitan within a week. Mexlight dropped charges against the tannery almost immediately. However, it did not drop charges against the factory's employees. The public prosecutor's office indicted Caffarel, Heinriche, and Granados three months later. The status differences in the tannery's vertical labor hierarchy would be revealed in these trials. Per his instructions from Caffarel, for about a year Granados had performed a special *maniobra* (maneuver) twenty to twenty-two times a month: he took out the cable connecting the transformer to the meter, preventing it from registering power usage. Paz Villagaña, a former employee, had carried out the task until he had fallen ill. He had managed to train Granados before his death—even leaving him his set of rubber gloves and electrical pliers. According to Granados, Paz Villagaña gave him a little black notebook to keep a record of usage. He claimed that he had misplaced it, so it was not entered as evidence. For these efforts Caffarel rewarded Granados with ten extra pesos a month.

Granados's honesty did not serve him well, and he quickly found himself alone. Juan Caffarel, a thirty-year-old French citizen and the tannery's manager since September 1905, denied any involvement or knowledge of the operation. He insisted that he learned of the fraud only when the police arrived at the factory. Even the gloves and pliers that Granados used, he said, did not belong to the company. Caffarel acknowledged the payment of ten pesos but maintained that this "custom" had preceded his arrival—it was a bonus or compensation for "cleaning the engines." Two other employees, Justino Gibaux and Eugenio Heinriche, also denied having any knowledge of the maneuver. French tanner Gibaux admitted that he had "heard, but did not remember from whom," of the missing connection between the meter and the transformer.[82] Like Caffarel he claimed that he did not learn the details of the operation and who executed it until the day of the inspection.

Likewise, Heinriche, a German citizen employed by the tannery for seven years—who was, strictly speaking, in charge of the machinery in the factory—also distanced himself. He could offer the police only broad generalities. The engines had previously run by steam, which had been replaced by electricity for the last three or four years. Gibaux oversaw the machinery used for manufacturing leather goods, while Granados was responsible for electric engines; neither reported directly to him. Heinriche claimed to be as dumbfounded as anyone in the matter. Prosecutors offered Granados the chance to ratify his original statement, but he reiterated that only two individuals knew of the maneuver: Caffarel and Andrés (an employee identified only by first name), who had left three months before. The testimony of Granados and Gibaux absolved Heinriche. Despite Granados's statements, the court also ordered the release of Caffarel, declaring that his responsibility "had not been proven."[83]

From there, matters only worsened for Granados. Left as the scapegoat, he was easily convicted in absentia by the court. He had fallen ill with pneumonia and was not present at the indictment hearing or at the rescheduled hearing. Court personnel arrived at his tenement house on June 29. A messenger told the bedridden Granados that a sentencing hearing would be held. Granados's lawyer had already entered a motion for acquittal, claiming that the evidence was too scanty for a conviction: his client, with no judicial record, had not benefited personally from the alleged theft. The judge denied the motion and found Granados guilty of illegally taking electric power to benefit the tannery by preventing the meter's proper function.

Once again, as in the guilt cast upon motoristas after accidents, employees were left to take the blame for the company. In addition to article 413 (fraud against private property), the judge applied article 432, which penalized unspecified fraud and set a fine of 25 percent of the damages, not to exceed 1,000 pesos. Granados was convicted of the second charge and sentenced to 100 days in jail or a fine of 1,000 pesos. There was no payment; Granados died a day later from pulmonary tuberculosis.[84] His family laid his body in a sixth-class grave at the Panteón Dolores, the city's largest public cemetery. Coincidentally, on June 29—the day of Granados's trial—a newspaper reported Caffarel's death.[85] He would be buried at Panteón Francés, however, the more exclusive cemetery for Mexico City's well-to-do citizens.

The malfeasance at La Velocitan may appear exceptional (why, after all, would a dominant manufacturer that already enjoyed several advantages over its competition engage in theft?), but sizable factories and workshops were theft-prone. The reason was simple: because they, unlike smaller competitors, could. The introduction of electrical machinery not only altered the nature of work and production but, especially in the large factories, created specialized positions to oversee electrical machines and engines. These employees became familiar with, even expert in, electrical matters. Such was the case of Villagaña and Granados and perhaps others at the tannery. Curtailing power consumption seems to have been part of their job or at least a special task rewarded with a bonus. Granados's training, the tracking of consumption levels via the little black book, and the regularity of the practice indicate that the practice was not impromptu: it was an essential part of the production process, developed in hopes of deceiving the electrical company and escaping the inspectors' attentive eyes. Yet, as in the case of the motoristas with schedules that they had to keep or face fine and firing, how free were workers like Granados to say no? For an ordinary worker, ten pesos more a month could constitute a substantial raise. Irrespective of the amount of pressure on employees, that money would be hard to turn away.

Act II: Molineros

Commensurate with their size, smaller businesses like cantinas, bakeries, *molinos* (maize mills), and various commercial houses carried out less audacious thefts than large factories like La Velocitan. Use of electricity at businesses like these was far below consumption in factories and large establishments, typically powering one or two engines or a small circuit of incandescent lamps. Consequently, most lacked skilled personnel to execute more ambitious theft. Still, they wrestled with the cost of electrical consumption while struggling to remain afloat during an economic downturn in the first decade of the twentieth century and an armed struggle in the second decade. With varying success, they contracted with outside individuals to help them get out from under the weight of electric bills.

A month after uncovering fraud at La Velocitan, Paliza heard from an unnamed source that a molino on San Antonio Abad Street had installed a mechanism that prevented its meter from functioning properly. He secured the arrest of the *molinero* (mill owner), Roberto del Villar; his niece, who actually managed the mill, Dolores del Villar; and a business partner, Esteban Castillo, on charges of theft.[86] Roberto del Villar was the sole proprietor of the San Antonio Abad mill and co-owner with Castillo of a second on Chabital de Monserrate Street. Suspicious, Paliza decided to inspect the other mill and found two incandescent lamps not connected to the shop's meter. Paliza deserves credit for his tenacity rather than for having a brilliant hunch, for it was well known that owners who stole power in one establishment typically stole in others.[87] Despite official efforts to curb theft, the city's dailies claimed that it abounded among merchants and industrialists.[88] Commenting on del Villar's case, *El País* explained that the mill made greater use of electricity than recorded by the meter via a "special cable." Installation of the cable was so easy and ubiquitous, maintained the daily, that it had become commonly used among businesses.[89] In del Villar's case, a pin-sized hole in the upper part of the meter allowed for direct access to reverse the dials. However common commercial theft of electricity may have been, finding an expert skilled enough to make such a surgical incision was hard, so that kind of manipulation was rare.

Dolores del Villar, the manager responsible for overseeing the grinding process, explained that her uncle was an upright man who paid no less than 72 pesos for the electricity to power the mill every month. In fact, she claimed, the amount was frequently higher. A Mexlight employee typically stopped by once a month to read the meter, but that April an inspector arrived. Upon examining the meter, he showed her a cable that had been placed in it, which, he claimed, prevented it from recording the energy consumed. She assured the inspector that she did not know who could have placed it there: fearing an accident, she never got close to what she called the "measuring apparatus."[90]

Roberto del Villar was no stranger to device-regulation, having served as inspector of the Federal District Weights and Measures department since 1896. He initially did not admit any knowledge of the cable in question, but he changed his story before signing his declaration. About a month before the inspection, he explained, Castillo, who was also a mechanic, offered him "a service": a simple "fix" to the meter that could lower his electric bill. Castillo kept his fix a secret and del Villar never asked what it entailed. He did notice, however, that he had been paying much lower bills since then. Two character witnesses attested to his moral decency and honesty: he had a permanent home, was of "decent moral background," and led a "decent way of life" as a government employee and owner of several businesses.[91] In the eyes of the court, del Villar was no common thief whose whereabouts and ways of earning a living were questionable but rather a public figure whose business activities and residence were public knowledge. The court approved his request for bail. Unlike the owners of the La Velocitan, however, del Villar could not simply pay his fines and move on. The court wanted a guilty person to hang the crime upon. Del Villar, who clearly was not about to throw his niece under the bus to secure his own freedom, apparently found it more palatable to offer up Castillo. The court granted him, like Caffarel, the benefit of the doubt, accepting his claim that he had not been the perpetrator. Justice in Mexico was blind to all things except class.

Two years later, a second case against a molinero demonstrated the relationships that contributed to theft within the trade. At the San Antonio mill in Calle de Pueblita, owned by Tomás Sánchez, Inspector Alfonso Gómez found that the meter was not functioning because of a missing fuse.[92] Gómez believed it had been intentionally removed, but Sánchez proposed a simple explanation. Three days earlier, a carter hauling a sack of maize had accidentally banged against the device, knocking the fuse to the floor. No one had touched the meter since. Manuel Aviera, a Spanish citizen and friend of Sánchez, vouched for the story. Aviera had bought several maize sacks and asked Sánchez to store them because he had no room at his own establishment. Sánchez agreed and made the room that housed his mill's engine available. The pile of maize was enormous, almost touching the room's ceiling. A few days later, when a handyman and a cart driver came to pick it up, the room was so crowded that they knocked it into the meter.

Inspector Gómez was having none of it. Based on the mill's consumption history, he estimated that the loss came to 1,089 pesos. Prosecutors indicted Sánchez and sent him to the Belem jail, where he elaborated on his earlier testimony. Sánchez explained that for several months he had been a regular paying customer but had stopped paying for electric service only because he had not received a bill. However, he was more than willing to pay. His electric service had been interrupted a year before. After he talked to Mexlight's lawyer, who fined him ten pesos, service was restored a day later. He was more than happy to do the same now. He had not returned the fuse after the carter accidentally knocked it down, only because he had no idea that the meter needed a fuse or would improperly record his power usage. He also feared that he might install it incorrectly or risk an accident.

With time, Sánchez became more desperate and admitted to some wrongdoing but not electrical theft. He claimed that the difference in the current amount of power consumed by the mill was due to competition. Until the previous year, his mill had enjoyed a monopoly in the area. After rumors reached him that plans existed to establish mills nearby, he approached Mexlight to explain that providing electric power to future mills would be extremely detrimental for him. A Mexlight lawyer assured him that the company would not service those mills, but the promise was short-lived. Soon after their meeting, Mexlight offered service to whoever requested it, undermining Sánchez's monopoly. A total of four mills had opened in the area. As Sánchez had predicted, the competition hurt his business, leading to less production and, consequently, less need for electricity.[93]

Sánchez's bid to keep competition at bay might seem an odd and incriminating thing to admit, but it was a standard business practice. Mills controlled competition via restrained access to electric power. Although neither Gómez nor Paliza, who brought the case to court, commented on the supposed broken promise, such an agreement had likely occurred. The last years of the Porfiriato and the early phase of the revolution saw a dramatic increase in molinos, largely facilitated by corruption. On the one hand, bribes allowed access to specialized licenses required to open new mills within meters of successful existing ones, which, like Sánchez's, were often small and vulnerable.[94] On the other hand, as was also the case for Sánchez, connections between high-level government officials and company agents could keep vital electrical supplies out of competitors' hands. Over time, electricity provided one of the ways by which the two or three largest companies came to dominate smaller establishments and control the industry.[95] The Federal District had 72 mills in operation in 1913. Two years later, the number reached 130, but concentration of the market accompanied this growth. Compañía Mexicana Molinera de Nixtamal (Mexican Nixtamal Milling Company, a Spanish-owned business) alone owned 100 of the 130 established mills.[96] It is conceivable that Sánchez, seeing the writing on the wall, resorted to theft to keep his business afloat after his arrangement with the Mexlight failed.

Corruption and backroom dealing were part of everyday business for molinos, as Gómez experienced firsthand. According to the inspector, Francisco García, a Spanish citizen and owner of El León de Oro grocery store, after talking to Sánchez about the missing fuse, stopped by the mill and invited Gómez and two other Mexlight employees to a nearby cantina. Over drinks, García asked Gómez to dismiss Sánchez's case, handing him ten pesos with the promise that Sánchez would provide an additional five pesos. Gómez accepted the money but claimed to have notified Mexlight. García's friends denied Gómez's bribery accusation, but it probably occurred. For business owners who resorted to power theft and bribery to remain afloat, electric company employees occupied a privileged position. Given the ease with which García offered his bribe, not every inspector put Mexlight's interests first.

Sánchez's case also demonstrates the importance of *peritos eléctricos* (expert electrical witnesses). Seven months into the case, court personnel gathered at the San Antonio mill for a reconstruction. Neither Sánchez nor Viera could provide the name or whereabouts of the carter who bumped into the meter, so the judge ordered a reconstruction of the alleged incident to determine the force necessary to dislodge the fuse and whether the device was correctly protected.[97] The judge called upon expert witnesses to determine whether the fuse could indeed fall accidentally. Carlos Ortiz and Fernando Corona of the Ministry of Justice offered their opinion: although a fuse could accidentally fall, the company should not have needed to provide protection to ensure it would not fall, based on the type of fuse in question and the height of its location. Therefore, they concluded, neither side was responsible for what had happened. Juan Espejel and Guillermo Arrington, experts for the defense, disagreed, arguing that Mexlight had to protect fuses against both accidental and intentional disconnections. Expert objectivity would not be the solution that the court hoped it would be.

The court acquitted Sánchez, finding that the missing fuse was the consequence of "a fact solely accidental, of which the accused could not have been responsible, and not fraudulent machinations to defraud the Mexlight."[98] The company was not happy: after failed appeals, it changed course and filed a civil action demanding 1,089 pesos in monetary damages. When Mexlight learned that Sánchez was trying to sell his mill, his only known property, the company asked the court to place a lien on it to cover the outstanding amount. If the company could not have the satisfaction of a guilty verdict, it would still have its money. Either would serve as a warning to those who would deviate from scripted consumption. No record exists of how Mexlight's battle with Sánchez was resolved, but that was largely beside the point. What mattered most was demonstrating the lengths to which the company would go.

Mid-sized businesses trapped in cutthroat competition experienced the advantages and disadvantages of electrifying their shops. Access to electricity, particularly after Mexlight had a monopoly, could give a business the upper hand in a restricted market if it could work with Mexlight to deny electric services to competitors. However, competitors sought the same advantage. It was a ruthless environment where proprietors had to be creative—and even a touch devious—to survive.

Act III: Sin Dolo *(Lack of Criminal Intent)*

Inspector Paliza kept up his frantic pace into 1907, filling his days hunting down suspicious activities among the city's many smaller shops and businesses. That year he accused Jesús Paredes (twenty-two years old, owner of the La Suerte diner on Plazuela de la Palma) of theft.[99] An official electrical connection had been interrupted a few months earlier for nonpayment when the diner was under different management. As Paliza could clearly see, however, the building had electric lights but no date for clear restoration of service. Paredes claimed that he had opened the diner a week before. On opening day, he explained, while he was busy fixing an oil lamp, a stranger had approached him and suggested the use of electric lighting, offering to make an immediate connection. The stranger also promised to take care of the contract with the electric company for a fee of five pesos. Trusting the offer, Paredes agreed. Instead of the five pesos, he handed over two pesos for the reconnection plus a free meal. That was the last time Paredes saw the other man.[100]

Paredes's lights unequivocally consumed stolen electricity, but did he mean to steal? As part of the 1906 reforms to the 1871 Penal Code, lawmakers added provision 8 to article 416, which demanded evidence of *intención dolosa* (criminal intent) of power theft. Only after intent had been established did legal proceedings move on to consider the physical execution of the crime. Consequently, ascertaining the intention behind fraudulent installations became an essential component of any case. Paredes was fortunate; because he had accepted an offer from an anonymous electrician and did not carry out the installation himself, he did not meet the law's standard of intent, so the company filed no charges.[101]

Not everyone was so lucky. In many cases defendants who admitted carrying out illegal connections but tried to raise doubts as to their intent found the company unsympathetic. Torres, of the Ambos Rumbos Hotel, admitted to an illegal installation, but only to test electric lights in the hotel. He swore that he intended to contact the electric company afterward—something he forgot to do. Librada Lara López used electricity for an application not stipulated in her contract but claimed that she did so only once and for only a few hours due to "urgent need."[102] Whereas the company filed charges against the first, no record exists of a legal case against Lara López for this matter. In at least one case, a defendant tried to equate ignorance with a lack of criminal intent. Concepción Delgado asked Nicólas Sánchez to run a connection from his commercial establishment to the room she occupied so that she could use three electric lights for a get-together with friends. Delgado admitted that she ended up entertaining them with a single light: she had "no criminal intent" because she did not know it constituted a crime.[103]

The courts could be more forgiving than Mexlight and consider unfamiliarity and personal circumstances in their ruling. In the case against Vicenta Castañeda, who powered a lamp in her shop without a contract, Judge Adalberto Gómez Jareregue sentenced her to pay Mexlight fifteen pesos in damages. In his verdict, the judge cited both the lack of a prior criminal record (article 39, section 1) and the fact that the accused did not know beforehand that it would be a crime (article 51, section 3) as extenuating circumstances, proclaiming that "the humble class *still* lacked a clear concept of the crime [power theft], which had provoked digressions and grave controversies among jurists."[104]

Most often the accused, like Paredes, narrated scenarios in which anonymous individuals introduced themselves as independent electricians or employees of electric companies and offered connections for cash. Having been duped themselves, these power thieves believed that they had not committed a crime because they had no intent of breaking any law. Sometimes the accused were clients—just not of the electrical company. Cristobal Rubio claimed that Federico Brown, a German electrician, had set up an connection for him and that he had paid Brown directly every month for service.[105]

Crooked electricians like Brown proliferated in the period, most commonly taking cash but also frequently bartering for their services. Manuel Herrera y Olvera, owner of the drugstore La Salud in Tlalpan, had been a Mexlight client for over seven years when inspector Gabriel Abrego arrived unannounced in March 1913 and found a brincador.[106] Herrera y Olvera first claimed that his reduced electrical consumption resulted from a recent Zapatista assault, which had forced him to close the pharmacy's doors.[107] He later admitted that he had been called to the house of Enrique Gavito three or four years before, to look after his sick wife. Without money to cover the expense, Gavito offered to reduce his electric bill and explained that it entailed simply loosening some screws. Herrera y Olvera claimed he rejected the offer because he would never allow a crime like that and never saw Gavito again. It is not clear whether Gavito went ahead and tampered with the meter anyway or Herrera y Olvera tried to do so himself. In the end, Herrera y Olvera got off on a technicality: expert witnesses confirmed that there had been no fraud because the brincador was improperly connected, so the court dropped the charges.

In their efforts to curb and prosecute fraudulent connections, companies and the courts ran up against the realities of a city in flux. Not only was the city physically expanding, bringing in waves of new faces with it, but eléctricos facilitated journeys into and across the capital. With cheap mass transit, residents of adjacent villages could come and go daily: people could visit and work in different neighborhoods and be back by dinner. Population movement was high, turnover was frequent in cheap rentals, and informal employment was common. In this environment, it was not difficult for anonymous individuals not only to offer services of dubious legality but to pose as employees of electrical companies who could offer "fixes" or special installations.

Ignacio Vera and a few friends made money doing this. On August 24, 1905, Juan Esparza of the Compañía Mexicana de Gas y Luz Eléctrica (Mexican Gas and Electric Light Company) was investigating interruptions at a tailor's shop in the area of San Lorenzo when he identified a clandestine connection to a residential house. Following the line, he arrived in the room of Efrén Vilchis, where a party was in full swing under the warm glow of twenty-five electric lights. The connection was makeshift and had not been carried out by the company.[108] Esparza immediately notified the inspection department and the secretary at the police station. But by the time the authorities arrived, they found the installation destroyed and not a single light. The police still arrested Vilchis for having an illegal connection to his home. Later they also arrested Ignacio Vera, one of the invited guests at the party who, they alleged, had carried out and then destroyed the installation.

At the station, Vilchis confessed that, having insufficient light for his party, he had accepted an offer from Vera and a friend to install ten electric lights for him. Pleading his innocence, Vera claimed he did not directly carry out the installation but merely helped his friend—a man identified in court documents only by his first name, Jesús—facilitate the deal. Upon further questioning, Vilchis expanded on his original statement, both helping and hurting Vera's case. Vilchis explained that the ringleader, who offered the connection and determined its price, was not Vera but someone he knew only by the nickname "El Chato." However, both the ringleader and Vera claimed to be employees of the electric company and carried out the installation.[109] As electric thefts go, it was a small crime: experts set the defrauded amount at three pesos and thirty cents and the company lawyers demanded twenty-two pesos and seven cents in damages. While Vilchis paid his fine and served no time, however, the courts convicted Vera of carrying out the installation and sentenced him to over fourteen months in jail and a fine of six pesos and twenty-five cents.[110]

After the reforms of 1906, the law clearly established the criminality of stealing electricity, but the same reforms offered a loophole. If the accused could demonstrate that they acted *sin dolo* (without criminal intent), then they were innocent of breaking any law. In some cases the lack of intent was clear, but in many others the accused could only surround their actions with a vague cloud of legitimacy. Some were duped, while others appear to have embraced an almost willful ignorance. Some tried to be savvy, while others were driven by fear. But no matter what their motivation, those who acted without intent shared one important quality with those who acted with intent: they both desperately wanted to be a part of the modern world defined by electricity.

CONCLUSION

The desire to pay less—or nothing—for power was at the root of theft, but curiously those charged and brought to court rarely admitted this or questioned electricity rates. Salvador Ruelas acknowledged placing diablitos in his meter to "save four pesos."[111] Esteban Castillo, co-owner of the mill with Roberto del Villar, complained about Mexlight's exorbitant rates.[112] But they were exceptions. Even critical comments from the press were few and far between and often oblique. For instance, in an article on a theft case, *El País* conceded that Mexlight had a right to use inspectors to safeguard its interests but showed some sympathy with the company's customers—not because they paid too much, but because the company's service was so poor.[113]

For over two decades, criticism of electrical companies, their rates, and their service had rarely been aired. It must have been astonishing for capitalinos opening their copies of *El Demócrata* on October 27, 1924, to read the paper's harsh commentary on Mexlight. That year alone, the paper estimated, as many as 80 to 90 percent of energy subscribers had used illegal diablitos, but it did not criticize the morality or ethics of the citizens. It was Mexlight's "prohibitive rates" that lay behind this widespread practice. The newspaper claimed to have received countless letters demanding that the Ministry of Industry and Commerce and the Ministry of Agriculture and Development step in to reduce rates and denounced Mexlight's monopoly for charging too much for service. As a comparison, the paper retold the experience of a "distinguished individual" who traveled to St. Louis, Missouri, and was shocked to find that the electric company there charged rates considerably lower than those found in Mexican cities. Nothing could explain the difference. After all, the author reminded his readers, the Mexican government had given the hydrodynamic power of the "magnificent" Necaxa falls to Mexlight for a price so low that it amounted to free. All Mexicans got in return, according to the distinguished individual, was exorbitant rates.[114] The paper considered it fair to cheat a company that abused its subscribers.

According to the law, power thieves were criminals. This included not only the motley crowd of individuals who sold or bartered the technical skills required to make fraudulent connections but also the innovative individuals who manufactured "energy-saving" mechanisms designed to bypass the vigilance of electric meters. Courts sought to quash these practices and protect the interest of large electric companies, but, criminal or not, capitalinos were not going to be passive participants in the electrification of their city. Their thefts amounted to a demand to be included in the wonder of modern, electrified Mexico City and a form of protest, questioning who had the right to determine who enjoyed the comforts of modern life and who did not.

For electric companies, theft was not just lost revenue. The tenacity of their inspectors and their lawyers' demands for convictions even after fines had been paid speak to an injury that went beyond finances. If cultural norms had to be violated in order to secure redress of those injuries, then the companies were prepared to do so. Public utilities began to penetrate previously off-limit private spaces so that their scripted consumption was the only path people could follow. Companies like Mexlight effectively declared that Mexicans would get electricity under their rules or not at all, yet court documents and the testimony show that capitalinos often refused, went off script, and asserted their own prerogatives.

Theft trials offer a fruitful avenue to document the nuances of the capital's electrification, not merely to analyze the transgression per se but to "reconstruct the texture of crime as experienced in everyday life by those who formed the majority of offenders and victims."[115] As the testimonies of the accused reveal, the interpretation of the law was not the exclusive purview of legal experts. The accused framed statements in terms that lessened responsibility, so judicial records thus become valuable not merely for what witnesses said but for how they drafted narratives.[116] Michel de Certeau posits that in looking at the everyday we begin to "see how [lives] are lived and 'how sites are constructed in the interstices of the vast socio-economic systems.'" With the lens on the quotidian, we come across "evidence of occurrences of resistance in which ordinary people undermined the imposed relations of power"; although individuals "cannot escape the dominant cultural economy . . . they can adapt it to their own needs." The companies' disciplinarian efforts to regulate electricity consumption were thus "constantly deflected and resisted by those . . . caught in its 'nets.'"[117] Unlike the conversation surrounding eléctricos and motoristas, where the middle and upper classes invented lamentations about how the lower orders might derail the nation's race toward progress, the conversation about electrical theft was far more democratic. People from nearly every layer of society participated in it to justify and explain their actions. To one degree or another, they knew the script the electric companies sought to propagate, but they refused to follow it and challenged it in the process.

PART III

| FIVE |

Becoming Electro-Domésticas

ELECTRICAL APPLIANCES, MAIDS, AND MIDDLE-CLASS DOMESTICITY, 1930S–1950S

A vision of the modern Mexican home enhanced by scientific housekeeping began to be promoted in popular culture in the twentieth century, linking *electrodomésticos* (electrical appliances) with gender, class, and color.[1] In her book *Almanaque 1936: "La familia,"* María Aguilar de Carbia assured housewives that home decoration was a more reliable statement of a family's wealth than an income tax declaration. She warned her readers that guests would sleuth around their houses like Sherlock Holmes and examine floors, furniture, curtains, and rugs to assess a family's character, culture, and financial situation. In a chapter titled "My Kitchen and Pantry," she explained the particular importance of the kitchen, the room "without secrets," which she believed revealed clues to a housewife's taste and personality.[2] She had solutions for all the anxious housewives reading her book, and her vast audience gobbled them up.

Beginning in the 1930s, under her nickname "Marichu," Aguilar de Carbia penned culinary columns for women's magazines and authored some of the decade's best-selling cookbooks.[3] The *Almanaque* crowned her collaboration with *La Familia*, a leading women's magazine. Marichu commiserated with her audience that model kitchens reproduced in magazines were beautiful but difficult to replicate. Rather than ambitiously try to have a kitchen that replicated the kitchens in magazines only to wind up with a mess, Marichu advised homemakers to take suggestions from each model to achieve a kitchen that was comfortable, beautiful, and within their means. She championed an accessible middle-class kitchen like her own:

> The electric range occupied the center space on the main wall. Over it, a hood hover crowned the alcove. Next to it was a sink with a chest of

> drawers to store cleaning supplies, an opening for a trashcan, and exposed plumbing pipes to facilitate repairs. A small spice rack with shelves on the sides for flowerpots and a small table covered with oilcloth for washing dishes and a drawer for spoons, knives, and other kitchen utensils completed the section. On the opposite side, recessed into the wall, a cabinet divided by shelves. The top shelf holds the ovenware. Housed under it, everything needed to strain, whip, grate, and press. Dishes for the domestic servants were kept on the shelf underneath. On the shelf below, items for boiling: teapot, kettle, pots, and pans of varying sizes. The last shelf to the bottom stored the metate, *molcajete*, grinder, and cutting boards, tools for grinding and mincing.[4] Under the window, a large table covered with granite. A broad chest of drawers on the side stored pastry-making devices. The "Westinghouse" refrigerator sits on the pantry. Everything in my little kitchen and pantry is red and white.[5]

Marichu's careful description of her own "room without secrets" testifies to the values and expectations of a middle-class, mid-1930s kitchen. Anchoring the room's technification were the electric range and refrigerator. The former stood where the *brasero* (brazier) once reigned supreme. Cooking with electricity meant the absence of soot, however, which in turn made white cabinets a reasonable choice. The Westinghouse refrigerator alluded to the family's discretionary spending not just on the appliance itself but also on the quantity and diversity of food stored inside it. Even nonelectrical elements of the kitchen had meaning. Indoor plumbing spoke not just of potable water and sewage services but of health and cleanliness generally. The modern kitchen, like the homemaker who ruled it, was disciplined. The spatial arrangement and rigid organization of the items on shelves and the categorizing of utensils emphasized efficiency. Storage space for stone implements indicates both that they remained useful and that the kitchen was—again, like the woman who used it—a liminal space where tradition and modernity met. Segregated dishware for the help suggests the presence of live-in maids. In model kitchens and Marichu's writing, middle-class domesticity manifested itself through commodities that were not just instruments to make life easier but social markers that revealed the character, taste, and personality of the woman who reigned there.

Three decades later, the famed poet, essayist, and chronicler Salvador Novo reflected on the revolution's impact on the capital's gastronomy. He mourned that yesteryear's submissive domestic servants, pulque, and siestas were mere relics. Braseros, along with metates, molcajetes, clay pots, wood spoons, and *molinillos* (wooden whisks for hot drinks) "had been swept away, vanquished, outdone." Yet he questioned whether what took their place was really an improvement on what had come before. People cooked with gas, on aluminum cookware, with the help of blenders, but was it really a better way to cook? After all, Novo claimed, the blenders that everyone was so excited about "indiscriminately grind grains, peels, fruits, and skins." Refrigerators reminded him of morgues, sheltering "preserved chicken corpses and fish mummies in hibernation." Who worked in the kitchen had also changed. The young girls who had once filled the ranks of respectable maids now found their financial freedom as either *ranchero* (ballad) singers or go-go dancers.[6]

In the first half of the twentieth century, the transformation of Mexico was mirrored in the transformation of the nation's kitchens. Strictly speaking, the occasional lamp or string of bulbs was the first introduction of electricity into Mexican homes, but the electricscape truly entered the home through the kitchen and transformed everyday behaviors. In the center of this room stood the figure of the housewife and mother. The kitchen is the place where the electricscape fused modernity and domesticity. Marichu's and Novo's descriptions attest to the new ideal of domesticity that emerged in the aftermath of the revolution, which prescribed the transformation of domestic spaces and cast the housewife as an instrumental actor in the country's physical and moral reconstruction.

Promoted by revolutionary governments and labor unions, this cult of domesticity found its voice in women's magazines, advice literature, newspapers, fashion guides, catalogs, and works of fiction that reached millions. Its proponents advanced new, fundamentally middle-class attitudes about work and family in which homemakers were to employ the latest domestic technology to build a modern nation one house at a time.[7] If some aspects of modernity allegedly disrupted the family as indicated by new forms of entertainment (like movies and radio), women working outside the home, and fashion, the homemaker and her tidy, efficient, stable house was at least one aspect of modernity that would strengthen the family. The middle-class wife would be everything that the poorer masses roaming Mexico City's streets and the dangerous motoristas were not: safe, moral, disciplined, and encouraging, facilitating modernity rather than endangering it.[8] The nation's kitchens and homes alone would be a site of unmitigated success in advancing technological progress with moral improvement. This was always more an ideal than a reality: in order to live up to this ideal, some middle-class homes stole electricity (see chapter 4). But it was a powerful narrative that shaped the lives of nearly two generations of Mexicans. As production expanded abroad and prices dropped, electrical appliances, which had been luxury items in the early twentieth century, became the sine qua non of a modern, middle-class home in Mexico City by mid-century.[9]

Though the desire for technical devices in the home was widespread, not everyone had them. Cost was one obvious barrier to acquiring the trappings of a middle-class life, but so was the availability of services. Of the 547,477 dwellings that housed the 2,832,133 capitalinos, 37,851 did not have access to piped water at home in 1960, while 295,597 lacked sewage service.[10] However, electricity—the necessary service in order to become consumers of electrical gadgets—appears to have been more readily available. By 1970, the first year the national census recorded whether a home had access to electrical service, 548,915 out of the 563,844 dwellings had access to electricity.[11] That is, electric power reached over 97 percent of the city's homes.[12] The census does not explain what it meant by "access."

With electricity so available, electrical devices, unlike other modern appliances and improvements to the home, were more widely attainable and were actively promoted by advertisers. They marketed bright home electrical appliances as being like the ideal middle-class wife: clean, respectable, and light-skinned. In the first section of this chapter, I discuss early advertisements for electric lighting and electrical devices, which were presented as luxurious and embodying universal values of science, hygiene, and modernity. Appliances promised not just to ease work (ironically, mostly for maids) but to express status.

In the second section, I shift away from manufacturers' and advertisers' use of aspirational ideals to promote the consumption of electrical devices and discuss the work of cookbook writer Josefina Velázquez de León and her efforts to promote scientific housekeeping. Advertisers promoted new electrical devices as time-saving and labor-saving appliances, placing users in sync with a broader scientific ideal that privileged the sacrosanctity of time and efficiency. But it was figures like Velázquez de León who mediated the everyday experiences of Mexican women and the cultural ideals of the time. She lent a hand in the domestication of kitchen appliances, shifting technology from being solely the domain of experts and making it more democratic.

In the third section of this chapter, I show how all these elements came together in an issue that manipulated people's anxieties: the maid problem. Direct and indirect advertising of appliances not only classed and gendered them as female, converting metal, glass, and plastic into accessories of a modern, middle-class woman, but also racialized them by crafting a distance between the fair-skinned modern users of electrical devices and the backward dark-skinned maids using stone and wood implements. Since the

1930s, advertisements, advice literature, and women's magazines had spoken of a crisis in securing domestic servants. Unable to employ someone to undertake housework, homemakers faced the daunting prospect—in terms of time, energy, and status—of undertaking it themselves. Advertisements both deployed and contributed to this so-called maid problem by presenting electrical appliances as the ideal answer. Simultaneously, advice columnists—some of them working with manufacturers—offered smart tips to meet class expectations despite diminished or nonexistent help. In so doing, they popularized the changing notion of what it meant to work in domestic spaces, the kind of labor—even the kind of other activities—that respectable women did, and to what degree being respectable demanded being lighter skinned.

SELLING THE ELECTRIC LIFE

At the beginning of the twentieth century, advertisements promoted electricity as safe, hygienic, and modern power. Even as late as 1910, however, newspaper advertisements for even utilitarian domestic electrical appliances and electric lighting presented these items as luxury objects, unattainable by most of the city's population.[13] Along with electric water pumps and elevators for large businesses, the first ads for electrical technology typically promoted electric lights by showcasing the type of attractive homes they would be in. Electric lights were cleaner than oil lamps, so homes filled with valuable items deserved this new, supposedly hygienic technology. Decorations, tapestries, and paintings—all items that a respectable upper-class home might have—should be illuminated by electric light.[14]

This approach implied that only more modest, less well-appointed domiciles might still make use of the older lighting. In addition, electric lights were promoted as safer. A wind gust could safely toss an incandescent lamp against fabrics without danger, but an open-flame lamp in the same circumstances could have grave consequences. The message was clear: people who have nice things want to keep them safe. (Notably, the ads ignored the fact that the risk of fire from faulty wiring surpassed the danger from wind.) Just as wealthier neighborhoods increasingly had better street lighting, so did wealthier homes. Within a decade home illumination became an obvious social marker. While the lower classes zealously guarded single candles, the wealthy enjoyed profuse lamps and candelabras. Indeed, a new term entered capital's lexicon: *derroches de luz* (ostentatious use of lights).[15]

The attributes of electric bulbs—marketed as clean, bright, odorless, sootless, smokeless, hygienic, flameless, convenient, and always ready—spoke of science, safety, and human control. According to one ad, all artificial lighting except electric lights consumed oxygen out of the air, "stealing the most essential and vitalizing element of the air that your family's lungs breathe."[16] The advertisement claimed that an open-flame consumed as much oxygen as four adults. Another ad echoed this sentiment, specifying that open flames not only "corrupted the property that gave life to the air we breathe" but, even more threatening, "left behind only a poisonous carbonic gas that rendered [the] home impure."[17]

Marketers understood people's embrace of science in the period—their faith that it was always marching forward to greater and better things—and worked to promote electric lights on that basis. Mexlight offered free expert advice for illuminating businesses and residences, claiming that modern research had demonstrated that illumination was an exact science.[18] An illumination engineer would show interested parties the right way and the wrong way of using lights and design a complete scheme to light a store, residence, or shop window based on modern and scientific principles. Another ad deployed the rhetoric of progress, asserting that kerosene and older incompetent illuminants were a thing of the past. People had considered burning bundles of wood sufficient until the arrival of tallow dips.[19] Kerosene lamps, although a significant improvement over tallow dips, were destined to be pushed aside by the advance of science and the greatest of its illuminants: electricity.

Electric lighting extended daylight. At least in the eyes of advertisers, however, the ways in which this added time should be used differed significantly by sex. Images of well-dressed gentlemen sitting on high-backed chairs absorbed in reading were common. In figure 5.1, the gentleman sits in front of fully stacked bookshelves—likely his home library.[20] Incandescent electric lamps nearby allow him to read from the comfort of his chair: extending daylight for him translated to male leisure time and intellectual improvement. Ads depicting scenes exclusively peopled by women were also conventional. Unlike gentlemen absorbed in their books, women were generally portrayed in more highly gendered activities. An Edison Mazda light ad depicts an oversized bulb in the foreground (fig. 5.2).[21] At such a close range, the bulb's inner workings are visible to the viewer. The monumental bulb is flanked by two women in lavish nightgowns smilingly engaged in genteel needlework. They can carry out such a delicate activity thanks to a lamp with a Mazda bulb sitting on the table. Two homes on the bottom brightly glow in the night. The message is clear: an activity previously performed during daylight hours now could be done at night.

FIGURE 5.1. *Mazda electric light ad, selling comfort for men.* El País, *September 4, 1910. Hemeroteca Nacional Digital de México.*

When electrical ads did not promise leisure to their audience, they promised ease. Life could and should be less of a burden. Take the demanding, uncomfortable chore of ironing, for example, a burdensome task for nearly all women. Even upper-class housewives with maids might periodically iron: as a Mexlight ad recognized, they had "lace and fine fabrics that one could not entrust to the maid or the laundress."[22] In a world where clothing still signified class and status, wealthier Mexicans made essential investments in clothing and linens and sought to conserve them.[23] A Mexlight ad claimed that ironing by electricity made the chore pleasant by "saving many steps, troubles, lousy temperament, time, and sweat."[24]

This was true even when potential owners and users were not the same. Advertisements for electric irons, for example, emphasized control and efficiency, promising to lessen the burden of one of the most time-consuming housekeeping chores. To iron clothes before that time, several irons—four to six to get the job done quickly—had to lie on top of a hot stove for extended periods.[25] Essentially just large blocks of metal with a handle, irons cooled down with even limited use, so others had to be ready to keep a

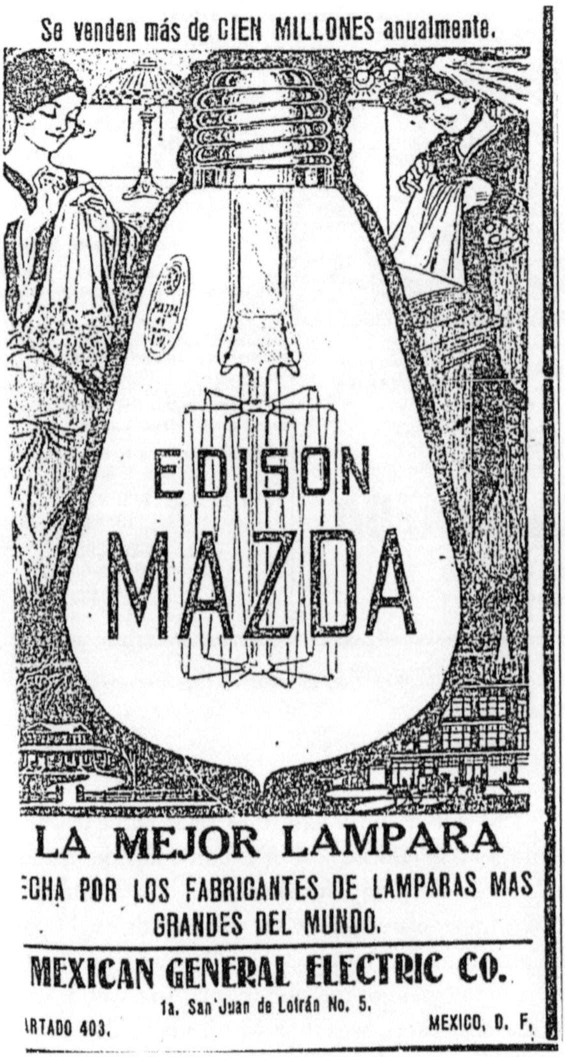

FIGURE 5.2. *Edison Mazda electric light ad, selling comfort for women.* El Imparcial, *October 2, 1913. Hemeroteca Nacional Digital de México.*

steady work pace. It was uncomfortable work. The person doing the ironing typically commandeered a small table to serve as an ironing station and moved it close to the stove or brasero to keep the warm irons in reach—but not too close or else they would be unbearably hot. The electric iron eliminated this layout. Any room with a light socket allowed for ironing. At the turn of a switch, the iron "was ready at any time, day or night." Besides praising this flexibility, ads extolled the uniform temperature. The construction of an electric iron "was entirely scientific. All heat uniformly distributed on the flat surface, remaining hot as long as you need it."[26] Maids, who did the lion's share of ironing in middle- and upper-class homes, would be grateful that the electric iron freed them to perform other chores.[27]

The ordinary housewife also would be spared the indignity of suffering through the sweaty drudgery. In another ad, a sudden storm has struck, creating an emergency in which a housewife must do the ironing. The accompanying illustration depicts her at a table conversing with two female friends as she works. The housewife's clothing does not clearly establish her social class, but the large broad-brimmed hats of her friends indicate their affluence. At least according to this ad, electricity transformed ironing from a sweaty, awkward chore to a genteel pastime that could be shared with friends.[28]

The defined user for these appliances is often not a housewife but her maid. Ads for electrodomésticos promised to do away with the most tedious chores in the kitchen. Instead of highlighting the benefits for housewives, they emphasized the advantages for maids and, through their work, the entire home. Electric ranges allegedly used heat and time hygienically and efficiently, with no fire to light, no smoke or soot, and no heat escaping from the stove. During the summer months, the temperature in the kitchen would remain pleasant and comfortable for those in charge of cooking. For a housewife who did her own cooking, the benefits were obvious, but life would also be better for a woman with a maid, because "a happy maid contributes significantly to the home's well-being."[29]

Early electrical devices promised to create new forms of domestic sociability, like the woman chatting with her friends while she irons. At the forefront of advertisers' efforts was tabletop cooking. Toaster-oven ads described them as a pleasant, cost-efficient, and elegant way to enjoy a meal. Instead of cooking on a hot stove, meal preparation occurred at the table. Like the modern world in general, the new art of tabletop cooking was proclaimed to be more rational and efficient than what had come before. Kitchen cooking not only produced unbearable heat but resulted in wasted leftovers. Now cooks could prepare the exact amount of food and not waste money.[30] Tabletop cooking was ideal when cooking for two. Meal preparation remained a woman's duty, but this new technology allowed her to enjoy her husband's company as she prepared a meal.

The real pleasure of a meal, the ads promised, lay in serving it. According to one ad, endless wait times ruined countless painstakingly prepared meals, but electric toaster-stoves allowed housewives to start cooking after the family sat down at the table. Each plate could be served steaming hot "directly off the grill."[31] To cook on an electric toaster-oven at the table meant preparing an exquisite meal in the midst of the family.[32]

Illustrations for toaster-ovens worked hard to sell the idea that a simple appliance could harmonize the home and improve people. The husband sits at a table, reading a newspaper, while the wife attends to a coffee percolator and a toaster-oven. As in the previous ads for electric lighting, the percolator and toaster-oven added to the husband's leisure time, spent catching up on the news and contemporary politics, while the wife engages in domestic tasks. Now they can lovingly pursue their respective responsibilities together. As the caption reads, the elegant couple represent "A 20th-Century Refinement—Electrical Cookery."[33]

Unlike the image of the harried, distracted housewife of old, the modern woman could be calm and focused, but only if she would "allow electricity to come to her aid."[34] An electric tea service meant that a wife could have her friends over for an afternoon without effort or the assistance of domestic servants. Her attention could be where it was supposed to be: on her guests. The General Electric Company (GE) claimed that guests would be delighted with meals cooked on a chafing dish.[35]

Ads and illustrations planted the idea of what it might mean to own electric appliances, but they were not enough. People wanted to see for themselves what this expanding new front of the capital's electricscape could offer. Companies obliged with window displays, model kitchens, and salesrooms in the downtown area. GE appliances were available for purchase at the offices of Mexican General Electric Company (MGEC) on San Juan de Letrán Street, at the American Grocery Company (AGC) on Avenida 16 de Septiembre, and at Mexlight's sales department at the intersection of Cinco de Mayo and Isabel La Católica. All three companies ran newspaper ads enticing readers to visit their showrooms for firsthand experience of the delight of electrical cookery. The AGC set up an entire model electric kitchen for customers, complete with coffee percolators, teapots, toasters, grills, flat irons, disc stoves, and water heaters.[36] The company held a culinary exhibition in the summer of 1912 to demonstrate that electric cooking was clean, cheap, and so simple that anyone could easily prepare any type of dish. Boasting that "seeing is believing," the model sought to prove to customers that "Edison's dream [has] at last been realized [and that] within a decade, we all be cooking by electricity and would be today were it not for a false notion of the cost." More than a curiosity, electric cookery in these showrooms was cooking of the future. Given its low prices, MGEC promised that this future was coming sooner rather than later.[37]

Mexlight was right on the tail of the MGEC and sought to attract customers through its sales department. It proclaimed a "greatly increased demand for electrical heating and cooking devices," which were no longer experiments and were now guaranteed by manufacturers.[38] Customers knew that, with the simple turn of a switch, they could enjoy the comfort of electricity in their domestic space: heating the baby's milk, toasting bread at the table, and making a delicious cup of coffee during a meal—all performed outside the kitchen. If by chance they did not know this, the sales department offered to send representatives to their homes. In the 1910s Mexlight's sales department established the Sala de Electricidad (Electric Showroom) for the demonstration and sale of household appliances. The showroom offered capitalinos a safe, respectable space in which to peruse the new technology and even take an appliance home for a 30-day trial. Rather than sending customers home unprepared, the showroom offered advice from a cooking and pastry expert.[39] For no charge, an expert taught customers how to cook with electrical appliances either at the company's showroom or at the customer's home upon purchase.

From early on, local retailers sought to familiarize potential customers with the advantages of electricity in the home. El Abastecedor Eléctrico, which boasted that it had the best selection of electrical appliances in the capital, joined AGC and Mexlight in displaying GE products in action. It invited the public to visit the store in 1912, having come up with an ingenious marketing device. The company had a nine-year-old girl use every machine in the store, proving to even the most skeptical that electrical devices were safe in the house.[40]

Mexico was not alone in associating electrical devices—whether large electric ranges or a hand-held curlers—with hygiene, science, and class aspirations. This universal language of modernity meant that marketing messages could be borrowed and even transplanted from one national market to another without much consideration of the new context. Mexican advertisers routinely reached over the northern border to borrow the rhetoric employed in US campaigns. Two ads for electric irons, one from the Idaho-Oregon Light and Power Company and the other from Mexlight, include an identical image of a cartoon housewife at her ironing station.[41] Such cartoon characters, consistently depicted as fair skinned, were reproduced in other ads: whether in America or Mexico, modernity was white.

During the decade of revolutionary struggle, advertisers and commercial stores in the city continued to deploy the universal language of science to sell their products. The symbolic transformation of the kitchen, and the home at large, involved the ongoing availability of maids. As members of the darker-skinned lower classes, they stood in the way of glamorizing housework and the consumer goods that made it possible. The maid problem and the racialization of appliances would serve to reinforce the idea that the ownership of modern technology constituted privileged racial whiteness. Advice literature became one avenue to craft this view.

JOSEFINA'S RECIPES FOR THE HOUSEWIFE: DOMESTICATE MODERNITY, MODERNIZE TRADITION

The construction of the middle-class housewife, her house, and the advertisement of electrical appliances came together in the 1930s cookbooks. Advice literature experienced a transformation in that decade as female contributors came to dominate the field. Male chefs had dominated Mexican culinary history until then, but women had recently emerged from professional culinary training. Manufacturers, eager to reach out to a nation of housewives, were ready to employ them.[42] As national and international manufacturers, local retailers, and Mexlight looked to sponsor cooking classes and insert ads in locally published cookbooks, they turned to women to write them.

As a historical source, cookbooks register the minutiae of everyday life rarely found in other sources. They provide insight into the genderized role of women and the construction of female domesticity, as well as the creation of the ideal modern kitchen, and give clues to class issues through the responsibilities of domestic workers. Even the recipes offered are fundamentally products of their time, place, and the individuals who crafted them. At the same time, through comments, references, and historical capsules printed alongside recipes, they expose not just top-down notions—author to audience—but the ways in which people understood and responded to these ideals.

No twentieth-century figure looms so large over a genre in Mexico as Josefina Velázquez de León does over cookbooks. She was the nation's Julia Child, only more influential: a family cook with no formal training who wrote over 150 cookbooks for middle-class Mexican wives. She ran her own school from the 1930s onward and wrote columns for women's magazines that catered to socially privileged women.[43] Her career, however, began with the modern advertising campaigns of electrical device manufacturers. Early in her career, Velázquez de León secured the backing of GE, which she leveraged in the mid-1930s into a culinary school called Academia Calrod. It was named after the Hotpoint Company's flattop cooking coil, the Calrod, which had revolutionized electric cookery. Located in Ontario, California, the Hotpoint Company specialized in the manufacturing of electric flatirons and table appliances, when GE bought it in the 1910s. Velázquez de León described her academy as "French cuisine through the General Electric System." In 1936 she published her first cookbook, *Manual práctico de cocina y repostería* (Practical manual of cooking and baking), which featured lengthy inserts introducing the GE-Hotpoint range and refrigerator. The thick *Manual* also included a photograph of her academy's kitchen, including the GE range on one side of the wall and the refrigerator at the other end. In order to "cook efficiently on the GE-Hotpoint range," she also included time and temperature tables for baking, roasting, and cooking on the hotplates.[44]

Public health officials and social and educational reformers shared Velázquez de León's view of women as the pillars of the home and conduits for the government's goals to modernize the nation. When the government turned to women as a stabilizing force after the revolution, women, especially middle- and upper-class women, embraced the role. For them, Velázquez de León's cookbooks were practically how-to manuals. They helped in articulating the construction of "ideal motherhood" while making sure the modernity served not to "emancipate women and children but to subordinate the household to the interest of national development."[45] Similar efforts to rationalize the domestic sphere elsewhere advocated that mothers should be educated in scientific hygienic household management and child-raising to produce "healthy, efficient, patriotic citizen-workers."[46] Discipline, efficiency, and modernity were to be inculcated at home.

Female domestic economy teachers, usually social workers, joined cultural missions deployed to rural areas to teach housewives what authorities believed was "a better way of life," by improving hygiene habits, exploiting the region's resources, and sewing and cooking, among other skills.[47] Capitalino reformers armed with the science of nutrition and carrying the gospel of middle-class domesticity also penetrated the domestic sphere to transform the working class's eating habits to subordinate the household to the interest of national development.[48] Having earned degrees in areas such as education, nursing, and social work, women like Ana María Hernández increasingly found professional work enacting the state's welfare and education policies. Born and raised in the central state of Querétaro, Hernández moved to the capital from 1917 to 1921 to attend the Escuela Comercial Miguel Lerdo de Tejada and Escuela Hogar Gabriela Mistral, where she received commercial and vocational training.[49] She would later teach domestic science courses through the Department of Technical Education, one of the many departments under the umbrella of the Ministry of Education, and work as an inspector for the Ministry of Labor. Complementing Velázquez de León's domestic advice, Hernández's cookbooks, directed at urban female workers and peasants, combined her passion for domestic science with her activism. She was a prototypical Mexican feminist, exalting so-called feminine qualities such as abnegation and self-sacrifice.[50]

Authorities, reformers, domestic advisors, and advertisers worked to normalize middle-class ideals of domesticity. Drawing a direct link between the home and the nation, housewives were held responsible for upholding domestic felicity and family unity, the bonds through which they would become productive consumers.[51] Publicists cast housewives as the family's consumers par excellence. Consequently, campaigns targeted female consumers through dramas that made women responsible for fortifying the family's unity and happiness through consumption choices. A Listerine ad, for instance, depicted a little girl devastated after her father left her mother due to negligence in hygiene. Her bad breath drove the husband away, broke the family apart, and brought unhappiness to its members.[52]

This updated cult of domesticity was in part a response to the anxiety over the increasing presence of women in the city's workforce. Beginning in the 1920s, the expansion of public and private commercial education spread rapidly across the capital. Equipped with the necessary tools, middle-class women took on office work alongside their male counterparts. By the 1940s their sizable presence among federal employees merited the label *burocracia femenina* (bureaucracy of women).[53] Women were also important contingents in the areas of education and welfare.[54]

Increased visibility of women after the revolution added to concerns over menaces to womanliness. The rise of alternative models of female behavior such as those exemplified by outspoken revue actresses—with their overt sexuality—sparked heated commentary on the threats of metropolitan modernity.[55] To defuse these controversies, social commentators reasserted the domestic sphere as women's natural domain. Vocational schools such as Gabriela Mistral's trained and educated women, but only for occupations compatible with housework. Mistral declared categorically that whether "a professional, worker, peasant or simply a lady, [a woman's] only raison d'être in the world is maternity, both physical and spiritual, or the latter for women, like me, without children."[56] Thus, the most patriotic education for a woman was one that accentuated family as the reason for living. Like the cautionary tales of the silver screen that stressed the devastating consequences that the corrupt metropolis could have on women, home advice literature offered guidance on preserving domestic felicity and family unity.

Velázquez de León shared her contemporaries' anxieties about the growing presence of women in the workforce and its effect upon the family. She lamented that young women failed to gain the essential knowledge of how to run a home while receiving a commercial education. In the past, she held, respectable women had trained to become perfect housewives. Even then, however, they spent little time learning how to cook because they had servants to do it for them. Now, even more than then, housewives needed help in the kitchen, for they could not rely upon a maid to make up for this culinary ignorance. Their years of education, followed by work in an office, meant that the modern Mexican woman came into her marriage and home wholly disoriented, without the slightest idea of how to select products at the market based on quality and price, unable to plan menus, and, ultimately, how to prepare and serve dishes.

Like the shifting, ambiguous notions of womanhood and domesticity of the times in which she wrote, Velázquez de León saw herself as a mediator between the traditional and the modern. She invited housewives to study the old, not to remain traditional, with the understanding that "the old is always new when enriched with experience and modern science."[57] She offered the nation's middle-class homemakers (and, equally notable, homemakers who aspired to the middle class) a way to navigate a democratic, industrialized Mexico while still conforming to the ideals of being a "proper" wife and mother.

As women became distanced from the domestic sphere—yet still nominally responsible for it—Velázquez de León saw a market for her academy and her literature. Her academy was nothing new: by the end of World War I, culinary training had a solid history in the capital, reaching back into private culinary schools for middle-class women in the late nineteenth century. Young single women from the Santa María La Ribera neighborhood flooded Alejandro Pardo Landa's Academia de Alta Cocina (Academy of Haute Cuisine) at the turn of the twentieth century.[58] In the following decades, Velázquez de León's famous Academia de Alta Cocina, Repostería y Decorado (Academy of Haute Cuisine, Pastry, and Decoration) shared the spotlight with the academy of Aguilar de Carbia, the cookbook writer discussed at the beginning of this chapter, and other schools such as Lucía T. de Madrigal's Academia Hogar (Home Academy), which offered cooking classes to middle-class single women.

Given her influence in defining domesticity, it was almost inevitable that Velázquez de León would offer book packages as wedding gifts by the 1940s. *Los 30 menus* (30 menus), *La cocina económica* (The economic kitchen), and *Manual práctico de cocina y repostería* (A practical manual of cooking and baking) were among her first books, available in a washable, white Keratol cover with the bride's name engraved in golden print, all placed neatly inside a cardboard box decorated with orange blossoms and ribbons and wrapped in cellophane.[59] Luxury packages, with leather bound cookbooks placed in satin-covered cases, were also available. Adapting the author's personal motto *saber cocinar es base de la economía* (knowing how to cook is the basis of economy), an ad for the books promised that *la economía es la base de un hogar feliz* (economy is the basis of a happy home) and assured customers that this present would "contribute to the bride's happiness." The modern marriage, as in so many things in modern life, required expert advice.[60]

The wife was solely responsible for the happiness of her home's happiness, which, in Velázquez de León's view, centered on securing her husband's contentment. The surest way to do that, she confided to her female readers, was to satisfy his belly: she advised each wife "always to receive him in a lively spirit, with a kiss on the mouth and a savory and well-seasoned dish in her hand." Attention to cooking was thus not only a practical matter of physical sustenance but also a way to satisfy emotional needs, ensure a peaceful home, and secure a husband's faithfulness. Velázquez de León argued that men were home-loving by nature: and if a husband went out searching for entertainment, blame fell on his wife for failing to provide it at home.[61] After all, no husband could resist the charm of a comfortable home, a smiling wife, and a well-set table. She captured this sentiment in an illustration in *La cocina de la recién casada* (Cookbook for newlyweds) (fig. 5.3). As a husband steps out the door to greet friends, an apron-wearing housewife in the background, visible through a window, gives the final touches to the dinner table with a cake, a bottle, and a dish on it. The husband has two options: enjoy a family dinner with his dedicated wife or go into the city with friends. As if to heighten the tension, among those seeking to persuade him is a well-dressed woman, who, in stark contrast with the modest wife, demands her friend's response. The husband answers, "If my wife did not know how to cook, I would gladly join you for dinner."[62] The message is clear: the wife's culinary talent keeps the enticing, corruptive elements of the city at bay.

Velázquez de León did not singlehandedly invent the postrevolutionary cult of domesticity, but she believed in it and played a large role in broadcasting it. Its influence is best demonstrated by her insistence that the modern wife could not rely merely on old knowledge and traditions passed down orally. The world and the demands placed on the home had changed too much. The food that a man's mother or grandmother had made for

FIGURE 5.3. *The wife who can cook keeps her husband at home. Velázquez de León,* La cocina de la recién casada, *51.*

him, cooked by ancient techniques and with implements often just as old, was no longer sufficient. A wife who prided herself on being a good housewife had to dedicate time to the study and improvement of the culinary art. Velázquez de León assured readers that her book was not only a compilation of recipes but an educational tool. Whoever studied it carefully and put it into practice could dominate the art of cooking and, in so doing, develop the skills and temperament to master all the arts of being a successful modern wife and mother. After all, "any woman, regardless of her social position and activities, must have in mind that marriage constituted her destiny's natural goal and should not forget all the training that her future family required."[63]

Velázquez de León's prolific pen and entrepreneurial spirit helped her earn a prominent position in the commodification of culinary knowledge and the construction of the professional housewife. She had used her cooking school as a platform to build "a virtual publishing empire" by the 1940s. Her publishing company, Ediciones J. Velázquez de León, printed all her cookbooks for the next two decades.[64] Questions poured in from throughout the country and even from the southwestern United States. Women considered her monthly culinary magazine, *El Arte de Cocinar*, to be a bible on how to organize their own homes and meals. But her audience was even broader than that. She penned cooking columns for women's magazines like *Negro y Blanco* and *Socia* and some of the city's leading dailies, including *El Universal* and *Excélsior*. For those who devoured her words but could not afford to come to her culinary academy in the city, Velázquez de León offered a correspondence certificate program. Each month for two years, students received a package that included several lessons with detailed instructions and photographs. She promised inexperienced housewives that they would certainly succeed as long as they carefully followed instructions. At the end of the course, students who felt prepared could travel to the capital to take an exam administered by a judging committee. Successful candidates received a diploma that accredited them as "Culinary Art Experts."[65]

A businesswoman at heart, Velázquez de León was only halfheartedly critical of what industrialization had done to foodstuffs. As a recognized culinary expert, packaged food companies and government agencies approached her to develop booklets with recipes based on their products.[66] Among her first sponsors were Productos Oro, a baking products company, and Empacadoras de Calidad, which sold concentrated tomato puree.[67] One particularly lucrative collaboration was with Fuertomate canned tomatoes. Tomatoes occupy a central place in Mexico's cuisine, but Velázquez de León assured her massive audience that the taste and quality of Fuertomate's Etiqueta Negra would meet their standards. Moreover, she claimed that the product would save housewives, who were always busy, effort and time in "selecting, washing, peeling, crushing, straining, and cooking [tomatoes]."[68] Other corporate sponsors followed suit. Nestlé approached her to serve as a judge for a nationwide contest in which housewives mailed in original recipes using Nestlé products.[69] In the demonstration classes at her cooking school, company representatives served cups of Nescafé coffee and gave Milo Milk samples to attendees.

Despite the increased number of middle-class female professionals, Velázquez de León's presumed readers were full-time housewives. Although completely committed to handling the needs of her family, each reader was still a modern woman with social commitments that took her away from the kitchen. In two of her most popular cookbooks, *Cocina instantánea* (Instant cooking) and *Cocina rápida para la mujer moderna* (Quick cooking for the modern woman), Velázquez de León catered to this modern housewife's fast and busy lifestyle. *Cocina instantánea* offered a range of meals based on canned and packaged products that promised to save time, labor, and servants. If her readers had to go out or had a commitment, they could to select dishes that were easy and quick. Contemporary wives, Velázquez de León wrote, may envy their grandmothers who had servants to cook for them; given the diversity of products now at their disposal, she concluded, "Perhaps they would be the ones to envy us."[70] *Cocina instantánea*, she clarified, was not merely a cookbook but also a source of suggestions, practical advice, and simple ideas on how to prepare dishes in a few minutes for those rushed occasions when there was not enough time to cook. It offered instructions not just on how to prepare food but on how, even in the crush of modern life, to present dishes and serve them with salsa and other condiments to help mask the taste of the processed foods used to make the meal. It was, in its modest way, a testament that modern life did not demand the sacrifice of civility and that the housewife need not feel ashamed to serve industrialized food.

Yet there was a limit to how much even Velázquez de León would accommodate modernity. Far from being a substitute for traditional cooking, her recipes and tips on fast cooking were to be used only for emergencies when a homemaker could not prepare a proper meal of several courses. After all, Velázquez de León cautioned, "our traditions would not adapt to such a diet." With the help of these new food products, modern appliances, and some goodwill, she argued, "We could easily live up to our responsibilities as housewives both in the home and in society." She promised Mexican women that they could fulfill social commitments, attend to their families, and—though it was less obvious in her cookbooks—have a job without being any less of a wife or mother. As her sales attest, people welcomed the message.[71]

Velázquez de León's *Como cocinar en los aparatos modernos* (How to cook with modern appliances) marked a change in her culinary writing career in the late 1940s: she explicitly embraced cooking with modern appliances (fig. 5.4).[72] In her foreword, she claimed that many housewives approached her with a dilemma. Women in Mexico City invested in acquiring pressure cookers, blenders, mixers, Miracle molds, and other devices. After bringing their exciting purchases into their homes, Velázquez de León claimed, they realized that "they did not know how to use them."[73] Accompanying manuals offered instructions to prepare foreign-style dishes but were of little help to homemakers wanting to cook the Mexican-style dishes that their families expected. Recognizing the potential for expanding her culinary grasp, Velázquez de León set out to solve her pupils' quandary.

For two years, Velázquez de León assiduously studied modern appliances, a program of "practical work" to develop "a pattern of use" that was culturally acceptable and could become part of daily routines. As part of this work, she took two hundred of the country's most popular recipes and modified them so that they could be created with modern devices. The Birtman blender, for example, promised to turn avocado and tuna into molded salads appealing to the eye and pumpkin flowers, corn kernels, and milk into a Mexican soup pleasing to demanding palates. The book included not only helpful illustrations with easy and detailed instructions on how to use

FIGURE 5.4. *Cover of Velázquez de León,* Como cocinar en los aparatos modernos.

each appliance but also advice on how to wash and take care of them. Calling itself the "housewife's best counselor," the book promised that if she took care of the devices, they would take care of her.[74]

Como cocinar en los aparatos modernos was only part of Velázquez de León's effort to integrate these appliances into everyday cooking. Her cooking academy also held free demonstration classes for those seeking hands-on learning, where housewives learned "to cook in the way present times demanded."[75] The Burpee bain-marie, the Birtman blender, and the Lamex oven, among other items, were used in the classes and available for sale at the academy. Indeed, anyone who purchased any one of these products also got tickets to four demonstration classes. Those who bought them but resided outside the capital received a free copy of the cookbook. Velázquez de León offered dedicated classes on Fridays for those who purchased a portable Lamex oven, where they could learn how to make the most of it. Her connections to the parent corporation did not end there. On Tuesdays and Saturdays she also sponsored "Lamex radio concerts" on the XEW radio station.[76]

Of the multiple devices that she researched, Velázquez de León believed that the Birtman ("the blender of the Mexican home") offered the most help to housewives. The Birtman could mix chilies, spices, and other ingredients for the preparation of favorite dishes such as mole and tamale fillings and ground cooked maize for fine atoles (cornflour drinks). Velázquez de León held that "the old is always new when enriched with experience and modern science."[77] To demonstrate her point, among recipes in *Como cocinar* that called for the use of a blender, she inserted four illustrations of the Birtman. The first illustration depicts a kitchen scene with a woman wearing a white apron and chef's hat, standing before a large kitchen island, like a scientist in a lab. Recipe in hand, the woman is absorbed in reading the instructions. As in other illustrations in Velázquez de León's cookbooks, the woman does not have a maternal figure or a maid alongside to help her. She is attractive, neatly dressed, but alone in her kitchen, with just her instructions in hand and modern and traditional cooking implements scattered across her working station. The Birtman blender stands in the forefront. A large clay pan filled with pieces of fruit sits on its left and a wooden spatula on its right.

The recipe metaphorically connects the young woman with her female ancestors. As if emerging from the recipe on paper, the blurry profile of an elderly woman wearing eyeglasses holds a note labeled "old desserts." With her left hand, she passes down two more notes to the younger woman, suggesting the transmission of culinary knowledge from one generation to another. The elderly woman, perhaps the grandmother of the housewife, hands recipes to her descendant, who can replicate the desserts savored by her ancestors with the aid of the Birtman. To drive this home, the label "sweets" sits inside the Birtman's glass jar. The triangle is complete: knowledge passed down from previous generations in written form serves as the basis for the preparation of desserts with the blender's help. The Birtman blender, according to this ad, is not a rupture with the past but a tool that enables the new generation to connect with those who came before.

A second illustration connects electrical appliances to the country's religious heritage (fig. 5.5). As Jeffrey Pilcher shows, Velázquez de León leaned heavily upon salient religious associations in defining the national cuisine, thus offering a Mexican identity distinguished "from the more secular versions of government propagandists and male intellectuals."[78] In another kitchen setting, a similar female chef converses with a nun garbed in a traditional habit standing behind a large table. The nun's posture suggests that

FIGURE 5.5. *Valuing the culinary heritage. Velázquez de León,* Como cocinar en los aparatos modernos, *284.*

she is in the midst of explaining something to the attentive young housewife. Two nicely decorated dishes sit next to a Birtman blender, and the caption reads, "[O]ld desserts from the convents, a Mexican pastry heritage, easily re-created in the Birtman blender."[79] New appliances not only could fit within traditional foodways but could have religious blessings too.[80]

The last two illustrations are reflective of the postrevolutionary period's contradictory views of indigenous people and how they and their customs could both be romanticized and be seen as a threat to the modern nation. The first invites readers to enjoy the exquisite cured pulques made in a Birtman. Pulque, the preferred beverage of the capital's lower class for decades, had been a controversial subject for authorities and reformers through much of the twentieth century. By the 1940s, however, it could be strategically rehabilitated into a source of pride connecting the countryside to the modern kitchen.[81] The recipe for almond-cured pulque called for thoroughly blending peeled almonds, sugar, orange juice, and a half-liter of pulque, adding the other half-liter of pulque, and setting the mixture aside to rest in a cool place for two hours before serving.[82]

The accompanying illustration (fig. 5.6) is set in a kitchen that metaphorically connects a modern home to a field of agave through an open door. An indigenous man dressed in manta carries a pulque jar into the kitchen. In the background, an *aguamielero* (individual who extracts the mead) kneels before a maguey plant. The image visually connects the two spaces. A housewife wearing an apron and a chef's hat supervises a happy indigenous man in her modern kitchen who is pouring the pulque—as if he is a child. She is ultimately the one who will manage the blender. He brings the raw material; she processes it via the technology.

The fourth illustration reinforces the racial, ethnic, and classed overtones of the third one (fig. 5.7). The caption triumphantly announces, "The Birtman blender has simplified a lot of housework!" Unlike the scenarios above, this image juxtaposes two different activities, counterpoising modern blending (mixer) and traditional kneading (by hand). On the left side, a short, stout indigenous woman with a sad expression kneels on the floor behind a large clay casserole in which her right hand appears to be kneading dough. Her left hand scratches her head, indicating her poor personal hygiene or even the presence of lice, and offers a miserable picture of unsanitary food preparation. Readers can only imagine that the left hand would eventually join her right in mixing the dough. The right side of the illustration presents an idealized image of scientific cooking, personified by a tall, slim, and

FIGURE 5.6. *Preparing a mixed pulque drink for modern entertaining. Velázquez de León,* Como cocinar en los aparatos modernos, *319.*

FIGURE 5.7. *Untouched mixing and kneading. Velázquez de León,* Como cocinar en los aparatos modernos, *319.*

fair-complexioned woman standing behind a counter. Dressed in an apron and chef's hat, she happily supervises a Birtman mixer. Her hands never come into contact with the dough, ensuring the hygienic handling of food.

Whereas the first three illustrations reconnect the modern housewife to a vibrant culinary past, the last one visually imposes a polarity, distancing the housewife from a nonmechanical past. The dissonance between the two women could not be greater, as if they were representatives of two diverging periods in the country's culinary history. The two technologies and bodies stand as mutually co-constitutive. The indigenous woman here is not a source of pride and inspiration—like the pulque carrier in the previous illustration—but an unhygienic and backward-looking representative of rudimentary technology of the past. The present appears bright: in fact, better said, white. The scientific housewife showcases how little effort goes into contemporary cooking. At the push of a button, metal blades work for her while she stands in the idealized working posture: upright behind a counter. Her counterpart sits on the ground, muscling her way through a pile of dough.

The contradictory representation of indigenous figures in the last two illustrations, far from being isolated, is representative of what historian Rick López has called the "fluidity and contested nature of [the] embrace of indigenousness as part of the national identity, even as . . . prejudice against indigenous people was continually reconstituted" in the post-1920s period. While nationalists embraced some parts of indigenous culture, he explains, "they denigrated other aspects and even the people themselves." In disparaging indigenous people, López argues that nationalists "claimed that such things as poor hygiene, squalid living conditions, and illiteracy, rather than being merely products of poverty and racism, had become integral aspects of indigenous culture." Furthermore, the derogatory portrayal of the woman kneading echoed typical contemporary representations of indigenous women as "hunched, black-eyed, disheveled, graceless, filthy, and with thick red drooping lips."[83]

Juxtaposing traditional and modern cooking devices became a common feature of ads for electrical devices. Traditional utensils made from wood, clay, or stone were most reduced in size and often employed by indigenous women in traditional garb with long braids; they appear to be the targets or foils of the ads. *Cocina instantánea* included an image of a woman crying while chopping onions. For no particular reason, a molcajete and metate sit on the counter beside her. The woman's tears are caused by the onions, not by the stone instruments. Including them hinted at a dreadful past when a woman's time in the kitchen was full of toil.[84]

Concurrent with the visualization of the modern Mexican home and scientific housekeeping was the association between electrodomésticos and whiteness. Increasing images of fair-skinned housewives operating electrical appliances "signified, reflected, and reinforced the link between ownership of [electrodomésticos] and privileged racial witness."[85] At the same time, the nation's postrevolutionary visual economy transformed stone, clay, and wood kitchen utensils—which might have been perfectly ordinary before the turn of the century—into signifiers of Indianness. Viewers could connect the Indianness of cooks and washerwomen with unhygienic handling of food and laundry, making them unredeemable, without a place within modern homes. Such symbolic "expulsion" from the modern home contradicted the actual presence of hired domestic help—often poor and indigenous—in middle-class homes.

Historians of technology have used the term "domestication" to describe the processes that technology goes through before being integrated into daily life. This integration involves "a taming of the wild and a cultivation of the tame."[86] Through domestication, technologies go from "unfamiliar, exciting, and possible threatening things into familiar objects embedded in the culture of society and the practices and routines of everyday life."[87] Historically, and independent of place, scholars have noted that domestication occurs on three levels. The first involves symbolic work, "in which people create symbolic meanings of artifacts and adopt or transform the meanings inscribed in the technology."[88] The second, practical work, occurs when "users develop a pattern of use to integrate artifacts into their daily routines." Finally, learning about artifacts is termed cognitive work: is experimental work with devices.[89] In her writing and classes, Velázquez de León traversed each level, making electric technology and the fruits of mass production safe, respectable, and (most important) practical. Velázquez de León's writings and the ads that accompanied them addressed the first of these most directly. As she sold recipes, she was selling the idea of what a modern home should be and what consequence this would have not just in a family's life but in the national culture as well. An electric blender or a tabletop stove made life not just easier but better.

THE "MAID PROBLEM": WASHED-OUT MAIDS

Ideas of linking appliances, housewives, and the nation began to gestate in the 1920s. By the 1930s, advertisers were showing their awareness that the kitchen had traditionally been a lower-class female domain. Not just in Mexico, but in the United States and Europe as well, advertisers explained to their initially middle- and upper-class customers why the new electronic appliances were so necessary by referencing the "maid problem." This particular narrative took on strong racial connotations in Mexico: middle-class families saw themselves as whiter, and therefore better, than the darker-skinned working class and rural poor. Many middle- and upper-class families imagined that the class from which their maids came was racially darker and hence less respectable.

Beginning in the late nineteenth century, industrial societies on both sides of the North Atlantic decried the servant shortage, "the difficulty in finding competent servants, and rising costs." Attempts to address this problem ranged from setting up programs to train servants to the establishment of cooperative kitchens and even the proposed "construction of large blocs of kitchenless apartments . . . served by central kitchens run by professionals."[90] In some respects, these fears were real, for the ongoing decline in immigration and the rise in factory jobs in the aftermath of World War I translated into a reduced pool of domestic servants. French bourgeois writers in the early 1920s bemoaned "the crisis of domestics," which they attributed to industrial war jobs that offered domestic servants better wages and a sense of power.[91] After working at these jobs, the people who traditionally made up the domestic labor pool demanded higher pay and refused to offer the deference that their employers expected. Live-in servants had become inaccessible to many middle-class families in England after the war: as one historian has noted, "most women preferred factory, shop, or office work to the long hours, low pay, and restricted freedom of the live-in servant."[92] But, as in Mexico, wherever the perceived problem of a lack of servants arose, appliance manufacturers offered their products as the solution.

By the 1940s, the "servant problem" had become a common trope that marketers employed to pitch the need for labor-saving appliances, though it was more a rhetorical device than a reality in Mexico City.[93] The city entered a period of rapid urbanization in the post-1920s that only accelerated in the 1940s. This demographic explosion was partially driven by the arrival of poor women from the countryside, who not only replenished but enlarged the pool of *sirvientas* (servants), *cocineras* (cooks), and *lavanderas* (washerwomen). Historians of Mexico generally agree that evidence for any decline in the domestic labor pool is lacking. The so-called maid problem, in Mexico at least, was less about conditions in the labor market and more about how technology transformed work in the home: who would labor and how? Although manufacturers and advertisers borrowed marketing tools developed abroad, they had to reflect sociocultural values, beliefs, attitudes, desires, and anxieties in Mexico to find an audience there. Those tools had to be adapted and translated into local cultural codes. One of the most powerful of these codes in Mexican society was race. Beginning in the 1940s, ads for electric appliances increasingly portrayed maids as backward, ignorant, and unhygienic, especially when compared to the machines that they were selling. The simplest, most compelling way to do this in Mexico was to make maids darker skinned and indigenous.

Domestic service in Latin America has been a central component of female employment since the sixteenth century, but the makeup of this labor force had undergone many changes by the eve of independence.[94] During the early colonial period domestic laborers came from a variety of backgrounds, including mestizas, criollas, freed male and female slaves, and white women, but most were Indian women. In the sixteenth and seventeenth centuries, as legislation was passed that protected indigenous people from abuse, the importance of Black laborers grew, creating a racial transformation that associated "domestic service with the lower end of the class/caste/color system." That in turn translated into "a gradual alienation between employers and servants, [and] a loss of status for the occupation of domestic service."[95]

Independence did not change laboring women's economic opportunities. Domestic service remained the most widely available source of employment for women of all ages from all over the country.[96] Over half of the females working in Mexico City were domestic servants. The 1811 census revealed that 70 percent of the young and single women who migrated to the capital entered domestic service, giving rise to the stereotype that maids were single young Indian women recently arrived from rural villages. The numbers of migrant women working as domestic servants—and the stereotypes that surrounded them—changed little in the twentieth century. From the 1850s to the early twentieth century, over 40 percent of the female workers were employed in salaried housework. During this period, domestic service not only employed female migrants but also played a crucial role in their acculturation into lower-class urban life.[97]

Women became servants in more than one way: most frequently they sought the jobs directly, but often poverty and the state forced women into service. The Hospicio de Pobres (Mexico City's poorhouse) in the late colonial period began to house and reform individuals arrested for begging through religious instruction and work training.[98] The institution aimed to transform its wards—especially female wards—into pious, industrious, and modest workers.[99] The confiscation of church property after the Reform Laws undermined institutions like the Hospicio de Pobres. Its conversion into an orphanage for a juvenile and mostly female population in 1884 "reflected broader trends in urban women's economic vulnerability as well as the feminization of concepts of dependence and assistance." The Hospicio then trained the girls in its charge almost exclusively in housekeeping, in keeping with the condescending notion, popular even among female philanthropists, that they needed to "maintain each one at her limit, without creating unrealizable aspirations." Institutions like the Hospicio churned out servants who were then adopted by their employers, for liberal views on adoption, codified in the 1868 regulations, had "explicitly favored the adoption of girls as servants."[100]

During the Porfiriato, dissatisfaction with domestic servants abounded, for a new reason: secularization. Newspapers reported that the cry *¡Ya no hay criados!* (Servants no longer exist!) could be heard throughout the city. The Catholic dailies *Voz de México* and *Tribuno Católico* nostalgically looked back to happier times, when individuals lived more comfortably, with their needs readily met, despite the absence of electrical appliances. In the newspapers' opinion, the servant crisis arose from the lack of religious education. As the papers explained, "there were no servants because there were no longer masters." Previously, Catholic education had taught women to attend to the family and provide maternal care to those in their homes, following Saint Peter's admonition that those "who abandon their homes and families are worse than the unfaithful" (paraphrase of 1 Timothy 5:8). A true Catholic housewife and mother was to offer loving Christian words to her servants, comfort them in their sorrows, celebrate their joys with them, and look after them in sickness, as "children of God."[101]

The maternalistic relationship that had once prevailed and made domestic service tolerable, the article claimed, had become obsolete in the postrevolutionary period. Instead of teaching women to do their duty as the angels of the home, modern education dismissed religious training. Women no longer treated their servants well.[102] Desiring equality with men, women ignored housework and consecrated themselves to study. They preferred to hold a pen to write about history or literature than a needle to sew their husband's and children's clothing. Instead of seeing the heavens through prayers and faith, they looked through a telescope, making them forget their families and abuse their servants. Treated like machines, discarded when no longer useful, servants were no longer comforted in their sorrows. Their employers no longer looked after their morality or thought of them as individuals with souls. Consequently, as *amos* (masters/mistresses) came to an end, so too did *criados* (servants).

Blaming employers for the maid crisis was rare. It was much more common to blame the maids themselves. The superiority that had long characterized the master-servant relationship began to be legitimated by racial theories in the late nineteenth century. The middle classes claimed that servants came from dark-skinned, indigenous backgrounds, were imprudent and unsubmissive by nature, and refused domestication. Supposedly objective researchers who studied the nutrition of workers agreed, concluding that their traditional diet left them unable to perform consistent meticulous work.[103] Employers called for ecclesiastical intervention to convince "domestics to recognize their place." If that failed, they threatened an employers' strike or, if need be, "the importation of more docile workers from Asia or Africa."[104]

The racial background of maids and servants was not going to change anytime soon, and mass importation of workers was both politically and economically unfeasible. Electrical devices entered the fray. Home advisor María Luisa, who directed *El Imparcial*'s section "Two Pages for the Ladies," penned an article on fashionable New York apartments that housed a plenitude of appliances in 1910. Electricity powered a long list of "marvelous inventions," simplifying housework to the point of almost eliminating the need for domestic service.[105] The automatic kitchen consisted of a revolving table to sit down and chop meat, grind coffee, knead and bake bread, fry eggs, and peel potatoes. All housework was done electrically except for making the bed—an omission that inventors were busy addressing, as Luisa assured her readers. Best of all, the machines cost only one cent per hour. American women lived in an Eden, freed from housework, with time to study or go for a stroll. The building also offered maid service by the hour.

Electrically run homes remained an unfulfilled dream in most of the world for decades to come. Live-in and part-time maids remained a common marker of middle- and upper-class homes in the Mexican capital. Despite their numerical importance, the state excluded domestic workers from the improvements ushered in for the working classes. A revision of the federal labor code in 1928 had a devastating consequence for salaried domestics. An early draft had defined domestics as "laborers and not workers, allocating [them] to a status like apprentices and day laborers." It gave them protections against sexual harassment as well as unemployment and vacation benefits. The final draft dropped these provisions, however, demanding only that employers treat the domestic workers "with due consideration, abstaining from bad treatment either by word or by actions."[106] The law confined domestics to a position like minors "learning a trade, [who] received room and board in partial payment and like casual laborers lacked formal contracts."[107]

Lax protections coincided with an overall expansion of the number of female workers involved in paid housework. This was an unintended consequence of article 123 of the 1917 constitution, which dealt with labor, and increased female migration to the city. The article was among the most progressive in the hemisphere: it created an eight-hour workday limit, set minimum salaries, granted maternity leave and two thirty-minute breaks for nursing, and protected women against dangerous or unhealthy jobs as well as night shifts. In practice, however, the protective legislation had unexpected results.[108] Rather than comply with the spirit of the law, employers simply removed women from factory night shifts and dismissed them when they became pregnant.[109] This unexpected consequence pushed more women into the informal economy at a time when increasing numbers of female migrants were arriving in the city, expanding the pool of domestic servants. The city had 61,226 servants, 9,989 men and 51,237 women, in 1910. By 1940, when the city's population reached 1,757,530, it had 82,745 servants, 9,787 men and 72,958 women.[110]

The social view of housework as menial, unskilled labor undertaken by recently arrived female migrants to the city reinforced the racialization of domestics as indigenous. The concept of the racialized maid would serve to distance the *clasemedieros* (middle class) in the postrevolutionary period from the women who labored in their homes.

The "maid problem" was fundamentally a concern of the middle class, which had historically been the largest employer of servants. Now middle-class employers complained that too few domestic workers were available (even while changes to the law increased their numbers)—at least, not enough efficient and respectable ones. It was also members of the middle class who would respond to manufacturers' and advertisers' pitch to solve the "maid problem" with electric machines. The degree to which the electricscape would penetrate the private homes of Mexicans depended on the size of the Mexican middle class and its disposable income. Successive postrevolutionary presidencies did all they could to make sure that happened. They privileged the middle class, ensuring that it grew in number and prospered in wealth.

The membership and relative size of the middle class in Mexico have been a contentious issue.[111] James Wilkie and Paul Wilkins found substantial social mobility: the middle class represented 16 percent of the population in 1940, 20 percent in 1950, 22 percent in 1960, and 29.1 percent in

1970. The authors argued that several researchers and commentators before their study "underestimate[d] the growth of the middle classes." Their miscalculated conclusion served the political Left's argument that the revolution had not resulted in substantial social mobility.[112]

Louise Walker finds a reciprocal relationship between the official party, the Partido Revolucionario Institucional (better known as the PRI), and the middle class.[113] After the revolution, successive presidencies sought to increase the middle class and privileged its interests by making it the main beneficiary of the state's educational opportunities and bureaucratic jobs. Professionals, intellectual workers (university professors, teachers, and artists), white-collar workers (administrators, secretaries, and clerks), and technical workers as well as small business owners, merchants, and supervisors benefited from the midcentury boom (1940–1970). The country's gross domestic product (GDP) grew at a 6 percent annual rate. These individuals were mostly engaged in "non-manual or technical work, and lived in urban environments." Recent studies, however, have made a case for going beyond the usual combination of income, occupation, and education to define the middle class and to include a state of mind (identity, lifestyle, and cultural world) and political discourse. This perspective privileges purchasing power, patterns of consumption, and diet as essential elements of middle-class identity.[114]

One good reason to privilege culture and consumption over income alone is that Mexico's middle-class identity emerged during World War II and the Cold War. During the war, the United States established the Office of Inter-American Affairs (OIAA), a propaganda agency meant to cultivate sympathy for the Allied cause via "radio programs and films . . . [that] promoted a middle-class version of the American way of life."[115] Propaganda did not stop with the end of the war. For decades Mexicans consumed large dosages of US advertising, creating a longing for consumer goods that epitomized the American lifestyle. The United States offered an appealing consumer society that "mesmerized [the middle class] with the quantity of things the metropolitan culture offered through advertisements of automobiles, clothing, electrical appliances, movie series, and trips."[116]

Not all were moved by the advertising. The poet José Joaquín Blanco, for one, sneered that after 1940 "[Mexicans] were taught to dream a pricey delirium of lavish, leisure, and pretentious affluence." The middle class risked becoming "a covetous class of whatever and above all, what it least needs—useless consumption, such as appliances or clothing of little use, eating up a great part of the middle-class expenses."[117]

Blanco may have kept Mexico's growing middle class at arm's length, but US advertisers reached out to it. In a piece on world advertising in 1947, Don Thorburn, manager of the J. Walter Thompson advertising agency (JWT) in Mexico City, wrote, "One of the brighter spots in the U.S. export horizon, Mexico . . . shows what U.S. manufacturers can accomplish abroad when they really try . . . the average Mexican of the prospective-buyer class has two essential things in common with its U.S. counterpart: the desire to buy; the power to buy." The peasant class, in Thorburn's estimation, still represented two-thirds of the population, but the upper-middle-class, though small in number, had the spending power "on about the same level as ours—in the metropolitan centers." Mexico was a potential market for "low-cost essentials which improve living standards, such as household appliances, ready-to-eat cereals, packaged foods, cigarettes—all things which appeal to the Mexican taste and can fit into the family budget." Mexico imported most of these items. In Thorburn's words, the country's industrialization had yet to reach "the point where national manufacturing can supply the increasing demands of a people who are daily becoming more exacting in their requirements of consumer goods."[118]

An advertiser himself, Thorburn boasted that the middle class's discernment did not arise on its own but rather because "the Mexican has been educated to the realization that the best products are brought from the U.S. or from abroad." Two years later, in a self-congratulatory tone, he bombastically claimed that "in the past five years, advertising has proved . . . a prime factor in elevating the standard of living in Mexico."[119]

Ads agreed. A 1929 Mexlight ad from the Sala de Electricidad showroom called for the laundress to be replaced with a *lavadora* (washing machine; see fig. 5.8).[120] Titled "Reflections of a Housewife," the ad praised a sleek, light-skinned housewife who compared the cost of the appliance to the cost of her laundress: though the washing machine was more expensive initially, she would save far more money over time if she bought one.

An essential element of the middle-class ideal, *economía* (economy) depended upon the thriftiness of the housewife, which, many argued, should be assured through training. Postrevolutionary governments advocated for mothers to be educated in scientific hygienic household management and child-raising. Following the lead of Porfirian home economists, Marichu and Velázquez de León included in their cookbooks techniques and tips for considering economía when attending to the family needs.[121] Indeed, it was a popular theme for cookbook writers and home advisors. Alejandro Pardo Landa, in his *Los 30 menús del mes* (30 menus of the month) for instance, offered recipes explicitly with economy in mind, which, he instructed his reader, was "well recognized as the base for order and happiness."[122]

FIGURE 5.8. *Sala de Electricidad,* "*Reflections of a Housewife*" *ad. D. M. V. de R.,* Manual de cocina práctica, *no page.*

Economía meant thinking like a businessperson, like the housewife in the lavadora ad: seeing the benefit of a long-term investment. Installment payments facilitated this investment: even a modest middle-class family could make purchases that were not just financially sound but contributed to the family's happiness by creating less work and cleaner clothes.

Besides saving pesos, the machine offered reliability, and hygiene, two additional markers of bourgeois identity. Successive Mexican governments since the late nineteenth century had quested after improvements in personal and public hygiene, for a recognizably modern city entailed not only public works but also campaigns to reduce the health-endangering practices of its inhabitants. Anxiety about reliability and hygiene arose from the fact that most laundresses, unlike others employed in housekeeping work, did not reside with employers but earned a living by washing and ironing laundry for multiple clients. Laundresses took linens and clothes—precious class markers—back to their humble homes, where, one ad proclaimed, "they used washboards and sinks in patios of tenement houses to wash, communal clotheslines to dry, and their floors, or perhaps kitchen tables, to iron clothes and linens."[123] Others would use public fountains for washing.

The handling of laundry from several households in communal settings generated anxiety over exposure to unhygienic conditions as well as the risk that the lower-class lavandera might use cloth items as collateral in pawnshops.[124] Although not visually present in the ad, the human laundress is present in the absence of her faults. The washing machine's efficiency and hygiene stand in contrast to its unpunctual and unhygienic human counterpart. Technologically regulated, the lavadora offered hygienic cleansing of the family's clothing, the middle class's first and foremost public symbol of respectability.

Revolutionary governments in the 1920s and 1930s developed health policies and programs to inculcate the gospel of health.[125] Domestic economy courses, for instance, taught mothers and daughters the importance of improving and safeguarding the hygiene of their home and family.[126] Washboards were unhygienic, for example, and mothers should strive to have washing machines in their homes. A 1933 Sala de Electricidad ad promised that a washing machine would provide "clean clothes daily . . . washed and ironed at home without the danger of being destroyed or contaminated by being mixed with other clothing or washing with impure water."[127] The ad included two contrasting domestic scenes to mark the distance between yesteryear's laundering and modern laundering. The first is chaotic: a mother busy ironing and a daughter hunched over the washboard surrounded by several water buckets next to a brasero with several pots and irons over it and clouds of smoke engulfing clothing on the clothesline nearby. By contrast, the middle-class home has a dedicated laundry room where a relaxed housewife in high heels holds a freshly ironed and folded man's shirt, while her daughter—who in yesteryear would likely have been helping her mother—plays with a ragdoll. The irony in the contrast between the two images was clear. The housewife in the second might be a "working" woman who needs appliances like washing machines, but she could still uphold the ideals of femininity because her work was not back-breaking manual labor, while the woman doing the strenuous work is not a "real" modern woman.[128]

According to these advertisements, the attributes of middle-class life, far from being exhausting or demanding, allow the homemaker to live up to the languid ideal of ease and beauty. The postwar era had witnessed the refashioning of corporeality and ways of discerning physical beauty. Ageeth Sluis, in her work on the intersection of urban reform, public-works projects, and gender norms in Mexico City, uses the concept of the "Deco body" as the period's new female ideal, which "stressed length, height, and androgyny." The Deco body's attributes paralleled those of the machine age, in which the body, as "physical machinery," reflected ideas about health, hygiene, and standardized lines. The female body, a metaphorical machine in the ad illustrations, stands comfortably next to the laundry machine and helps to elicit similarly "desired qualities such as precision, productivity, and discipline."[129]

When the feminine form became delineated and made to reflect the ideal characteristics of modern machinery, as was the case with the Deco body, it became "not only the commodity's emblem but its lure [operating] as a conduit and mirror of desire, reciprocally intensifying and reflecting the commodity's allure."[130] Like the elongated Bataclán (a burlesque show) dancers who adorned much of the 1920s visual economy, the housewife in the ad echoes those "sophisticated, independent, and stylish women" who peopled the black-and-white imagery of the day. Indeed, she closely resembled the drawings by artist and cartoonist Ernesto García Cabral. As in his drawings, the body of the modern housewife is rendered "languid, confident, and relaxed," with her gaze boldly engaging the viewer. The posture of the woman in the 1929 Mexlight ad from the Sala de Electricidad, at once confident and defiant, invites housewives to reflect on what the shiny washing machine offers. Her hand firmly resting on the machine, she has already made up her mind: "I no longer worry about my laundress's informality."[131]

The black-and-white figure of the Deco body appeared regularly in the advertising artwork throughout the 1920s. These women were feminine in a new, modern way: simple, rational, but powerful—like the machines they used. A 1929 ad (fig. 5.9) for an electric range not only shows its massive size and height (it is nearly as tall as the two stylized women next to it) but also expounds on its labor-saving promise. The range makes cooking simple and fast, freeing the two women to go out, move about, and demonstrate their grace and sophistication—that is, their modernity.

FIGURE 5.9. *"Cleaner kitchens, better food, and less work," electric stove ad. D. M. V. de R.,* Manual de cocina práctica, *no page.*

The modern woman's embrace of the Deco body was, on some level, threatening to the postrevolutionary ideals of the family. It "ran directly counter to . . . ideas of acceptable feminine roles and clashed with the mainstay of revolutionary programs for women, which allowed women a measure of power only as mothers."[132] Previous images of women had emphasized obvious markers of biological reproductivity (accentuated hip and bust lines) as well as other figures—husband and children—around her. The Deco woman had none of these things. She was alone, confident, and androgynous: not tied to her body, which was nondescript, but bound only by her attitude and what she could do. This ran counter to what postrevolutionary governments said women should aspire to. Ads for products that were themselves clearly targeted to a woman in the home, taking care of a husband and family, nonetheless incorporated chicas modernas. This otherwise radical figure of femininity became as domesticated as the appliances being marketed, and her subversive nature was tamed through images that brought her back into the home.[133] The Deco body was so safely domesticated that in 1934 Carbia used the image for the cover of her cookbook *Marichu va a la cocina y recibe con distinción* (Marichu goes to the kitchen and hosts with style).

Velázquez de León also helped to tame the modern woman. On the cover of her book *Cocina rápida para la mujer moderna* (Quick cooking for the modern woman, ca. 1950), two crossed spotlights highlight four modern activities for the "modern woman"—who was of course a wife and ideally a mother. In the top-left corner, a modern wife confidently sits behind the wheel of a car conversing with her female companion. It is not just any automobile but the period's quintessential symbol of freedom, independence, and even danger: a convertible. In the top-right corner, a woman in a swimsuit is water skiing, with one hand securely gripping the rope and the other waving joyously in the air. In the bottom left, she is depicted with cards in hand, sitting around a game table with female friends. In the image on the right, decked in a red gown, she dances the night away in her husband's arms. Her busy social life was proof of her modernity, but even her modern role had its limits. No matter how busy she was, she still had her traditional responsibilities to her family. Though her packed schedule regularly kept her away from home, Velázquez de León had a cooking style to suit her: *Cocina rápida*. At the intersection of the two spotlights sits a perfectly prepared dish.

The chica moderna's embrace of novelty and individuality might have been a radical break with previous notions of womanhood. Through efforts like this, however, it became less a break with past ideals of womanhood and more a new expression of a reliably familiar figure. Like the Deco body's elongated, androgynous physique, the look of a modern woman had notable variations. But writers and advertisers assured Mexican audiences that this new woman's grandmother might have recognized herself in the core work that constituted her day.

The chica moderna had her counterpart in the ads for electrical appliances: the figure that women were supposed to fear becoming or (equally dangerous) being seen as. Whereas the modern woman was elongated and notable, this woman was short and squat. The modern woman wore high heels, while this woman often went barefoot. The modern woman had short hair or at least hair done in the latest style, while her opposite had long hair, often in braids (to keep out of the way of her constant work). The modern woman had features that made her indistinguishable from women in New York or Paris, but her counterpart was both unmistakably poor and Indian.

In one ad GE used this unmodern woman to sell a laundry machine: a short, shoeless woman whose height barely surpasses the stone laundry washboard vigorously wrings a piece of cloth. With the woman's anonymity preserved by having her back turned, the viewer is left looking at two long dark braids cascading down her back. Facial characteristics were not necessary to place her in the country's racial hierarchy. Dirty and barefooted, she is clearly intended to be Indian: the antithesis of the Deco body. The negative associations—short, stern, and dark—served as visual reminders of the perceived unredeemable characteristics of the country's indigenous population.[134] Anxieties about indigenous people's potential to assimilate in modern Mexico and embrace consumer culture ran high as late as 1950. A census section dedicated to material well-being asked whether the individual being surveyed wore shoes or went barefoot. The middle-class readers of the GE ad needed few words to warn them. The presence of an indigenous woman spoke volumes, and the ad simply warned of the "nest of microbes" that lay in the "barbarous laundry washboard."[135]

By contrast, in the foreground of the ad lay modern society's savior: machine washing. The appliance's whiteness distanced it visually from its brown-skinned human counterpart. The machine stood taller and had rubber wheels on the bottom of its legs, allowing the housewife to move it around with little effort. This model included a wringer, which allowed users to further distance themselves from the process of cleaning clothes. Once the wash cycle came to a halt, soaked clothes were manually fed piece by piece through the rollers on the top of the washer to press out all water. The entire process repeated after the rinse cycle. The Indian woman in the background had to scrub and wring out her clothes by hand, but not the modern housewife. She was nothing like her drudge-like counterpart, who was less an expressive, accomplished individual than just a source of household labor.

By the 1940s, ads presented indigenous figures, particularly women, as antiquated objects for modern consumers to symbolically reject. The modern domestic spaces constructed by advertisers had little room for maids, let alone indigenous ones, and the anonymous lavandera in the GE ad enters the picture only to be conceptually discarded—destined to perish alongside the indigenous cocineras who had tended the braseros.

Foreign manufacturers, which until then had dominated the electrical appliance market in Mexico, responded to the emerging postwar protectionist landscape by establishing local manufacturing plants for their products and by shaping their advertising campaigns to fit changing attitudes. The ease of Mexico–United States relations during World War II gave way to increasing tensions in the postwar years and, for Mexico, growing nationalism. As the United States pushed for more access to Mexico's markets and enforced controversial loan conditions, Mexico responded with a "surge of postwar economic nationalism in defense of [its] autonomy."[136]

Under the campaign "Hecho en Mexico" (Made in Mexico), import substitution became government policy: Mexico would look to itself to supply the conveniences of modern life. It would also look inward to find figures that sold the idea of modern Mexico—ironically, in many cases, by looking to the past.

US manufacturers who only decades before had celebrated a figure of the modern, universal female consumer found themselves now crafting ads "with rhetoric that 'Mexicanizes' American products." This was the strategy that Industria Eléctrica de Mexico (IEM), a subsidiary of Westinghouse, followed in order to emphasize its independence from its American origins. As the historian Julio Moreno argues, through patriotic and nationalistic rhetoric, IEM underlined how "it contributed to industrial development and the progress of Mexico . . . while presenting its alliance with Westinghouse as the key to the company's success and Mexico's progress."[137] The company was not American but was happy to leverage its US connections to advance the interests of the Mexican appliance market.

The new white porcelain electrical maids had made not only the ethnic maid obsolete in these ads. By the mid-1950s the ads extolled the reduced interaction between the user and the machine as another step in the housewife's liberation. This liberation from domestic chores translated into siesta time and added comfort and happiness for the entire family. These elements found their expression in a 1955 ad for an IEM washer called a Laviem (fig. 5.10), a play on words, combining *lavar* (wash) and the company's name. The ad announced in simple, straightforward language what the machine promised: "She washes . . . and you rest!" In the foreground, a nicely dressed housewife lies down, fanning herself, while the washer runs. In the background, a caricatured maid in uniform shouts, "Undoubtedly better!"[138] This, the IEM ad suggests, is the new and better servant for the overworked housewife, which offers profound satisfaction and relaxation. Just in case the message was not clear enough, the add offered a litany of adjectives: faithful, obedient, tireless, quiet, skilled, and economical.

These terms popped up routinely in other ads. In one ad IEM applauded Laviem's reliability, praising it as a "beautiful washer that washed and washed the family laundry without getting tired, without the most minimal mishap, removing stains leaving it radiant as white as snow and, above else, without damaging delicate fine fabrics." Another opened with the question, "Do you wash your clothes? Or the Laviem?" It described the machine as a "modern servant, obedient, tireless, which did the family's laundry fast, well, cleanly, and without damaging fine fabrics." Lava-

FIGURE 5.10. *"She washes . . . . and you rest." Laviem ad.* Almanaque de la felicidad, *1955, 119.*

doras washed, rinsed, wrung, and almost dried, leaving laundry so clean and white that the clothing was a pleasure to look at. They saved homemakers from two unpleasant chores: "washing without getting one's hands wet" and having to "deal" with laundresses.[139]

In creating a market for electrodomésticos, the maid problem symbolically vacated domestic spaces traditionally occupied by lower-class women. Depicted as unhygienic, backward, and lacking in deference as well as discipline, the dark-skinned maid served as a counterpoint to the emerging ideal of middle-class womanhood that commentators, writers, state officials, and even middle-class women themselves held. The women who had historically embodied the productive spaces of middle-class homes were symbolically shown the door so that white, slender, hygienic, and modern housewives—aided by technology—could reclaim them. New appliances were not particularly revolutionary: class and racial hierarchies replicated themselves in this expanded electricscape. When it came to who should be celebrated and who should be scorned, modernity—for all its lights and switches—looked much like the past.

CONCLUSION: ELECTRICITY, LABOR, AND THE HOME ECONOMY

The opening paragraph of the edited volume *How Users Matter* reminds us never to take for granted not only the use of a technology but the users too, for owning and using are two different things.[140] Advertisers, domestic advisors, and middle-class women in Mexico City crafted a visual racial script in the symbolic domestication of household technologies, a process that inscribed race into not only artifacts but also the spaces that housed them. This script combined the applied value of domestic technology with the subjective view that worked to racialize and redefine middle-class womanhood. As David Arnold demonstrates in his study of India, "who sold a machine to whom, who used that technology or was displaced by it, who gained prestige or lost status by its use [are] questions in which issues of race repeatedly asserted themselves."[141] Race shaped how electrical appliances acquired social meaning in the Mexican context by standing as a signifier of technological aptitude in the reimagination of domestic spaces. Users were symbolically constructed as white middle-class housewives, but a gendered middle-class identity was not crafted solely in relation to technology use. Middle-class respectability had long been associated with having domestic servants, which is still true today. Maids, who historically served to mark class status, subsisted alongside the new modern appliances.

The electricscape had made significant headway into the homes of capitalinos by the late 1950s, particularly the upper and middle classes. Besides serving a practical purpose, manufacturers, advertisers, and domestic advisors claimed, appliances also announced that the home they resided in was a modern home, inhabited by people who cared about hygiene, efficiency, and style: progressive people of substance. Electrical appliances had become signifiers of class not only for the middle class: the market for used appliances meant that even poorer capitalinos could obtain smaller devices. In his acclaimed study of the Sánchez family in the capital, anthropologist Oscar Lewis found that 79 percent of the renters in the Casa Grande tenement had radios, 46 percent had sewing machines, 22 percent had electric blenders, and 21 percent had television sets.[142] Lewis found that even tenants under an often dilapidated common roof employed electrical appliances as social markers to claim status. This is how the electricscape expanded not only practically into the intimate spaces where people lived but also subjectively into Mexicans' sense of self and place within society. The ads for electrical appliances promoted not simply their practical aspects but what it meant to have and use them.

Not all the promises were true. The "maid problem" never materialized, at least not in the form of lack of workers. Through the first half of the century, maids remained prominent in census data, literature, cinema, and even floor plans of middle-class dwellings with maids' rooms.[143] A telling 1946 report by Don Thorburn from JWT Mexico City noted that one of the problems of the supermarket chain Super Mercados was that "most of the shopping here is done by cooks and housemaids—'criadas,'" who did not care to go into the new store. The markets were "so elegant that they are too embarrassed to go inside; they don't get the essential gossip from the contacts with the curbstone market 'marchantes' [vendors]"; and "Super Mercados won't let them manhandle the meat, fish, vegetables, and fresh fruit, so how can they tell whether it's good?"[144]

Zeroing in on lived experiences—who used what within domestic spaces—is a difficult undertaking.[145] We know that middle-class employers continued to employ maids despite the acquisition of electrical appliances. But who used these appliances? Were maids allowed to use them? Were they taught how to do so or did housewives decline to teach them based on their so-called technological ineptitude? It is possible that some maids resisted by feigning ignorance to avoid additional tasks or to guard themselves from a malfunctioning appliance (and, worse, having to pay for a broken one). Perhaps experiences varied house-to-house. We are thus left in a quandary as to who the users were. If middle-class housewives were considered the end users for household technologies, maids, although physically present, were made invisible, washed out.

| SIX |

The People, Their Electricscape, and the Vanguard of Labor, 1930s–1960

The February 1956 cover of *Lux*, published by the Sindicato Mexicano de Electricistas (SME; Mexican Electricians Union), pulled no punches (fig. 6.1). The magazine had a message for the government: economic growth had to be shared by all. A woman reaches out to one of the many shop windows in the commercial heart of Mexico City, frowning directly at the viewer. The glass in the window stops her hand; the object of her desire is so clear and close yet categorically out of reach. Signs announce the cost of the inaccessible appliances. A man—presumably her husband—caresses the woman's forearm in a consoling gesture. Unlike the cookbooks of the postwar period, this family scene does not celebrate the achievements of the nation's expanding electricscape. The cover contains no self-congratulatory images. It showcases the pain and alienation of consumerism denied. An editorial note explains the "involuntary and cruel irony" of the ubiquity of advertisements for electrical appliances: people saw them as essential for decent living, but the items themselves were priced so high that an average family could not afford them.[1]

An article inside the magazine explained that electrical appliances were no longer a luxury but an indispensable condition of a modest and comfortable modern life, yet inexplicably out of reach for the working classes. When daily wages fluctuated between five and twenty pesos (barely enough to cover food), how could a working-class family find the money for a refrigerator or a washing machine? These same families were subject to the constant siren calls of advertisers: only people who had these appliances could be hygienic, attractive, and modern. For people like the couple on the cover, calls to electrify came across as "a bloody jibe, and sometime rather criminal."[2]

The *Lux* cover gave voice to a working-class complaint: although state-

FIGURE 6.1. *Expensive appliances in a store window. Cover of* Lux, *February 1956. Courtesy of the Sindicato Mexicano de Electricistas.*

led industrialization in the postwar period had blessed the nation with unprecedented economic growth, too much of it had come at their expense.[3] The low wages of these workers made the nation's industries profitable but left them unable to enjoy the fruits of their work. Instead, they charged, only white-collar workers or government officials could enjoy the trappings of a modern life. The SME took it upon itself to address this paradox. Light and power workers emerged as one of the most powerful unions in the country in the early postrevolutionary period. The state's postwar industrialization—brought to life by electricity—only strengthened the political influence of electricistas.[4]

By the mid-1950s, electricistas had led a decades-old vociferous (if intermittent) campaign against Mexlight, the foreign-owned company that by then controlled power service in the country's capital and central region. The SME characterized the company as a rapacious foreign monster that undermined electricistas' purchasing power by refusing to raise wages.

Despite stellar economic growth—annual GDP expanded 6 percent in the period—the SME's attacks found a larger audience among the capital's industrial workers.[5] Real industrial wages had dropped "sharply after 1939, remained exceedingly low until 1952, and were not to recover their 1939 level until 1968."[6]

It is a testament to how rapidly and deeply Mexico's electricscape had expanded that electricity was considered as essential to modern life as shelter, food, and clean water by the early 1950s. What had started with a string of uncomfortably bright lights along the Zócalo nearly a century before was now found in nearly every home and business. Electricity cooked food and brought entertainment. It saved lives and kept people comfortable. Electricity, and the machinery it ran, was such an essential part of everyday living that it became in its own way a right. Unsurprisingly, the people who built and maintained the infrastructure for the electricscape promoted the idea that all should have some stake in it. Electricistas cast themselves as a new generation of revolutionaries carrying on the ideals of fairness and distribution that had been at the heart of the revolution, but now in a struggle against a foreign company. They reached back into the revolution's rhetoric to frame a discourse of nationalism, industrialization, and working-class masculinity particular to the industry: *masculinidad energética* (power masculinity).

The electrical workers held one revolutionary principle particularly close: Mexicans should own and control the country's natural resources. Nationalization of the rail system (1937) and the oil industry (1938) set a strong precedent that offered a model and bargaining chip to labor. The union combined these grand political principles with more practical and pragmatic attacks, claiming that Mexlight's inability to maintain adequate service and develop infrastructure was a looming threat to the nation's industrial development and, ultimately, its strength. The nation, the SME claimed, faced an *atraso eléctrico* (electric backwardness) because Mexlight continued to repatriate profits. In the face of rising demand for electric power, the company offered only a tepid response, preferring to keep its profits rather than invest and expand generation and distribution facilities. The SME claimed that nationalization—which the union preferred to call Mexicanization—was necessary to secure the country's industrial future and would serve as the final step in its economic independence.

The SME employed revolutionary rhetoric, historical narratives, and pictorial devices to persuade people to Mexicanize the electricscape via nationalization, but Mexlight fought back. As the 1950s drew to a close and the threat of nationalization loomed ever larger, the company rolled out a public relations campaign praising its role in modernizing the country. With the help of the advertising firm J. Walter Thompson de México, Mexlight embarked on a campaign to educate the public about electricity in Mexico, its history and importance, and how Mexlight was at the center of it all. Through diverse media from speeches and campaigns to comics and documentaries, the company celebrated its contribution to the nation's progress.

The competing narratives of the SME and Mexlight raised a fundamental question: Who could lay claim to the electricscape that not just capitalinos but all Mexicans sought? Was it the achievement of all Mexicans: the workers who built it and made the machinery that gave it life and not just investors and owners? Or was the electrification a visionary, entrepreneurial act, carried forward with boldness by a few who had the temerity not to surrender in the face of challenges and adversity? Behind the often technical discussions of increasing purchasing power, rates of return, and capital investment, the real issue was who deserved credit. The answer to that question would determine who set wages and investment. To address these intertwined questions, in this chapter I look at the 1930s–1960, a period that witnessed the strengthening of the SME and the growing intervention of the state in the electrical industry, in terms of regulating the industry as well as direct generation of electricity, which culminated in the industry's nationalization in 1960. A more assertive union became paramount in a post–World War II Mexican political context that privileged industrial development and increasingly sought to rein in organized labor.[7] By emphasizing the industry's role in the country's industrialization and the nation's right to control its resources, the electricistas sought to legitimize their group demands against capital.

POWERING INDUSTRIAL DREAMS

With the election of President Manuel Ávila Camacho in 1940, Mexico shifted away from the redistributive policies—particularly the agrarian reforms—that had characterized postrevolutionary governments and toward policies that favored industrial development. During World War II, Ávila Camacho envisioned an economic future beyond supplying agricultural and extractive resources for his wartime allies. His government employed the global conflict and wartime propaganda to sell industrialization both as the inevitable product of the nation's revolution and as a means for Mexico to participate in and support worldwide democracy. Mexican industrialists welcomed this shift and jumped on board, leading an aggressive press campaign to convince the public of the need to set industrialization as an economic policy priority.[8]

The path to industrialization proved to be a rocky road, and the president tripped in his first steps. Established in 1943, the US-Mexico joint Mexican American Commission for Economic Cooperation (MACEC) set out to plan a long-range program that would encourage Mexican economic growth.[9] Among other goals, the commission recommended the expansion of public works industries and the establishment of new manufacturing enterprises. The Ávila Camacho administration also moved to protect these businesses via broad tariff increases but recanted in the face of diplomatic pressure from US ambassador George S. Messersmith and the precedent of reciprocal trade agreements. In the end, not only did the tariffs put up by Ávila Camacho come crashing down: broader protectionist trade barriers did too. At the same time, wartime production restrictions prevented the acquisition of equipment and materials essential for new manufacturing projects from the United States.

Although Mexican politicians hoped that these restrictions would recede after the war, they still faced opposition from their American counterparts, who feared increased tariffs and the threat of new industries south of the border. Dudley Bonsal (deputy director of the Office of American Republic Affairs), for instance, after conversations with Messersmith and the embassy's commercial attaché, wrote to his director, Lawrence Duggan, that the MACEC should "not submit for the approval and support of the Department and of other agencies in Washington any projects or programs which might conflict with the general commercial policy of the United States (i.e., lowering trade barriers)." It was equally important politically and commercially, he declared, for the commission to avoid facilitating "broad or long-range industrial projects" in Mexico.[10] Facing growing opposition from US interests, the commission disintegrated in 1945.

During the war, the United States had made promises to assist Mexican industry after the end of the armed conflict. As the years went by, however, little help would come from the north. Instead, as the United States pushed to dominate and direct trade between the two countries as well as interpret loan agreements in its favor, a surge of economic nationalism swelled up in Mexico.[11] Politicians shared industrialists' and technocrats' concerns that the United States pressure for Mexico to commit to free trade and raw material production really meant a commitment to continued dependence. Electric workers were not alone in charging that American corporations benefited from this dependency at the expense of Mexican consumers.

Mexico's state-led economic modernization plan mirrored the plans of other Latin American countries. Building on the experience of Argentina, Brazil, and Chile, the Mexico plan called for a postwar model based on protectionism, industrialization, and the state's increasing intervention in the economy. The electric industry, one of the most critical industries in the economy, was dominated by foreign-owned companies.[12] For years, economists had argued that shortages of electrical power were a major cause of low productivity in the Mexican economy and that wartime constraints on imports of machinery and fuel prevented the expansion of generating capacity.[13] Power supply shortages threatened Ávila Camacho's postwar plans. His ambitious goals for industrialization demanded a large, reliable supply of power, but in the postwar years he faced a severe shortage of fuels (petroleum and coal) and an extraordinary drought, which resulted in hydraulic resources going to irrigation rather than power generation. With the wartime restrictions coming to an end, the government began to talk about an emergency mobilization of the nation's electric industry.[14] Electricity had become the lifeblood of industry, a national necessity.

Mexico's experience was not unique. A global consensus emerged after World War II that markets were incapable of providing the means for recovery, reconstruction, and economic development and that government intervention was necessary. In terms of the electrical industry, this meant that the state should develop or extend a reliable national electric grid in order to facilitate and secure economic growth. Electric companies in Mexico and beyond felt the pressure to raise production levels and enlarge distribution networks to meet increased demand. Companies—especially foreign ones—swore that they were ready to funnel significant investments into power production as long as they could guarantee adequate returns and the security of their holdings.[15] Companies like Mexlight explained that massive capital investment was not simply a matter of paying for imports from abroad; they often had to confront inflation or depreciating currencies. The issue of rates, or how much profit and security electrical companies deserved, quickly became a contentious political issue: companies lobbied for higher rates, but the public wanted lower rates.

An assessment of the industry and the drafting of a suggested course of action fell into the hands of an experienced group of foreign power consultants in 1946.[16] Their diagnosis was succinct: Mexico, a nation dependent upon hydroelectric power, suffered from an inadequate power supply, with the risk that shortage of water would soon bring further power rationing. A steady expansion program and permanent additions were imperative. Without an ample electric energy service, Mexico would not reach a balance of industrial, commercial, and agricultural development. The report noted that the scarcity of electrical energy was already stagnating the operations of numerous factories and "asphyxiating the development of new industries," because they had no assurance or defined date for the availability of electric power. Bureaucracy was not the problem. According to the commission, the country's legislation was not only favorable to private enterprise but even generous, conceding benefits and privileges that other countries, with more "advanced" political and economic conditions, rarely granted. The fundamental problem was shortages created by the war. The government should spare no effort in obtaining portable power plants and all types of electrical equipment and systems wherever possible.[17]

The consultants considered it vital to follow a two-pronged coordinated program to stimulate private companies while the Comisión Federal de Electricidad (CFE; Federal Power Commission) embarked on a comprehensive national electrification program. The federal government had established the CFE in 1933 as a state-sponsored power producer as part of its effort to develop the electric-power industry, because of its "inherent social interest for the national economy." The CFE had a broad mandate and broad powers, including the development of programs for the "formation of a national system of electric-power generation, transportation and distribution, composed of semi-official enterprises and consumer cooperatives."[18] Its derisory budget, however, left the country's electrical landscape effectively in the hands of foreign private companies. It would take many years for the CFE to become an important electrical producer rivaling the giant foreign-owned companies that dominated the industry.

Although the consultants saw opportunities for electrical development all around the nation, their focus remained on the capital and the central region of Mexico.[19] This focus upon and privileging of Mexico City was both a response to and a cause of industrial centralization. Half of the country's manufacturing was concentrated in the nation's central region by 1930, with 27 percent in the Federal District and 25 percent in the central states of México, Querétaro, Hidalgo, Tlaxcala, Puebla, and Morelos.[20] These percentages reached 38 and 18, respectively, over the next two decades, concentrating nearly 60 percent of nation's industrial footprint.[21] In addition to channeling a disproportionate percentage of investment in public works, infrastructure, utilities, and communication to Mexico City, the federal government subsidized the distribution of foodstuffs and other factors of production so that prices favored the Federal District, encouraging even further industrial expansion in the area.[22]

As industry congregated in the capital, so too did labor and business organizations, which were eager to stay close to the center of political power in the nation. The Cámara Nacional de la Industria de la Transformación (CANACINTRA; National Chamber of the Manufacturing Industry), for instance, arose in the capital in 1941, and 4,585 of its 6,925 constituents—all manufacturers—were located in the capital by 1945.[23] Much of the electricity feeding this growth came from the abundant hydraulic power offered by the eastern mountain range between Veracruz and Puebla (the site of the largest hydraulic complex). The Federal District and bordering states accounted for 80 percent of the 500,000 HP of hydroelectric power generated in the nation in 1934.[24]

The 1946 commission of foreign consultants blamed wartime constraints for a lack of investment in electrical production, but the issue started decades before. Electric companies, which had enjoyed enormous power over their workers and consumers, began to face growing state regulation, labor-friendly administrations, and the nationalization of the railroad and oil industry in the 1930s. The electric industry responded in 1936 by not risking further investment and remained stagnant until 1945, with an increase just under 100,000 kW (kilowatts). The CFE accounted for 40,000 kW, while private utility companies contributed less than a third (30,000 kW). Although the lack of industrial equipment from abroad factored in this standstill, "the real cause was the cold war waged between the State and the companies," according to the historian Miguel Wionczek.[25]

Even before the issue of stagnation began to take root, the rates that electric companies could charge were a source of contention. Leading economists, foreign experts, public officials, and scientific bodies in the 1920s, such as the Mexican Society of Geography and Statistics, attributed the standstill of the Mexican industry to the excessive and discriminatory rates of electrical power producers. A report by the National Commission of Motive Force in 1926 concluded that electric rates had been "practically immutable" for nearly thirty years, despite numerous technical developments in the field that reduced production costs. Improvements in the generation and transmission of electricity reduced operation costs, so the company enjoyed increased profits without raising rates. Differential rates for electricity hindered the development of small shops. For example, whereas large industrial consumers paid 100 pesos per horsepower in 1907, the median rate for small shops was 477 pesos.[26]

The radical nationalism of the Lázaro Cárdenas administration (1934–1940) became a real threat to foreign companies. The former general came to power with a populist agenda that soon backed industrial unions and responded to their economic demands in hope of consolidating the power of the state.[27] His presidency marked the beginning of an interventionist state that would move rather quickly to control and direct the national economy. The electric industry felt this quite early. Electric companies, including Mexlight, grudgingly rolled out price reductions in 1934–1935 after various government reports concluded that rates were too high. The only voluntary price reduction recorded before then took place in 1906, when Mexlight inaugurated the Necaxa hydroelectric plant, slashing rates by 50 percent. Ten years later, however, the company was authorized to raise rates and effectively remained free of any state rate regulation until 1934.[28] Legally, the state had no jurisdiction over the electric industry. Despite an early effort to regulate the industry in 1926 via the National Electric Code and legislation in 1929 to strengthen the government's oversight of national waters, authorities had no clear precepts on how to regulate and control the industry.[29] The two laws were enough to make other electric companies agree to a downward revision of rates in late 1932, but Mexlight resisted, claiming that both laws were unconstitutional.

Congress did not address the issue until late 1933, when article 73 of the 1917 Constitution was reformed to include the electric industry. It had previously been limited to economic matters in mining, commerce, credit institutions, postal services, general means of communications, and water resources. Before the reform, the federal government's only leverage with the electric industry was its control over concessions for the use of water resources for the generation of electricity.

The 1934–1935 reductions resulting from legislative reform were a watershed in state regulation of the industry and a triumph on several fronts. They not only broke a thirty-year-old tradition of untouchable rates and established the principle of revising rates every five years but also proved that rate readjustments would not ruin electric producers. Mexlight calculated that the rate reduction—from 30 to 25 cents per kilowatt hour (kWh, the energy unit that measures electrical energy used over time) for lighting and similar adjustments for heat and power—would translate into 4 to 6.8 million pesos in lost earnings. The government sought to assuage the company's anxieties by allowing it to collect on bimester fixed-rate electricity plans in advance, to employ rigorous measures for the collection of metered services, and to install devices that would prevent power theft. Perhaps even more important for Mexlight's bottom line, the government authorized it to set differential rates for large power consumers, with only a vague warning against favoritism and unfair pricing.

For the masses of Mexicans who had railed against the electric companies, the law pulled back the curtain to reveal the confusing situation that reigned in the matter of rates. Popular discontent over rates had been rising for decades and picked up steam in the late 1920s when small domestic manufacturers felt particularly resentful of the high rates they paid.[30] According to an editorial in late 1928, "We are in a swamp of permanent crisis. Production costs are high, given the price of electricity. It is a matter of life and death for the national economy. The power company has become an enemy of progress." More and more residential consumers gradually joined the dissident ranks: tired of the state's inability to protect them, they demanded rate revisions.[31] Leagues of consumers organized throughout the country to pressure local and state government officials to examine the business of electric companies and regulate them.

A League of Consumers in the state of San Luis Potosí, north of Mexlight's radius of operations, turned to militant actions in September 1932.[32] Claiming to represent residential users, small manufacturers, shopkeepers, and merchants, the league organized a strike of 5,000 out of the 6,000 light and power customers in the state capital.[33] Strikers stocked up on petroleum and candles and asked the company to disconnect service. Besides suspending payments on electric bills, the league began a blacklist of those reluctant to strike and published a small newspaper titled *Luz y Justicia* to air complaints and petitions. It issued a stirring call in October to those whose "condemnable bourgeois sybaritism" prevented them from going without electricity. Their reluctance to join the strike was "proof of servility toward the powerful monopoly." Perhaps knowing that this rousing appeal would go unheeded, the league printed a boldfaced list of "selfish sybarites" in the same edition, so that "their disgrace, their [names] would be stamped as an eternal stain over the historical awakening of a weak and exploited people, who with all courage struggle against a bloodthirsty octopus: the Foreign Electric Light and Power Company."[34]

Inspired by the Consumer League in San Luis Potosí, other consumer movements sent delegates to Mexico City in late 1932 to establish a Liga Nacional para la Defensa de Servicios Públicos (National Defense League of Public Utilities). The newly established league not only applauded the militant actions in San Luis Potosí, but, in a symbolic gesture, called for nationalization of the industry. The league put its hopes in the government, but the state was in no position to delve into the business of the electric companies. It had no technical, financial, or administrative capacity to do what the league called for. The government lacked not only reliable, precise data but also competent personnel to carry out an accurate valuation of the companies' capital assets.[35] But the league and its demands were a wake-up call.

President Cárdenas's brand of economic nationalism, commonly associated with the nationalizations of the railroad and oil industries, also reached the electric industry. The state finally began listening to the pleas of the people and their complaints against the electric companies in the 1930s. The growing electricscape was important enough that the government saw a role for itself in regulating it, though the role laid out in the 1930s was weaker than many in the consumer leagues might have hoped. Mexlight began work at Valle de Bravo to connect twelve power-producing divisions in 1935 to better supply the market. To recoup the project's significant cost, the company proposed a substantial rate increase, but the Cárdenas administration opposed it, worried that an increase would be harmful for industry. Mexlight responded by abandoning the project of creating a vast interconnected and reliable power system. The government could do nothing about the situation. No new investment or advancement of the project occurred for three years. Mexlight briefly recommenced the project but severely scaled it down a year later.[36]

From the mid-1930s onward, as the government heeded the calls of consumers, regulators and electric companies became immersed in a "cold war" in which any expansion of electrical production became linked to rate evaluations.[37] If electrical companies wanted to expand their production—and hence market—the government asserted its right to determine a fair rate. The annual growth rate was hardly over 1 percent, which led to a "critical power shortage" by the end of World War II. Oscar R. Enríquez, an engineer serving as director of the Bureau of Electricity in the Ministry of the National Economy, found that many of the nation's plants and installed equipment had severely deteriorated by 1946, having been in operation (mostly uninterrupted) for over thirty years. Maintenance had been so spotty that most electric plants were obsolete from a technical and economic point of view. This troublesome assessment continued into the immediate postwar period, as requests for much-needed machinery and equipment for the electrical sector and other industries vital to the domestic economy went unanswered.[38] Foreign power consultants would concur with the dire conclusion.

For those struggling to get by while the economy grew, such hubris on the part of the electrical companies was frustrating. They wanted to be full participants in the electricscape but found themselves priced out. Even those who could still afford power and appliances were bitter at their dependency upon electrical monopolies. They now needed electricity in their daily lives but had to acquiesce to the demands of companies like Mexlight, which could claim financial hardship even while its books were closed to the public. Pushed too hard, as the example of Valle de Bravo project proved, the companies could make service worse. With regulation of rates a mixed success at best, people were open to other strategies: calls for nationalization did not seem terribly radical.

A NEW REVOLUTIONARY FRONT

The year 1936 was a harbinger of a broader, more empowered labor movement, including the union of the Cárdenas presidency and the SME. Even the Communist Party, through its Popular Front strategy, became an ally of the president.[39] The new political landscape would see the legalization of the Communist Left, the encouragement of labor unity, the sanctioning of strikes, and the forced exile of Calles. For the next four years, with Cárdenas in office, labor would find a friendly ear in the Presidential Palace.

The Consumer League strike of 1932, one of many strikes to sweep over Mexico in the decade, shared a common enemy with the electricistas: foreign electric companies. Unable to reach a compromise, electricistas represented by the SME led a major strike in 1936 and emerged as one of the most aggressive unions in Mexico. Its story actually begins with a strike the year before by the capital telephone workers and the reaction of former president Plutarco Elías Calles. Calles, who had been out of office since 1928, nevertheless remained the real power behind three consecutive short presidencies. He raged that current president Lázaro Cárdenas was being too lenient with the union, calling the strike "not only ungrateful but treasonous." When the former president's words reached the public, incensed workers responded energetically. The SME took the lead in inciting other unions to back Cárdenas in the struggle to protect workers' rights. Their efforts culminated in the establishment of the Comité Nacional de Defensa Proletaria (CNDP; National Committee of Proletarian Defense), an organization determined—in its own words—to fight against "the possible implantation of a fascist regime in Mexico" and "maintain class unity." Because of its roots in many different unions, the CNDP's greatest weapon was the threat of a general strike.[40]

Within the year, the CNDP developed into the Confederación de Trabajadores Mexicanos (CTM; Confederation of Mexican Workers). Under the leadership of Vicente Lombardo Toledano, the CTM became the nation's largest labor federation and a staunch ally of Cárdenas. Cárdenas publicly acknowledged the strike's legitimacy and subsequently alienated Calles, thereby earning the backing of a now more belligerent SME.

The SME took advantage of the shift in the political winds and demanded legislation that addressed a wide range of demands on economic and social issues important to workers, in particular increasing the union's membership. Besides a wage increase and compensation for work injuries, the SME demanded that no worker be laid off because of the introduction of new machinery and that technical workers could join the union, an important change. Lombardo Toledano, leader of the CTM, spoke on the issue, saying that he believed Mexlight's refusal to allow such union membership revealed its objective "to convert the SME into a group of unskilled workers, [by] establishing a deep divide between engineers and technical personnel and the company's manual workers and office employees."[41]

The struggle to bring engineers into the union was years in the making. When the Lerma-Tepuxtepec hydroelectric plant was completed in late 1931, the union moved to protect a group of engineers from layoffs.[42] Once they were allowed to affiliate with the SME, engineers responded by joining the union in droves. Soon they rose to leadership positions, catalyzing a transformation in the union. The engineer Luis Espinoza Casanova was elected to the central committee in 1933 and subsequently rose to secretary-general. This marked the beginning of the "Age of the Engineers." For the next seven years, engineers Francisco Breña Alvírez, Manuel Paulín, and David Roldán served on the central committee as secretary-general or in top positions. Their hold on the SME would bring about significant transformations in the union's relationships with Mexlight, organized labor, and the government.[43]

Mexlight responded to the SME's energized agitation in 1936 by either stonewalling or declaring itself indisposed to satisfy the union's demands. The SME extended the deadline for negotiations, but that too passed without settlement. Seeing the writing on the wall, the union voted to strike—agreeing not to return to work even if the government declared the strike illegal.[44] Union leaders had reason to worry: Genaro Vázquez, head of the Labor Department, notified the SME that its demand went beyond what the Federal Labor Law specified. Realizing they could not take on both Mexlight and the government, electricistas did all they could to make sure that the strike would not alienate the government. The union made special preparations so that the president's office and other government offices, police stations, and hospitals would have power and made arrangements to deliver power when it heard complaints of "shortages of vital goods and services."[45] Mexlight, in contrast, was not so politically adroit. It refused even the suggestion of a government investigation into its finances or demands for information that might inform subsequent negotiations. This was a miscalculation. With the strike looming, the company asked the president to intervene, but he ignored the request, referring the company to his personal secretary.

On July 16, 1936, over three thousand striking electrical workers brought the city to a standstill. Assembled in the patio of the Labor Department building, a mass of idle workers broke into a roar when the Federal Conciliation and Arbitration Board ruled that the collective contract could be legitimately reviewed because of an imbalance in the factors of production. For the first time, the inner workings of the electricscape would be open to scrutiny and critique.

The CTM, as the largest labor organization, initially sponsored the strike—adding bitter polemics against British and American imperialism—and offered to serve as a mediator. It lost SME support, however, as CTM grew anxious over thousands of workers standing idle. The entire electricscape was affected, but industry was particularly vulnerable. As factories stood idle, employers refused to pay salaries. In response, a four-member committee of SME electricistas met with President Cárdenas and offered to submit their collective contract to him directly. He refused, arguing that resorting to arbitration in the midst of the strike would set a bad precedent. This move reassured the union of the government's noninterference in the strike. When Cárdenas asked his personal secretary to stay out of the dispute, the union knew that Mexlight would be its only opponent. However, the government's neutrality was universal. When the union floated the possibility of confiscating the foreign company, Cárdenas dismissed the idea: not only was it illegal, but it would set an unwanted precedent.[46] The two sides would have to work it out on their own.

It took ten days, but they did so. The SME came out of the strike reinvigorated. With ninety-five of its demands met and the other twelve scheduled for discussion, the union could claim a near-total victory. The SME's techniques—maintaining a direct dialogue with the government, repeatedly asking for Cárdenas's opinion, and taking the necessary measures to supply power to vital areas of the city—had been central to its success. More importantly, its determination to push for far-reaching demands and its refusal to compromise on vital points established its leadership within the organized labor movement. The struggle had demonstrated that the benefits mandated by the Federal Labor Law were not the top limit but only the base of what workers rightfully deserved. After the strike, one historian has claimed, the union had "expanded the definition of Mexican workers' rights."[47]

President Cárdenas also emerged out of the conflict strengthened, having capitalized on the SME's aggressive actions. The combination of the union's militancy and its deference to the president reasserted his role in labor-employer relations. By simply not bowing to a powerful company he had implicitly sided with labor and became seen as a force that could defy the dictates of foreign capital. The experience left him immensely popular and set a precedent that he would call on a few years later when the oil workers launched their own famous strike.

The electrical workers' strike flipped on a switch and showed labor in a new light. Lombardo Toledano considered the strike a "beautiful testament" to the proletariat's organization and discipline, revealing what workers could achieve. However, the strike also shone a light on something painful to Mexicans: their economic dependency on foreign companies. Lombardo Toledano celebrated labor's success but also pointed out the discrepancy. While workers and residents went without essential services, stockholders in London and Brussels likely received news "drinking a wine pleasing to the palate, comfortably seated in the terrace of a summer hotel from which they could contemplate, satisfied and tranquil, the calm July ocean." For him, the stoppage highlighted one of the country's fundamental problems. "As long as we do not surpass the current colonial state," he declared, "it will be impossible to improve or achieve steady economic progress."[48]

For all its labor militancy, the SME was not ready to take up Lombardo Toledano's call with real enthusiasm. For that to happen it would need a contemporary crisis that brought its members back to Mexico's revolutionary ideal. That crisis emerged at the end of 1936, when generals Francisco Franco and Emilio Mola led the Nationalist coup that sparked the Spanish Civil War. The far-right movement that the generals led targeted all working-class groups and organizations. SME members roused themselves to defend their rights in case right-wing elements in Mexico tried to replicate their Spanish counterparts' actions.[49] In a November SME general meeting, Agustín Salvat, the secretary of foreign affairs, addressed the assembly with a proposal that any electricistas so inclined would receive military training to defend against a fascist plot. The assembly responded in fervent applause. However, not everyone was swept up in the call to arms. After the rumble calmed down, Espinoza Casanova openly challenged the idea. The ensuing debate grew so heated that the proposal had to be pushed to the next assembly.

The proposal for training in arms was pushed back but not stopped. Once it was announced, the cause of labor became infused with a call to patriotism: a revolutionary fervor swept through the SME. This was not exceptional, for by the 1930s generations of Mexicans had come to see themselves as revolutionaries, whether or not they fought in the Mexican Revolution. This perspective was informed by the received understanding of the revolution, which one seminal historian has described as "a product of collective memory, mythmaking, and history writing."[50] The notion of *la Revolución*, a form of collective memory, offered noncombatants a way to wrap themselves in the mantle of the nation's most transformative period. Unlike later twentieth-century revolutions, Mexico had no state-sanctioned narrative or official history that was ruthlessly policed. A diverse group of individuals, groups, and institutions—only one of which was the state—appropriated "the promise of revolutionary transformation" and elaborated upon it by weaving together memory, history, and myth. As a political myth, the Mexican Revolution offered a malleable story with "an emotive past . . . [whose] implications for the present and future" could be deployed to mobilize support and provide a mantle of legitimacy.[51]

Organized labor generally—and the SME specifically—had made its own contributions to the concept of la Revolución. The SME existed during the revolution but took little direct part in it. Unlike the *rieleros* (railway workers), electricistas, as a group, did little fighting.[52] Despite exaggerated coverage of such episodes abroad, all factions worked to avoid attacking foreign-owned interests like the electrical infrastructure, so the electric sector emerged from the civil war without much physical damage.[53] Moreover, the union had only recently been founded and its members at the time were among the better-paid workers in the nation and were more conservative than others.[54] They responded poorly to radical calls and utopian anarchism. The SME gave rhetorical backing to other groups but largely eschewed direct participation.[55]

In light of this history, the 1936 call for military training was a way for the union to burnish its revolutionary credentials. Unlike that earlier generation of electricistas, those who emerged victorious from the 1936 strike would fight to protect the promises of the revolution. They were further pushed to do so by the fact that Lombardo Toledano and the CTM were already lobbying the government to arm workers. Such training had been discussed earlier, when Lombardo Toledano encouraged the government to arm workers within weeks of the Spanish conflict "to defend the conquests of the Revolution." Cárdenas rejected the proposal, perhaps anticipating the uproar it would cause among his critics, who interpreted his 1935 decision to arm peasants against landowner-backed militias as a looming "dictatorship of the proletariat." Lombardo Toledano was not discouraged: despite a lack of government support, the labor leader "ordered all CTM unions to form militias . . . and throughout much of 1938 electrical workers met on weekends to train under the supervision of leftist elements of the military."[56]

Calls for preparation continued for as long as the Spanish Civil War raged, along with the idea that some right-wing fascist or imperialist threat loomed over the nation. In March 1938, days after Cárdenas nationalized the Mexico's oil fields, an army major told the SME general assembly of the urgent need for military training. The nation might need them to face problems "such as the case of the oil workers, which were not entirely solved since the dispossessed imperialist companies would resort to all the mediums within their reach to recover what they had lost."[57] The major advised marching drills and exercises in order to form a disciplined and cohesive spirit as well as technical knowledge of the art of war. He and other left-leaning elements in the military recognized the SME's potential as a source of dedicated soldiers who had both an education and above-average skills.

Although it had been around for years, SME was reborn as a militant version of its previous self in 1936. That year became the key to the union's identity and the center of its calendar. On the fifth anniversary of the strike, for instance, *Lux* commemorated the occasion with a pledge to safeguard the group's militant history. In much the same way that la Revolución celebrated the mythology of the nation's revolutionary founding and ideals, the union tried to celebrate—even mythologize—the 1936 strike. The SME strove to "keep alive the memory and gratitude of our leaders" as it recounted the details of the union's emergence. *Lux* warned that a failure to do so threatened the organization itself. "Just as pueblos lose the physiognomy and characteristic traits of their history by not passing down the most outstanding events of their life and succumb in nothingness," organizations had to "feed the votive lamp for the eternal remembrance of outstanding men." Alongside an account of the timeline leading to the strike, the July 1941 issue included the narrative poem "La huelga eléctrica." While the disciplined army of three thousand "held their breath," through the city "the miraculous current was left prisoner and the silence of machinery began the feat [work stoppage]." After "ten endless days of storm and scorching sun," the *soldados de la luz* (soldiers of light) "picked up the gauntlet and won the duel."[58]

Lux had not always been so keen to push for a militant labor rights. Founded in 1928, it did not declare its aim "to develop the moral, intellectual, and physical progress of its members" until five years later, with the "Age of Engineers." The SME used art vigorously after 1933 as part of its efforts to shed its traditional apolitical orientation. That year, *Lux* announced it would use "art as an expression of [the union's] radicalized political culture."[59] Keeping this promise, artists associated with the Liga de Escritores y Artistas Revolucionarios (LEAR; League of Revolutionary Writers and

Artists) took over the artistic design of *Lux* right before the 1936 strike. Santos Balmori provided the first poststrike cover, a visual rendering of the results of the conflict: a colossal worker towering over a fallen capitalist.[60] Using common allegorical images, Balmori showed a fat, knocked-down capitalist wearing a top hat.[61] With his mouth half opened, the capitalist looks flabbergasted at the imposing electricista, who rivals an electrical pylon in height. With a firm grip on the power switch in his left hand, the victorious electricista energetically extends his right arm: written upon his palm is the word "HALT!" The capitalist, knocked to the ground and holding "sophistry" and "exploitation" in his two hands, has been dislodged and no longer has a hold over labor or the industry.

The liberated electricista of 1936 would remain part of the magazine's visual lexicon for decades. For *Lux*'s cover on the fifth anniversary of the strike, a colossal chiseled-bodied worker emerges with a red-and-black flag, having broken the chains that held him back—classic symbols of socialist iconography.[62] In the October 1941 issue—accompanying an invitation for workers to attend union-sponsored lectures, readings, and theater productions—is an image of a worker emerging from an opened book with broken chains flying from his wrists, with the caption "[L]et's break the chains of our ignorance!"[63]

Over and over again, the SME and *Lux* strove to show the liberated electrical worker as only one part of the invigorated, colossal electricista, who—endowed with the masculine traits of strength, independence, and firmness—dominates industrial landscapes: electricscapes. The consumer of the electricscape, when all was said and done, may have been rendered feminine by advertisers, but the source of the power was the masculine—independent yet affiliated—electricista. A 1941 cover features a solidly built man standing in the center of the frame dressed in overalls and a flat cap, essential working-class emblems.[64] It is his gloves, however, that confirm his electricista identity. Whether dealing with low- or high-voltage current, gloves were a symbol of the literal power that electrical workers generated and distributed across the city. The man's firm posture and facial expression demonstrate his confidence and boldness. Portrayed with an electrical wire in each of his hands, he embodies the connection that makes the electrical circuit possible. Behind him, white water cascades to meet a hydroelectric generator that powers the two electrical cables securely in his grip. Factories, water tanks, and exhaust pipes, symbols of modern industrial cityscapes, clutter the foreground.

The image was meant to portray how the hydroelectric potential of the Necaxa waterfall, represented by the white cascade, was captured only

FIGURE 6.2. *Colossal electrical worker. Cover of* Lux, *January 1953. Courtesy of the Sindicato Mexicano de Electricistas.*

by the work of electricistas. Through their efforts, generators harnessed the power and transmitted it to the industrial city via high-voltage cables. Dwarfing both the hydroelectric generator and the industrial city, the ginormous worker announces the central place of labor in the power system. Twelve years later, another *Lux* cover depicted another colossal electricista, this time *descamisado* (shirtless), towering over the cityscape (fig. 6.2).[65] The muscular electricista on this cover stands in blue trousers, with his left fist vigorously gripping an electric tower that supports overhead power lines feeding buildings along the bottom. His face optimistically and assertively looks outward, to what the New Year might have in store. He expresses the SME's confidence and security about the nation's industry: all modern industries require electrical power. Was not industry itself the final, most lucrative element of the electricscape? The man's right hand is clenched at his side, ready to confront challenges.

As the only union to organize Mexlight workers, the SME had understood its importance since its inception in 1914. When it expanded to encompass telephone workers and individual electricians, its influence only grew. From a small, conservative membership of just under 200 at its founding, the SME grew over the next decades, with over 14,000 members by the mid-twentieth century.[66] But the membership grew by leaps and bounds in the crucial years after 1936, when the SME cultivated an image of the electric industry and its workers as a constituent part of every industrial sector of Mexico. Their sweat, sacrifice, and struggles were not just part of a labor struggle but a patriotic defense of the nation's natural resources, the interests of the people, and the achievements of the revolution.

PICTURING THE ENEMY

Collectively the "soldiers of light" found in the colossal chiseled body of the electricista were not only a worthy representative of the masculine traits that workers aspired to but also a symbolic defender of the electric power consumers and the Mexican people at large. Labor organizations in the textile, railway, and automobile industries, among others, constructed a proletarian masculinity that privileged physical strength, honor, and unity. Individual unions developed their singular masculinity by adding traits and skills that were particular to their line of work. Labor syndicates such as the SME and the Sindicato de Trabajadores Ferrocarrileros de la República Mexicana (STFRM; Railroad Workers Union of the Mexican Republic) used masculinity to accrue power and prestige, constructing alternative and sometimes competing understandings of union masculinity.[67] The railroad works, for instance, developed a "heightened masculinity" particular to their line of work, "as they came to associate their manliness with the mastering of a mobile industrial experience critical to national development and international capitalism."[68]

In a similar vein, the electricistas developed their own version of masculinity particular to the electrical industry: they provided the muscle needed to bring the country's industrial dreams to fruition. In the construction of the concept of masculinidad energética and their gendered performances as men, electricistas emphasized not only honor, physical strength, and unity but also their technical skill, cultural sensibility, and strong commitment to revolutionary ideals. Their brand of proletarian masculinity pictured them as defenders of electricity—the lifeblood of the nation.

Lux and the SME liked to depict electricistas as strong and virile. In part this was idealization and flattery of its mostly male members, but it also emphasized their physical development. Echoing the discourse of social reformers on healthy bodies, nutrition, and respectable leisure activities, the SME advocated that workers exercise their bodies to construct a disciplined masculinity. From the magazine's earliest editions, the union promoted physical activities. The need to place the muscular system in motion was so natural, the magazine claimed, that "not only does man engage in physical exercise to preserve health and gain strength, but inferior animals exercise whenever possible, giving in to an irresistible instinct."[69] Physical development would transform not only their bodies but also their spirit. As a promotional ad urged, "Help yourself, comrade. Be physically fit and strong, and you will be strong in spirit. The gymnasium awaits you."[70] The union also drew a link between sports and patriotism, "To play sports," announced *Lux*, "is to be patriotic."[71] A similar association between patriotism and physical activity appeared in the 1949 cover, which features two lines of workers lifting weights standing behind a woman carrying the Mexican flag. Disciplined workers could be mobilized to provide the necessary muscle, both tangible and symbolic, to defend and support specific political agendas.

Lux did not always present electricistas as muscular, virile, monumental figures dominating electrical landscapes. These celebrations of union members often gave way to images of impoverished and exploited workers victimized by a rapacious foreign company that threatened their interests and those of the nation. The electricista-as-victim trope typically represented Mexlight as colossal hands hauling profits into oversized moneybags, stereotypes that tapped into a national visual treasure of worker-employer class animosities dating back into the late nineteenth century.[72] The SME typically portrayed its adversaries as top-hatted fat foreigners—typically American or British—mistreating the Mexican worker or, in *Lux*'s case, the electricista-as-victim. Class conflict and the "inequalities and abuses of industrialization" often found their simplest, but strongest, articulation in these images.[73]

Outside of mere caricature, *Lux* explained power, and especially fairness, to its audience via size. Its pictures often showed a foreign company as huge—nearly world encompassing—compared to the Lilliputian SME. Take, for instance, two images. In the first image, a genie in the background representing Mexlight appears out of the bulb of a modern electric lamp (fig. 6.3). Hardly a friendly, docile cartoon character, the hideously deformed genie with fangs and claws is obviously evil and not to be trusted.[74] Its top hat, adorned with British and American flags, signals its foreignness. The placement of the genie over the baby, in the foreground closer to the viewer, heightens its terror. The composition demands that the viewer share the perspective of the baby (that is, the SME), vulnerable and ready to be devoured. In the second image, the power differential between Mexlight

FIGURE 6.3.
Eloy García, "Cooperación." Lux, *April 1952, no page. Courtesy of the Sindicato Mexicano de Electricistas.*

and the union takes on class and racial tones. Sitting on the floor with his wares on display, a caricatured indigenous man in traditional garb, broad hat, and sarape engages a stout woman in a fur-lined coat.[75] The size difference between the two is emphasized by the woman towering over the man. The choice of an indigenous person to represent the electricistas emphasizes both the union's claim to be a true representative of the people and their vulnerability—particularly given the role of idealized indigenous figures in reimagining the Mexican nation in the postrevolutionary period. While Mexico embraced the image of indigenous identity after the revolution, everyone knew that the indigenous people were still powerless and marginalized in Mexican society. The man in the picture wears humble garb in contrast with the woman's fur-lined coat.

Yet size is not everything. Though the SME in both images could be interpreted at first glance as being a weak victim, that is not the case. Up close, the hand gestures of the baby SME and indigenous SME hold a different message. The baby, despite having a monster looming over it, stands confident, with its chest out and hands crossed behind its back. It is small but not fearful. Likewise, the indigenous SME figure does not rush to sell his inventory but instead negotiates patiently, calmly bargaining for a fair price.

The SME manipulated tropes expertly, as in its handling of the 1936 strike. The union rendered itself as an infant or as a colossal, muscular electricista, depending on the context. When it wanted to rally workers to keep at a difficult task—such as a strike—then *Lux* might depict adult men with male-associated attributes like force, strength, and even violence. If sympathy was its goal, the organization presented itself to the nation just as meek and innocent as an infant or as simple and frank as an Indian.

Besides caricature, the most common visual tool *Lux* relied upon was the union's emblem of itself: hands. The SME's logo of a clenched raised fist with lightning streaks radiating from the knuckles was meant to express the group's unity, power, defiance, and militancy. Hands were omnipresent in working-class literature and symbols of class conflict around the world and are still international symbols of communism today.[76] Images of hands were widely exploited in postrevolutionary Mexico. Murals and photography dedicated creative space to the construction of working-class aesthetics.[77] Artists visualized the agrarian remaking of the country in the form of campesino and indigenous hands shoveling the earth and sowing seeds. Industrialization was depicted with images of hands—virtually always male—grasping controls and levers to direct machinery or wielding tools to fix it.[78] The SME's iconography was a repertoire of hands firmly gripping tools and elements of electrical work, an allusion to the electricistas' control over the industry, expressed in a line from the SME anthem: "the force [electricity], the great nerve of the city, passes through our hands."[79]

Another vivid image that the SME and *Lux* relied upon was an exhausted worker slumped or laid out while capitalists—especially foreign capitalists—looked on dismissively. This image of oppression touched upon deeply Catholic sentiments in Mexico. One *Lux* cover, horizontally divided, uses the trope to mock the class injustices in the economy (fig. 6.4).[80] On the left side, a worker on a ladder, sweating profusely, rests on a water tank, exhausted from hammering and shocked by the question raised on the right side of the image: a cigar-smoking capitalist lying on a large pile of money asks the exhausted laborer, "Aren't you satisfied to see the fair distribution of the fruits of our labor?" In another image, an exhausted electricista leans against a structure while a fat man passes with a wheelbarrow overflowing with moneybags. The personified company mocks the worker: "[S]ee if you can perform better . . . this is no longer good business."[81] No exhausted individual worker could stand up to the greed of the corporation, the image implied, but perhaps the union could.

Workers throughout the economy were invited to see their financial problems in those of the electricistas, who, in turn, took on the strug-

FIGURE 6.4. *Worker and capitalist. Cover of* Lux, *October 1954. Courtesy of the Sindicato Mexicano de Electricistas.*

gles of the laboring poor. Another *Lux* cover features an obese, top-hatted man with cane in hand, followed by a chauffeur overburdened with gifts, carelessly strolling by shop windows while two barefoot children sleep on the cold winter ground.[82] Only the youngest child has a newspaper for a cover. Standing between the man and his chauffeur and the homeless children, a worker in overalls, with hands extended, asks, "Merry Christmas?" From firsthand experience, SME members knew the uneven reality of the so-called Mexican Miracle. The administration of Miguel Alemán Valdés (1946–1952) kept the minimum wage fixed despite persistent inflation. Economic historians largely agree that wages fell in real terms despite economic growth in the mid-1950s and living standards stagnated.[83] *Lux* explained to its readers that Mexico City workers had seen their cost of living rise almost 60 percent between 1951 and 1952, 21 percent between 1953 and 1954, over 75 percent the following year, and then 59.1 percent between 1956 and 1957.[84]

For the SME—and, indeed, electricistas everywhere in Mexico—the single biggest obstacle to decent wages was the nation's single largest electrical company, Mexlight. In an effort to reign in the company's power, the SME joined the clamor for protectionist policies, which had become, in the words of one notable historian, "the foundation for a new version of revolutionary nationalism rooted in statist, urban industrialism."[85] Artists, in the pages of *Lux*, attacked Mexlight not only for being rapacious owners but also for being foreign, as the cover for the October 1948 issue illustrated (fig. 6.5).[86] On the right side, Mexico is embodied by a woman dressed in the *china poblana* (a regional style of dress from the southern state of Puebla) along with a man in a traditional charro outfit.[87] The choice of dress had become synonymous in the postrevolutionary period with a general Mexican identity that had "little connection to race or ethnicity." The china poblana's style often included an embroidered Mexican eagle, a national symbol reserved for the flag, while the popular charro style included a broad hat and tight pants and was exploited by national cinema as a "masculine stereotype . . . rooted in the rural landowning elite, though in some circumstances it denoted the *revolucionario*."[88] The traditionally

FIGURE 6.5. *"Rates are a patriotic issue." Cover of* Lux, *October 1948. Courtesy of the Sindicato Mexicano de Electricistas.*

dressed woman and the charro exude dignity and confidence as they hand over moneybags to a teller behind a glass window. Uncle Sam and John Bull, the personifications of American and British companies, respectively, dominate the left side of the cover. They are withdrawing large profits in the form of gold bars and coins, bleeding the country financially. By highlighting Mexlight foreignness, *Lux* made the SME's conflict over labor's share of profits into a patriotic issue.

Another *Lux* cover depicts a capitalist hauling a massive sack labeled "profits" over the border, threatening not only the electricistas' livelihood but also the nation's.[89] This was a savvy ploy: even if some Mexicans were on the fence regarding the appropriate returns to labor versus capital, many would rally to a defense of the nation. The struggle over the SME's call for livable wages and regulation of Mexlight's rates was not merely an employee-employer conflict but a story of whether Mexico would continue to be abused. Would Mexico unjustly continue to be an economic colony of foreign powers—as if the revolution had never happened—or would it be the captain of its own destiny? For the SME the answer could only be the nation's full economic and electrical independence.

Electrical workers strove to position themselves as a heroic, patriotic force against the industry's greatest enemy: Mexlight. Increasingly after 1950, the pages of *Lux* sought to convince the nation that Mexlight was not only the union's adversary but an obstacle to the economic development of the entire country. Its exploitation of workers, coupled with profiteering based on its monopoly, made the—notably foreign—company an easy target. Electricistas "attempted to define [themselves as] positive heroes and create models for acceptable thought and action."[90] The electricistas' thinking merged a binary socialist conception of the world—the bourgeoisie versus the proletariat—with Mexico's equally binary postwar nationalist rhetoric—imperialist versus colonized.

DEFENDERS OF THE PUEBLO

The year 1948 was a critical political conjuncture not only in Mexico but throughout Latin America. This became clear at the Conference of Bogotá where the United States called for a "political-military unification of the continent as a part of its global strategy against communism." This change in inter-American discourse, according to historian Margarita López-Maya, frustrated the democratic tendencies in the region where the majority of popular governments, which had insisted on the necessity of economic collaboration, were soon replaced by military regimes.[91] As the democratic advance was mostly contained, the Left in general lost terrain with Communist parties throughout the region facing "proscription and severe repression." The unprecedented militancy of organized labor in the preceding years was drastically reversed as the state moved to discipline labor and bring it under closer control.[92] This conjuncture is best exemplified in Mexico by the infamous episode of the *charrazo*: the clash between two different views of industrialization that culminated in the 1948 overthrow of the rieleros' democratically elected labor leaders.[93] Unions energetically pushed back the state's effort for a tighter grip over organized labor.

Seeking to prove it was at the vanguard of the global labor movement, the SME set out to win the continent's best collective contract in 1950.[94] During his speech to the general assembly in April 1950, Roberto Ocampo González, the secretary of labor within the SME, stirred his fellow electricistas by evoking the Mexican Revolution: "*[F]ogatas* [campfires] had been set up and the stanzas of *La Adelita* could already be heard." His rhetorical allusions were not accidental. Members knew that the SME was *en pie de guerra* (up in arms) that April. It was not simply the movement of a given number of workers, but a movement *pro patria* (for the homeland). Ocampo González added that they did not need to announce their next step, which was already in everyone's mind: strike! Electricistas were ready, he promised: if conditions turned out to be against them, "just as Troy burned, all the foreign exploiters of Mexico will go up in flames," obviously including Mexlight.[95]

A month before the April general assembly, the union entered a contentious period of collective bargaining with Mexlight. Secretary general Juan José Rivera Rojas made clear at an early stage that, in addition to the usual demands to improve the electricistas' economic livelihood (a 65 percent salary raise, an increase of two pesos in daily housing subsidies, and one and a half peso increase in the transportation subsidy), the union wanted to "conserve whole the conquests won until now."[96] These gains included medical service, established with the money that Mexlight had given the union after the 1936 strike. Mexlight not only intended to end this service but wanted to nullify the workers' seniority rights and retirement pension.

Mexlight, the secretary general warned, "was set to eliminate the union's interference in the movement of personnel and in any other issues affecting workers' rights." The Mexlight directors wanted to "impose the same contractual conditions of US workers, who, although they enjoy a better standard of living, do not have the same social conditions of their Mexican counterparts, since our country is very progressive in labor matters, at the forefront worldwide, with the one exception of the Soviet Union." Rivera Rojas assured attendees of the early March meeting that the SME would not allow their rights to be truncated or denied and concluded by predicting that their upcoming contract was destined to become "the most important in the Republic, and the gains won will serve as a precedent for other Mexican workers' future contract revisions."[97]

As defenders of the pueblo—in this case, a nation of working people—SME leaders took on a nationalistic tone. They saw themselves as the heirs of radical revolutionary generals like Emiliano Zapata and Otilio Montaño, who had questioned ownership of natural resources, and, more recently, president Lázaro Cárdenas, who had nationalized the oil industry.

SME leaders felt confident that their financial demands in the 1950 collective contract revision were not just reasonable but fair. Mexlight, they pointed out, had increased its income by "100 million pesos per year" since the last contract revision in 1948, and power demand had indeed increased substantially across the nation in the previous eight years.[98] Despite this increased demand for power, however, Mexlight responded with only minor expansion plans. This was not new, as the Valle de Bravo episode had revealed years earlier. Power shortages had long been identified as the country's Achilles' heel, and the inability to supply the needs of industry tempered the nation's dreams of modernization. Former US ambassador George S. Messersmith, now chairman of the board of Mexlight, shared this sentiment in his address to the Mexico City Rotary Club in May 1948: "Unless this demand in Mexico . . . is adequately met . . . there will not only be restrictions on the present uses of power . . . but . . . a complete or partial arrest of any industrial development, and in measure, . . . any increase in agricultural production."[99]

Messersmith's forecast for the central plateau, "the heart of the country" and the center of Mexlight's business, was quite dire. The power industry had already achieved peak capacity. The backlog of domestic demand coupled with the rapidly increasing demand for industrial power not only warranted but demanded increasing productive capacity. A program for new plants had to be "one of considerable magnitude," but a lack of fuel handicapped any plans.[100] The country had no major coal sources, and the supply of petroleum was inadequate, in his view. Messersmith believed that water remained the best option, even though hydroelectric sites lay far from areas needing power most in many cases.

Besides fuel limitations, there were other shortages, not least of which was the lack of equipment necessary to produce electricity. Messersmith predicted that power demands in numerous countries would go unmet because the factories that produced electrical machinery could not meet global demands—many of these companies could not even satisfy their home markets. Equipment for either hydroelectric or thermoelectric plants was expensive, even by the standards of capital-intensive industries. Hydroelectric plants in particular not only were expensive—requiring building dams, tunnels, and other large-scale engineering works—but also took years to complete. Thermoelectric plants were faster and easier to build but still required vast pools of money. The problem of equipment still loomed over the nation's efforts, but Messersmith was optimistic. His discussions with manufacturers abroad gave reason to hope that equipment would indeed be available. Underlining this was important not just for the expansion of the nation's power capacity. The Mexican business community had grown critical of US reluctance to approve the exportation of machinery and equipment during and after the war. This fed speculation among Mexican officials and in the business community that "vertical trusts in the United States were controlling their access to modern machinery."[101]

Capital was another serious problem. A new plant, whether hydroelectric or thermoelectric, was costly and—Messersmith added—risky. Whether private companies like Mexlight would finance major expansions of electrical capacity depended on the treatment of foreign companies. In this regard, he noted, Mexico could learn a thing or two from its northern neighbors in the United States and Canada, where power industries enjoyed relatively high power consumption levels and were able to obtain capital from private sources. They also secured rate structures that yielded a reasonable return on capital investment, allowing amortization of the investment in due course and a reasonable return on their investment for company owners. Legal regulations in those countries allowed companies to sell bonds with a long maturity to the public. Banks, insurance companies, trust funds, and the general public favored bonds and stocks of power companies because of their security. Refusing to critique his adopted country directly, Messersmith diplomatically explained that private entities would not invest in public utilities "in many countries" where power companies did not receive "equitable and understanding treatment." In no countries would private investors "place . . . capital in an industry which does not offer a reasonable promise of a return on the investment."[102] However, the Mexican landscape was experiencing positive changes, in his opinion. The passage of the 1939 federal electricity law, which guaranteed a reasonable return on capital investment, was a step forward, but it did not go far enough. For the industry to have the sort of long-term stability needed for massive investment, the law needed changes and amendments.

That same year Messersmith and the Mexlight leadership unveiled a proposal to increase the company's capacity in the central region of Mexico. In coordination with the CFE, the company came up with a four-year plan for the power needs of the central plateau. As the principal producer and supplier of power in the Federal District and surrounding states (México, Hidalgo, Guerrero, Morelos, Michoacán, and Puebla), Mexlight's existing capacity included sixteen plants (fifteen hydroelectric and one thermoelectric plant at Nonoalco), which together reached 303,650 kilowatts—a substantial quantity, given that the entire country produced 963,000 kilowatts.[103] Mexlight's 1948 development program proposed, at a minimum, expansions to current plants, including an additional unit for the plant at Lerma and another unit for the Necaxa plant, as well as a new hydroelectric plant at Patla and a thermal plant at Lechería. For its part, the CFE was to complete a new plant at Santa Bárbara and establish another one at Durazno.

Securing capital, however, was another matter. The Necaxa and Lerma expansions and the new Patla and Lechería plants—along with their transmission, distribution, and accessory facilities—amounted to 47 million dollars. An additional 8 million dollars were needed for a plant then under construction in Tacubaya, which would supply power to the Federal District. Thus, at a minimum, Mexico needed 55 million dollars, a sum far too large to secure from domestic private sources. It would have to look abroad for financing. As connoisseurs on the subject, Messersmith added, attendees knew that there was "no real source outside of Mexico for private or government capital needs other than the United States."[104] US capital markets were the only source for Mexico and other countries in need of industrial development, but the private capital market north of the Río Grande was closed to Mexican public utilities. The reason was simple: Mexico's politics made investment unattractive. In such a state of affairs, the bulk of financing would come from only two sources in the world: the Export-Import Bank of Washington, an organ of the United States government, and the World Bank for Reconstruction and Development, an intergovernmental bank.

The World Bank was the more attractive option. When foreign private capital hesitated to continue investments in power, transportation, and communication networks after World War II, the World Bank stepped in and secured financing. The institution became a leading advocate of the view that investment in these areas, as well as other public utilities, "was a precondition for the development of the rest of the economy." During the early postwar period, the bank focused on electric power projects around the world, but especially in Latin America, where studies and plans for sites already existed. Two power projects in Mexico received $34.1 million from the bank in 1949. While $24.1 million went to finance equipment that the CFE needed to complete new power-generation stations and transmission lines, the second loan for $10 million went to the CFE and Nacional Financiera (NAFINSA), a semigovernmental Mexican financial institution that, in turn, loaned the money to Mexlight for its expansion program. For this latter loan, the bank requested from the Mexican government assurances that it would grant the Mexlight "reasonable rates." This had lasting consequences: by the 1950s, the bank routinely established "electric power rate covenants" as a condition for the approval of power loans around the globe.[105]

The Export-Import Bank was a more complicated source of capital politically. Created in the 1930s to fill a vacuum created in international capital markets, it extended secure loans to governments, their agencies, and private foreign companies.[106] On the one hand, because the bank was a US agency, accepting loans meant giving the United States leverage in the Mexican economy; on the other, the loans were readily available on generous terms. Between 1941 and 1945, the bank authorized credits to Mexico for 94 million dollars for highway construction, gasoline refinery equipment, railway equipment, steel mill equipment, and electrical equipment. That sum included 20 million dollars earmarked specifically for the purchase of US electrical equipment, materials, and services for the CFE and NAFINSA to construct "a broad electrification program in Mexico."[107] For the next decade the Export-Import Bank would be the main source of finance for Mexico's expanding electrification program. During that time, the nation secured $252 million or 16 percent of the $1.5 billion that the bank channeled to Latin America.[108] As had been the case in the first quarter of the twentieth century, the electricscape would expand because of foreign wealth.

During his 1948 Rotary Club presentation, Messersmith emphasized that, although both the Export-Import Bank and the World Bank continued to offer reasonable rates (from 3.25 to 5 percent), the two institutions extended loans only to sound projects and based on careful study. He was confident the Mexlight's application would be granted if the program was judged to be essential to the national economy and the bank determined that the company would be "able to repay the loan and to earn a reasonable return on the investment."[109] These principles, he added, would apply anywhere else and to any enterprise. Messersmith's request for stable rates had an imperative undertone, demanding assurance the company would not be nationalized.

Mexlight understood that expanding the country's electrical capacity meant including the Mexican government, NAFINSA, and—especially—private domestic capital along with foreign investment. Anyone familiar with the region, Messersmith added, "knows that it is not wise [for] any new industry to establish itself in any one of these American republics unless . . . associated with . . . a certain amount of local capital and, in many cases, as much local capital as can be secured." He believed that was the only way in which those industries could work on a sound basis. No matter how essential domestic capital was politically, the company would still need to reach out to foreign sources. Whatever capital funds the nation could not furnish "will have to come from the U.S. government or private banks and investors in Mexico and the United States."[110]

Upon taking his position, Messersmith encouraged strengthening the company's profile, starting with its board. Given the expansion plan in mind, Mexlight had to "get a very strong Board." He recommended adding "some good active American representation" as a gesture.[111] He also extended invitations to local high-ranking Mexican officials and business leaders, including Eduardo Suárez, formerly of the Ministry of the Treasury, and Guillermo Barroso, president of the Confederación de Cámaras Industriales de los Estados Unidos Mexicanos (CONCAMIN; Confederation of Industrial Chambers of the United States of Mexico). Messersmith did not stop there. He also assembled a sort of "junior board" called Mexlight's Mexican Advisory Committee, staffed with important private sector figures. The historian Stephen Niblo notes that the direct rewards to the board members were significant, including lavish expenses to attend board meetings as well as biannual payments for "fees and unspecified services" in exchange for "favorable policy" toward the company.[112] These payments would double to lure junior government officials to pursue policy friendly to the company in the hopes of one day becoming board members.

Messersmith was no stranger to representing foreign interests in Latin America, but it was a strange career path that led him there. He began in the Foreign Service as American counsel and then counsel general in Antwerp. There he established a lifetime friendship with Dannie Heineman, the American-born director of the Société Financière de Transports et d'Entreprises Industrielles (SOFINA), a leading global public utility holding company. Heineman, an electrical engineer, led the company to "assume an even larger significance in the international financing and development of electric utilities" in the 1920s. *Moody's Manual (Utilities) 1930* reported that SOFINA had 42 principal subsidiaries and participations. A number of these held additional public utilities, making the company one of the largest international players in the electric sector, with interests in all major markets.[113]

The Société Internationale d'Énergie Hydro-Électrique (SIDRO)—which held considerable Mexlight securities and obtained voting control in the company in 1924—got into financial trouble in 1928. SOFINA stepped in and took "effective control" of Mexlight.[114] From then on, Heineman had the final say in Mexlight matters. He demanded that the Mexlight chairmanship go to a man with established international contacts and a deep knowledge of the new global landscape after World War II. In 1947, after learning of Messersmith's retirement, Heineman reached out to his longtime friend. Having served as ambassador to Cuba, Mexico, and Argentina in the 1940s, Messersmith was Heineman's ideal candidate. He eagerly accepted the position, confiding in a letter that he found it appealing because he felt acquainted with country and had already decided to live in Mexico City after retirement.[115]

Only a year into his chairmanship, Messersmith wanted to make the most out of time before the Rotary Club. He struck a tone of urgency that evening, partially because he believed that the public remained unaware of the vital role that electricity played in their lives or the challenges that Mexlight faced, despite having improved relationships with political leaders. Upon accepting the chairmanship, he told René Brosens, chairman of the Compañía Argentina de Electricidad (CADE; Argentine Electric Company), that he considered the new government sincere in its support of expanding electrical production. "[T]he company is safe in undertaking this new program," but the president's support was not enough.[116] Mexlight would have to increase rates to recoup its investment and secure lines of credit, which Messersmith knew would be deeply unpopular. In order to get the public to swallow this bitter pill, it was paramount to educate people about the "real significance of power in the life of the country and of the people." As he put it, the public needed to comprehend "what a part electricity plays not only so far as their comfort and convenience are concerned, but also [in] their capacity to make a reasonable living."[117]

The electricscape had become such a fact of life by 1948—at least in the eyes of the chair of Mexlight—that ordinary people took it for granted. Although members of the public had some vague notion of the inconveniences that current limitations on domestic power created, they did not "appreciate to what degree [future] restrictions would affect the ordinary functions of our daily life . . . our comforts and conveniences." The effect on the national economy would be disastrous. The government and industrialists' program to develop the country's industrial capacity "would be *completely* arrested." New factories under construction, and those in the planning stage, "would be without power to move their wheels . . . the flow of goods so important to the country would be stopped in its tracks, and the source of greater employment and general wellbeing [*sic*] would be destroyed."[118]

Messersmith was likely wrong. People were vividly aware of how important electricity was in their lives. Just a year before his speech to the Rotary Club, a drought in the central region of Mexico in 1947 had led to widespread power rationing. To alleviate it, Mexlight broke ground for a 30,900 kW diesel-electric plant, the largest of its type, in the industrial suburb of Tacubaya.[119] The CFE secured a line of credit for 5,860,000 dollars from the Export-Import Bank. The federal government invited one of Mexlight's subsidiaries (Compañía Mexicana Meridional de Fuerza, SA) to buy machinery and electric equipment for the plant. Not everyone was mollified by the actions. Days before rationing and restrictions went into effect in mid-June, Jaime Ruiz Casas, a concerned engineer, wrote to President Alemán, accusing the CFE executive director Alejandro Páez Urquidi not only of failing to foresee the power crisis but of corruption and of misleading the president and the public with overly optimistic reports of a situation that had been deteriorating for some time.[120] The director had known for weeks that irregular rains had caused the Necaxa water tanks to drop but failed to warn the Ministry of the Economy. Thus the CFE took no emergency measures that might have lessened the blow. The CFE finally warned the ministry when it had enough water for only twelve days. By that time, the government's mandate to cut power consumption by 40 percent was already causing conflict: employers with no power to run their machines refused to pay their workers.

The SME stepped into the fray, criticizing Mexlight's habit of conveniently alluding to drought conditions whenever contract negotiations loomed and capitalizing on drought by reducing supply to request rate hikes when the company was partially responsible for the drop in supply in the first place. Mexlight, after all, had failed to undertake the necessary expansion program. In the SME's assessment, the General Electrical Administration, a department in the Ministry of the National Economy, enabled Mexlight by never pressuring it to expand production. The CFE also contributed by refusing to recognize and address the system's inefficiency.[121]

Labor unions denounced the burden that workers disproportionately had to carry. The Confederación General de Trabajadores (CGT; General Confederation of Workers), a far-left confederation of labor unions, criticized the Ministry of the National Economy for failing to properly implement the noble goals of the president's accord no. 1122, which was intended to save 40 percent of power consumption.[122] The confederation accused industrialists of taking advantage of the energy crisis to reduce work schedules, even though the many factory-owned generators could easily compensate for power losses due to rationing. The regulations drafted by the ministry resulted in a reduced workload, exacerbating the workers' deteriorating financial situation. The confederation claimed that employers were violating labor contracts and thus labor law. It called for the president to arrange a meeting between labor unions and the minister of the economy, the minister of labor, and the minister of social welfare to study measures to ensure that work hours remained at precrisis levels.

The Sindicato de Trabajadores de la Industria Textil y Similares de la República (Trade Union of Workers of Textile and Related Industries of the Republic of Mexico) also called on the president to intervene: their salaries had dropped by nearly 60 percent since the rationing.[123] Agustín Guzman, general secretary of the Miners Union, condemned the grave error of rationing and demanded that those responsible be held accountable. Nothing short of the dismissal of incompetent authorities, he declared, would prevent the "paralyzation of the mining industry, which would severely shatter the national economy."[124]

The only realistic way to make the electrical crisis of 1947 go away and stay away was to increase electrical production—a position that the SME and Mexlight agreed upon. However, they had a profound disagreement over who should bear the cost. During their 1950 collective bargaining, in addition to criticizing the pace and scope of Mexlight's expansion plans, the SME took aim at the company's December 1949 request for rate hikes. Representing itself as looking out for the interests of both its members and all working Mexicans, the SME printed an offer to the company on the cover of *Lux*. A tattered page in a medieval-style font read: "We electrical workers are willing to sacrifice our particular interests in defense of the Mexican people; we propose to withdraw our union's just wage raise demands if the rate hike request is dropped." The corresponding editorial inside explained that the union would seek wage increases and strike if need be, if Mexlight declined the offer: members of the union valued sacrifice and "always refused the improvement of their conditions in exchange for rate hikes or anything that harmed the interests of the people."[125] In claiming to be the guardian of the people, the SME was willing to put its money where its mouth was.

Both Mexlight and the industry's governmental regulatory body, Cámara Nacional de Electricidad (CNE; Electricity National Chamber), counterattacked in the press, calling the SME's claims a farce. In an article in the conservative daily *Excélsior*, Mexlight accused the union of requesting excessive wage increases while allowing no rate adjustment.[126] The article added that the secretary general had declared two years earlier that the SME enjoyed the best collective contract in Latin America. The CNE's response in the same issue was even more incendiary. The agency called out the union for creating a useless problem with serious consequences for the public. If, in reality, the SME's intention was to defend the public then it would withdraw its petitions. After all, the agency argued, every wage increase that the union secured translated in the end into a rate hike for consumers. The union's intentions were not to safeguard the interests of the people: "the only thing the union wants, in reality, . . . is to take the greatest advantage but appear at the same time to defend the people." It was the Mexican regulatory agencies that defended the public, not the SME leaders. The regulatory authorities were carrying out their mission through their "vigilance over the operation of all electric companies faithfully exercised." These companies were among the most essential for the country's economic health and consequently among the most carefully watched.

Rejecting the accusation that it was a greedy, deceptive union, the SME sought to prove that Mexlight was dishonest by exposing its argument for an upward rate revision: the 1949 Necaxa drought crisis. Drought in central Mexico had caused another electric shortage that made a rate hike of 56 percent for domestic consumers and 87 percent industrial consumers necessary.[127] The union and the company had exchanged heated words in the press in the closing months of 1949. Messersmith added the company's requested rate hike was caused not just by the drought but by the union's exorbitant demands for wage increases.[128] In an effort to appease the union, Mexlight had offered a 15 percent salary increase but demanded in return that contract revisions take place every four years, instead of every two. The company was willing to pay the price in order to avoid dealing with a biennial threat of strikes and interruption of service.

During the summer of 1949, James H. Heineman, son of Dannie Heineman, reported back to his father that among Messersmith's most important tasks in Mexico were trying to secure rates and appease "the Syndicate who have threatened slow-down strikes and demonstrations if our rates are increased."[129] In his enthusiasm to criticize the union, Messersmith went too far: the SME pounced, pointing out his lies and that no contract revision was taking place in September of that year.[130] Messersmith had to print a retraction, stating that the upward rate revision petition had no relation to the salaries or benefits enjoyed by the company's workers.[131]

The battle of accusations continued into the waning months of 1949. The union shifted to a new front, criticizing the company for its poor investment record and for enriching itself at the expense of the people. The ceaseless repatriation of profits out of Mexico and into the United States and Europe represented "a bleeding that diminished the potentiality and determining factors of our improvement, while maintaining the users' misery and sacrifice." Emphasizing repatriation was a powerful rhetorical weapon and pointed to an even more powerful action that SME could bring into its struggle with Mexlight: nationalization. An editorial in *Lux* asked readers to consider: "If our hydraulic systems generate electricity, with the force of Mexican workers, [and] locally produced fuel; if the Mexican pueblo consumes the electricity, there is no reason to pay American, English, and Canadian capitalists for it. The CFE's operation demonstrates to what extent we can and should nationalize the electric industry."[132]

It had been eighteen years since the Consumer Leagues first suggested nationalizing the electrical industry, but the SME's effort in 1950 was not merely symbolic—the union knew it had a chance.[133] One reason was that Mexlight had overplayed its hand in the 1949 rate hike. The SME published studies in April 1950 proving that the December 1949 rate hike increase had been nearly 100 percent. The "disastrous monopoly," the union argued, needed to be stopped for the sake of Mexican society. The country, it explained, was experiencing an "urgent need to nationalize the industry."[134] Secretary general Rivera Rojas argued that the SME needed to move past wages and working conditions and push for the industry to become the nation's property.

The SME delayed the call to strike from May 1 to May 16, so that it could use Labor Day commemorations to galvanize support.[135] The bet paid off. What could have been seen as simply a labor issue had become a patriotic one. Banners during the holiday's celebrations emphasized both working-class unity and the need to nationalize the industry, with slogans: "We fight for the electric industry's nationalization." "The electric industry is for everyone. Don't let yourself be extorted by Dracula Messersmith." "To expropriate Mexlight is another step in Mexico's economic independence." "We are with you, president. Nationalization of electricity." "Public enemy #1 of the Mexican people is Mexlight." "How to end monopolies in Mexico? Expropriate Mexlight!"[136] This support gave the SME greater leverage: it was no longer a mere union but an organization that described itself as having "an attitude purely patriotic and revolutionary."[137] The electricscape was too important—too valuable—to be left in the hands of foreigners who thought about only short-term profits. The nation would never achieve the sort of industrialization that postwar administrations had dreamed of if Mexico did not direct the growth and operation of its electric infrastructure.[138]

The May 16, 1950, strike lasted sixteen hours, ending when the SME and Mexlight reached a compromise. Salaries would increase by 15 percent, the December 1949 rate hikes would remain in place, and no additional rate increases could be requested that cited rising wages as their basis.[139] Although the strike had no clear victor—all sides got what they wanted—Rivera Rojas spoke of its results as a triumph for the union. The SME had succeeded in confronting one of the most powerful companies in the nation and its contract, unlike those of other labor groups, had remained unscathed. He boasted that "undoubtedly, if our Contract had been broken, practically the labor movement could be considered finished, as the SME is at the vanguard of the labor movement. If they manage to end our organization, unionism in Mexico would be liquidated." Though the rate hikes remained, electricistas could still claim a larger victory born of "a true patriotism." But their work was far from finished. "All of us gathered here," Rivera Rojas reaffirmed in his first speech after the strike, "would have liked that the money uprooted from the lifeblood of the pueblo, from the tears of the fatherland and economy itself, would not emigrate abroad; that the flow of gold that could be employed for Mexico's progress, would not go out of the country."[140]

SME had still accomplished something important. "We have moved away from the selfish situation that at most times animates men," Rivera Rojas claimed; instead, "we fight not only for the interest of electricistas but for those of the homeland itself." This defense of the people's interests would set a powerful precedent and would shape the history of the country's labor movement.[141] Later that year, in an editorial titled "The Electricistas and the Mexican Revolution," *Lux* broadly proclaimed that by defending the rights and successes of the union and the laboring classes in general, the SME had defended the revolution itself.[142] In fifty years, electricity had become so deeply woven into the fabric of Mexican life that the electricscape—like the family or ejidos—had been raised into the pantheon as an essential element of la Revolución.

The one clear victory in the 1950 collective bargaining that Rivera Rojas neglected to mention was that the strike had brought an open discussion of the industry's nationalization to public attention. The SME's proposal had enough momentum that President Alemán felt obliged to consider it, at least privately. In a meeting with the SME, the president confided that "expropriation did not scare him."[143] However, to nationalize the industry at the moment would bring a delay in the industry's development of about ten years, he estimated. Despite his initial hesitancy, the question of expropriation would hang over Mexlight-SME relations from 1950 on. The SME had planted the seed and would tend to it patiently until it bore fruit.

MEXICANIZING ELECTRICITY

As late as the mid-1950s, foreign-owned companies provided an estimated two-thirds of the electricity consumed in Latin America. Mexicans mirrored citizens around the world who widely and strongly felt that "foreign ownership of electrical utilities . . . an essential sector in the economy, was not tenable."[144] The ghost of nationalization loomed larger than ever, and electricistas masterfully deployed it in their struggles.

Self-identified as leaders of the labor movement and defenders of the pueblo, electricistas marched into the 1950s energized. *Lux* used the civic calendar to aid its message. Particular dates became like cardinal points on a compass: Labor Day on May 1, the Independence Day celebrations on September 16, and the November 20 remembrance of the revolution. *Lux* served as a platform for SME leaders to weave the electricistas into the national revolutionary fabric during those commemorations. Self-congratulatory articles, legendary speeches, and the ever-increasing use of art served as thread to stitch together the interests of the union and the nation.

The calendar date that mattered most to the SME in the 1950s was March 18, 1938, when President Cárdenas expropriated Mexico's oil industry from its foreign owners. If the SME's 1936 strike served as a foundation story for its militant self-image, the 1938 oil expropriation marked what the electricistas hoped to become. Cárdenas's move caused economic and political tremors that were felt beyond Mexico's borders, to the tip of the Southern Cone. Unions, politicians, and policy makers throughout Latin America lauded Cárdenas's decision as an example to be emulated. In Chile, for instance, students staged celebrations of the expropriation, parading with images of Cárdenas to rallies, shouting in unison "Long live Cárdenas!" Marmaduke Grove, Socialist Party leader of the Frente Popular, during his campaign discussed how the examples of the Mexican Revolution could be adopted in Chile's national struggles.[145] Cárdenas's decision offered countries in the region a model of how to reclaim national resources in the name of their pueblos.

The SME embraced the expropriation as an exemplary act. Manuel Paulín, SME secretary general, praised the decision as "of greater transcendence than the Cry of Independence since the expropriation could be considered the Cry for our economic independence."[146] In his estimation, Agustín Salvat alleged that the step was comparable only to the liberal government's appropriation of church wealth in the mid-nineteenth century. Paulín disagreed: the appropriation, which had only national repercussions, had not inspired the overthrow of the international capitalist class as the expropriation had.[147] The expropriation, and the labor struggle that led to it, provided labor with an important bargaining chip. The SME referred to the labor conflicts of the petroleros and rieleros in the biennial contract negotiations as early as the summer of 1938.

The image, legacy, and myth of Cárdenas became central to the union's identity throughout the 1940s and 1950s. The anniversary of the oil expropriation joined Cárdenas's birthday and, eventually, the date of his death as permanent presences on the SME calendar. His image and words provided the SME with a ready-made, widely recognized symbol that could easily be deployed not only to remind Mexlight of the real possibility of nationalization—a powerful bit of leverage in negotiations—but also to call upon presidents to act on the principles of the revolution as Cárdenas had.

During the 1950 contract negotiations with Mexlight, in a speech meant to assuage fears of what might happen if the industry was nationalized, Rivera Rojas declared that the SME would make sure there would be no sabotage or shortage of technical personnel. Unlike the *petroleros* (oil workers), who lacked technical personnel and faced sabotage in the aftermath of nationalization, Mexico's electricistas largely ran the industry already, had no interest in seeing any disturbance to power delivery, and—unlike Mexlight's foreign owners—were patriots who had the best interests of the pueblo at heart. Already, he added, the SME had undertaken management studies on how to administer the various divisions of the electric companies once they were nationalized. "We are going to show with facts that we are in effect the most *sabrosos* [genuine leaders] of the Mexican labor movement. Death to Mexlight!"[148]

Mexlight did not die during the 1950 negotiations. While it still enjoyed official support by the Mexican government, however, its position over the years grew weaker. In "Oil, Expropriation, and Electric Power," diplomat and politician Antonio Robles argued that expropriation had gone from being a slogan to rally the union into a real possibility and listed several reasons why. Because the nation's economy was stronger in 1958 than in 1938, any financial setbacks could be better handled. Just as the economy had grown, the electrical sector had matured. Mexican companies were now generating much of the country's capacity. Whereas in the past it had been hard to find Mexican technicians, there was now an abundance of domestic electrical technicians. Among all of these considerations, one remained most salient: "all the pueblo of Mexico suffered the weight of the elastic rates of the electric companies."[149]

This last point was perhaps the most salient. The electricscape had grown, reaching into people's homes and even into their notions of their status and self-image. As important as oil was in 1938—and it was undoubtedly essential to the nation's economy—it did not have the reach and immediacy that electricity did. Mexicans warmed and lit their homes with it, traveled and entertained themselves with it, and consequently paid close attention to its service and price. Electricity had become essential, the article concluded. If Mexlight could not expand and meet demand, or if it charged too much for something so valuable to so many, then the SME had the solution: Expropriate!

Electricistas strategically construed the revolution as on ongoing process that left room for the incorporation of new revolutionaries. In the article "Our Organization and Revolutionary Institutions," Fernando Casas Alemán, mayor of the Federal District, acknowledged the SME as one of the government's "most important factors within the progressive program of the Revolution."[150] Although he, like the electricistas of the 1950s, had been unable to fight in the Mexican Revolution, they were lucky to live in one of the most important periods of the revolution: fulfillment. The electricistas agreed completely with the mayor: they paid homage to the revolution in an editorial with an analysis of the electricistas' role in protecting the promises of the revolution. The article "The Electricistas and the Mexican Revolution," claimed the SME, had managed to coordinate their activities: "by defending the rights and conquests of our trade and the working classes, we defend the Revolution itself." Indeed, without entities such as the SME, the revolution "would lack a support base."[151]

By the forty-second anniversary of the Mexican Revolution, not all Mexicans believed, like Casas Alemán, that they were living in a period of fulfillment. Some even questioned whether the revolution had achieved anything. Responding to these doubters, a *Lux* article described the revolution as a composite of principles and people. While principles such as "Land and Liberty" and "The land belongs to those who work it with their own hands" had been the soul of the movement, individuals had been the implementers. The principles were "ideals, intangible, immutable," waiting to be materialized in the action of real, flesh and blood people. According to the writer, the question should be not whether people or the principles had failed but what people were willing to do to make the revolution real. "It is our responsibility," claimed the author, "to be the achievers of the Revolution, the ones to embody its spirit to convert it in something tangible."[152]

The SME capitalized on the construction of the revolution as incomplete. "Within our means," the article continued, "there is no one better qualified to carry out these noble principles than the SME." The syndicate had been born in 1914, "in the heat of the Revolution," and had contributed one of its principles to Mexico's great historical event of the twentieth century: respect for the rights of and justice for the worker. The defense of the country's natural resources provided a strong links for electricistas both to the revolution and to President Cárdenas. Beginning with the fifteenth anniversary of the oil expropriation, Cárdenas began to appear regularly in *Lux*'s March issues. The magazine dedicated far more covers to the former president than to any other political figure in the 1950s. Ironically, during his actual presidency, when the union was far more apolitical, Cárdenas had never once appeared on the magazine's cover. The SME had changed since then, along with its attitude toward Cárdenas.[153]

Lux covers in the 1950s showed Cárdenas as a monumental figure with an industrial landscape populated by oil wells, the national colors, and images of the eagle and the serpent (fig. 6.6).[154] The magazine placed an unusual emphasis on Cárdenas's eyes, suggesting that it shared his vision for the nation, but more commonly it presented Cárdenas as a nurturing father figure showing the nation the way. The cartoon "Gremito," based on and named after a Cantinflas-like character, dedicated an entire strip to Cárdenas's expropriation. It reminisced that the president, "with bravery and virile posture, said, 'Mexicans, the oil is ours!'" He was a patriotic example for the working classes and for those that worked "on behalf of Mexico, and for Mexicans."[155]

Repeatedly, *Lux* portrayed how Mexlight was not working on behalf of either the country or its citizens and how it ridiculed the country's revolutionary character. In one image, the central iron cupola of the Monument to the Revolution saturates the space. Along with the Torre Latinoamericana, the city's lone skyscraper of forty-five floors, the monument was one of the focal points in the capital's cityscape. Completed in 1938, the structure was a symbol of the new nation and doubled as a mausoleum for revolutionary heroes. The structure, originally proposed as the new Legis-

FIGURE 6.6. *"President Cárdenas shows the way forward." Cover of* Lux, *March 1953. Courtesy of the Sindicato Mexicano de Electricistas.*

lative Palace, a French-designed building that flaunted Italian marbles and Norwegian cladding, stood as an emblem of the Porfirian regime's reverence for European culture.[156] The cupola had been salvaged from destruction and converted into a monument. The visual syntax of the image indicated Mexlight's abuse. Appearing in the dominant position on top of the cupola is a capitalist on moneybags, a personification of Mexlight, visibly stepping on a group of sculptures of "strong, proud, unbowed peasants and proletarians, soldiers and citizens, men and women, unmistakably Mexican—Indian and mestizo."[157] The monument was supposed to celebrate "a people's revolution . . . described in the official initiative as a 'collective work' and 'the struggle of the people in the conquest of their rights,'" but the capitalists mocked these intentions. To add insult to injury, the caption mockingly states, "[A]nd yesterday, they [Mexicans] were revolutionaries!" Such mockery amalgamated the country's visual motifs to unify readers against the rapacious electric company.[158]

From the pantheon of national heroes, Cárdenas rose to become an international symbol of social and economic justice. Twenty thousand workers, from all sectors of the economy, gathered at Arena México in 1958 to pay homage to the expropriation and express solidarity with the SME in its quest to do the same with the electrical industry. Alejandro Carrillo, in the name of Cárdenas, claimed that the country was different after twenty years of the nationalized oil industry—stronger, more prosperous. After visiting the oil fields, he concluded that the president had "not only sown a seed of progress, but he had also sown his thinking for Mexico and his heart."[159] The SME's coverage of commemorations not only contributed to the ongoing mythmaking of General Cárdenas but, by stoking his popularity, created the impression that the former president endorsed the union's call to nationalize the electrical industry.[160] As the union entered its collective bargaining in 1958, it capitalized on the twentieth anniversary of the oil expropriation to raise its confrontational fervor.

Despite the foreign power companies' hesitation, electric power generation and distribution facilities experienced sustained expansion in the period between 1945 and the industry's 1960 nationalization. As Miguel Wionczek's study of the industry demonstrates, the installed capacity went from 700,000 kW to about 3 million, with the annual power generation increasing from 3.1 billion to 10.8 billion kWh.[161] This development reversed the foothold until then secured by foreign power companies vis-à-vis the Mexican government's Comisión Federal de Electricidad (CFE; Federal Power Commission). The CFE with a minuscule initial capacity of less than 50,000 kW added more than a million kilowatts. While Mexlight and the other major foreign-owned energy company in the country, American and Foreign Power, added 500,000 kW to the country's global installed capacity, the rest of the power producers in the country increased by 700,000 kW. At the beginning of the period, the two giant foreign companies held a combined 60 percent of the national installed capacity, the CFE a mere 5 percent, and other Mexican energy producers 35 percent. By 1960, however, the situation had dramatically shifted: the state company controlled 40 percent, while the two foreign companies dominated around 33 percent.

The SME regularly underlined that Mexico's plight was part of a broader struggle in Latin America. Contributions to *Lux* decried foreign intervention in the area. In a 1958 article titled "Revolutionary Struggle or

Subjugation to Capitalism," union leader Luis Alvarado Tello reminded electricistas that, when it came to Mexlight, the syndicate could not remain "mute and inactive in the face of provocations and robbery of this foreign imperialist company."[162] This anti-imperialist fervor was echoed a few pages later in an article on the situation in Guatemala. *Lux* had covered the coup d'état against Guatemalan president Jacobo Árbenz four years earlier and published on its cover Diego Rivera's mural *Glorious Victory*. The mural depicts John Foster Dulles (US secretary of state), his brother Allen Dulles (director of the Central Intelligence Agency [CIA]), and a bomb bearing the face of President Dwight D. Eisenhower standing over the bodies of lifeless Guatemalan children carpeting the foreground.[163]

Alvarado Tello's 1958 article was accompanied by Arturo García Bustos's print *Foreign Companies*, which denounced the exploitation and violence of foreign companies in the region (including Esso Standard Oil, US Steel, and United Fruit Company).[164] The text applauded the union's protests (probably after the coup in Guatemala). Along with the protest of the Mexican people, they had in the past "prevented the assassination of hundreds of political leaders, students, peasants, and patriots." Once again, the SME called for the reestablishment of a democratically elected government and respect for the fundamental rights of citizens. Both the cover and editorial a month later were dedicated to the Cuban situation. The editorial criticized the American press and their henchmen in Mexico, who were "whining over the summary trials that the Cuban people were carrying out." Although the SME did not wish "for useless bloodshed, the Cuban people had the right to punish their tormentor."[165]

REVOLUTIONARY FERVOR

The strong winds of 1959 stoked the revolutionary fire of the country and the SME. The deeds and words of Cuban revolutionaries came from the east to an already heated political landscape. Dissident railway workers had organized, fought, and won the right to elect their leaders democratically since 1958. Their militancy came under the national spotlight in February 1959 when the now independent union went on strike even after the Federal Conciliation and Arbitration Board ruled the strike illegal. The court's decision technically meant that the government could lawfully arrest striking workers for "engaging in a criminal conspiracy against the FNM" (Ferrocarriles Nacionales de México; National Railroad of Mexico). Surprisingly, the union's lawyer, Mario Pavón, managed to turn the tide when he was sent to negotiate personally with president López Mateos. The railroad workers struck gold, including important wage increases and the family medical and housing benefits that they had long sought.[166]

Their honeymoon was short, however. The national economy came to a sudden halt in late March as renewed strikes against privately owned firms paralyzed freight cars. The historian Robert Alegre notes that "in seemingly no time," as the government moved in to replace workers with soldiers, "rieleros and rieleras had gone from being hailed as pillars of the nation to being maligned as enemies of Mexico."[167] In the following months, the decapitation of the railroad workers' leadership struck a blow against the nation's organized labor. In Cold War rhetoric, state repression of unions, especially those that fiercely defended syndicate autonomy like those of the railroad, oil, mining, and electrical workers, was justified. Elisa Servín argues that a "belligerent anti-Communist discourse" served to legitimate state repression by accusing "union leaders of seeking to destabilize the country with the help of foreign agents as well as attempting to halt the [country's] march toward progress."[168] The excitement that the Cuban Revolution generated in Mexican soil exacerbated the mainstream media's anti-Communist line.

For the country's organized labor, particularly for electricistas, 1959 was important not only because of the federal government's about-face against railway workers but also because of the invigorated presence of former president Cárdenas on the national political scene and the Cuban Revolution's galvanizing effect on the region's anti-imperialist sentiments. Cárdenas, who had remained at the margins of the country's political life after concluding his presidential term, returned in 1957 as an essential and critical political voice. Through declarations and newspaper articles, Servín argues, "he became a constant presence in public opinion," who, among others, "called to strengthen the legacy of the Mexican revolution."[169] Cárdenas's support for the mobilization of organized labor was explicit throughout 1958, and he acted as an intermediary between president Adolfo Ruiz Cortines and the leadership of the railroad and teacher unions. His international tour through several Eastern European countries, the Soviet Union, and China was closely followed. Upon his return, Cárdenas spoke widely about the positive elements that he saw in these countries. It was his defense of jailed workers and the Cuban Revolution (the new cause of the progressive forces), however, that led Servín to conclude that Cárdenas's activities "contributed to the radicalization of the confrontation between the political groups aligned in a spectrum that had Cardenismo and Alemanismo in its extremes and, which at the same time, formed part of the Cold War's ideological dispute."[170]

The effect of the Cuban Revolution on Mexico's political affairs remains a source of historiographical contention. In her study of propaganda, Servín argues that the triumph of Castro and his troops generated a "gradual opening," which allowed Cardenismo and the Left to organize around the Movimiento de Liberación Nacional to gain ground in public opinion.[171] Sylvia Gómez Tagle notes that the Left benefited not only from the post–Cuban Revolution context but from the actions of the government of Adolfo López Mateos. Much in need of rebuilding popular alliances, he saw the opportunity once again to create a social base of support.[172] The historian Eric Zolov, without denying the president's agenda in appropriating aspects of revolutionary Cuba for his benefit, demonstrates how Cuba became "a mirror into which Mexicans old and young gazed, though not with common eyes."[173]

Electricistas remained carefully tuned in to Cuban affairs and were quick to integrate rhetorical devices and connect their demands to those of their sister country. *Lux* had run critical assessments of the state of the Mexican revolutionary program for years, sometimes going as far as to call it defunct. Having what Zolov calls "the romance of a vibrant revolutionary project" so close to home elicited fervor among the electricista leadership, as the expressions of support for the Cuban cause throughout 1959 attest.[174] The SME moved quickly to establish relations with the island's revolutionaries and connect to its radical narrative. *Lux*, for instance, provided detailed coverage of the union's packed event to honor and support the victory of the Cuban people at the syndicate's theater.[175] A mural depicting the embrace between electricistas and Cuban revolutionaries served as background to the presidium. In her speech, the Cuban intellectual Teresa Casuso Morín asked electrical workers to help the Cuban people by protesting media campaigns against the revolution and, perhaps equally important, to do whatever was necessary to prevent dictator Fulgencio Batista and his riffraff supporters from attaining political asylum in Mexico.

Electricistas were aptly aware of how mainstream media could misinterpret their rapprochement. After all, the Mexican press scaled up its anti-Cuban sentiment, denouncing Fidel Castro as a caudillo and describing events in Cuba as a "fratricidal war without precedent . . . a horrible nightmare."[176] Despite their enthusiasm for the Cuban delegation, the union leaders felt the need to underscore their solidarity with the goals of the Cuban Revolution while highlighting its distance. Electricistas had witnessed

FIGURE 6.7. *"A meeting of true revolutionaries." Cover of* Lux, *June 1960. Courtesy of the Sindicato Mexicano de Electricistas.*

in the recent past the publicity campaign that both the state and the FNM had carried out against rieleros, accusing the movement's leaders of being "Communists who manipulated the rank and file . . . promoting a foreign ideology and aiding a foreign state."[177] Electricistas themselves, as well as supporters of all other dissident movements, were characterized as Communists. It is thus unsurprising that coverage of the Cuban delegation's activities was visually and textually peppered with signposts. While one article described the two movements as "Mexico and Cuba united in the struggle for freedom," another marked the distance between them: "We will never be Communists: our doctrine is the Mexican Revolution."[178] Not even the increasing anti-Americanism in Mexico seems to have pushed the SME to take a bolder stance in supporting the Cuban struggle. *Lux*'s June 1960 issue (fig. 6.7) for instance, depicts an encounter between Mexican workers and Fidel Castro and a *guajiro* (peasant) fighter. Stoically facing each other, workers and revolutionaries, revolutionaries and workers look past each other toward the horizon, with only the national flags in the background touching.

By the time of Castro's appearance in the union's magazine, the city's and nation's electricscape had been radically transformed. The federal government bought out the US-controlled American and Foreign Power in April 1960 and was finalizing negotiations with Mexlight by the summer. Discussions of López Mateos's decision to move toward nationalization of the industry have credited the president, repeatedly leaving out the SME's historical calls and actions. Gómez Tagle argues that the Cuban Revolution, which came on the heels of the political crisis unleashed by the intense period of organized labor actions in 1958–1959, presented a "favorable conjunction to take measures against imperialism, which allowed [government officials] to call for national unity, without, fundamentally, modifying tendencies of capitalist accumulation." Nationalization, she adds without being specific, "had also been included in the programs of various leftist groups and parties as they considered it an indispensable battle against imperialism."[179]

In his in-depth study of the industry, Miguel Wionczek agrees with this "favorable conjunction" thesis, noting that the American and Foreign Power buyout was a "fairly spontaneous deal" in which the interests of the seller and the buyer "coincided for different, but mainly economic, reasons." It was not until the completion of that negotiation, he adds, that the López Mateos administration "became aware that it might be desirable, not only on economic but also on political grounds, to solve the remaining part of the problem by buying out the Mexlight properties." President López Mateos was caught between the cross fire of the US-Cuban conflict during the summer of 1960 and the Mexican Left's and PRI's view that the Castro regime was being persecuted "mainly because of its hostility to the powerful private foreign interests" disturbed by the seizure of oil refineries. Wionczek proposes that in that context the nationalization of the electric industry was "probably the only available issue" capable of restoring national unity.[180]

It is uncertain to what extent nationalization was a unifying factor. Rogelio Hernández Rodríguez argues that nationalization was not a source of unity, claiming that the Mexican Right associated "the Cuban issues with various social and economic measures that identified [López Mateos] with communism," including the move to nationalize the electric industry.[181] According to Hernández Rodríguez, the president's decision to re-

FIGURE 6.8. *"We support nationalization!" Cover of* Lux, *August 1960. Courtesy of the Sindicato Mexicano de Electricistas.*

cover the industry from foreign hands was not spontaneous but was instead included in his first speech as a presidential candidate and formally presented in late 1959.

Zolov also characterizes the nationalization not so much as a unifying issue but as a decision intended to reap personal benefits for the president himself. In a global context of anticolonialist struggle and an emergent "Third World" identity, López Mateos more "than any previous postrevolutionary president . . . proved anxious to counterbalance the preponderant financial and commercial influence of the United States on the country's economic sovereignty." To crown the triple commemorative significance of 1960—marking the fiftieth anniversary of the Mexican Revolution (1910–1960), the 100th anniversary of the liberal cause under President Benito Juárez (1860–1960), and the 150th anniversary of independence (1810–1960)—the president "unexpectedly announced the nationalization of the electricity industry." This action was framed as the final step in the recovery of the nation's resources (land, oil, electricity) and allowed López Mateos symbolically to inherit "the mantle of revolutionary nationalism."[182]

Whether or not the nationalization unified opposing groups in the political landscape, however momentarily, or whether the action was a response to López Mateos's desire to advance his revolutionary legacy remains unclear. It is clear, however, that the electricistas' long-established struggle for Mexicanization of the industry, which was finally achieved via an anticlimactic nationalization (indeed financed by foreign capital), went unrecognized. *Lux*'s cover for August 1960 flaunted an oversized flag enveloping two high-voltage pylons behind a line of mobilized workers (fig. 6.8). The merging of patria and syndicate reached an unprecedented fervor, perhaps as it would never again. An SME worker leads the file of workers in the foreground. His shirt with a rolled-up sleeve is a symbolic gesture of informality, camaraderie, and machismo. The union's standard-bearer, with a firm grip on the SME's flag in his right hand and a rolled-up collective contract in his extended left hand, reenforces the SME's posture of defense going forward.

CONCLUSION

Discussions of the state of the electrical industry included detailed examinations of the industry's obstacles, what these meant for the country's industrialization, and the role of the government in addressing them. It became commonplace for these studies to look back to the Cárdenas period and his handling of the oil industry as a model to be followed by the electrical industry. The idea of electric backwardness gained traction as it became identified with urban blackouts, high electric rates, and the hunger for electricity in the country. The industry's development was intertwined with the country's industrialization. The nation's industrial potential could be pursued only by addressing the shortfall in electricity.

As defenders of the nation, particularly the nation's natural resources, electricistas were a new generation of revolutionaries destined to end what the SME called the colonial condition of the country's economy—to finally achieve a Mexico for Mexicans. Electrical workers equated the fight for their own rights with their fight for the nation's resources. Their struggle was not an economic struggle but a patriotic one, in which they were defending and not demanding. The future of the electric industry was debated as the missing step toward full economic independence. Their call for the industry's nationalization was framed as completing the country's economic independence initiated by President Cárdenas. When president López Mateos moved to acquire the foreign-owned electrical companies, however, the narrative that cast electricistas as the protagonists fighting for the country's ultimate economic independence found no room in the chronicle that linked López Mateos to Cárdenas and past revolutionaries as the true defenders of the people.

Conclusion

¡LA ELECTRICIDAD ES NUESTRA! (ELECTRICITY IS OURS!)

A papier-mâché man hung from a crane outside the Presidential Palace in Mexico City on September 27, 1960. The effigy's pale face starkly contrasted with its black suit and top hat, but the large moneybag clutched in its hands drew the most attention. The bag, along with the figure's pants, was copiously covered with dollar signs as if his greed could not be overstated. Just in case the identity of the effigy was unclear, a sash draped over it on one side identified the "Mexican Light and Power Company": the other side said defiantly "¡No hasta luego, sino adiós para siempre!" (Not see you later, but goodbye forever!)

Thousands of labor union workers, in what seemed to be a never-ending stream, made their way to the Zócalo to applaud the nationalization of the electric industry. An atmosphere of jubilation filled the plaza and crowds spilled out into the surrounding streets. At the entrance of the Verónica Building, one of Mexlight's former properties, officials held a ceremony marking the transfer of Mexlight to the new public governing board. An enormous procession of union members made their way afterward into the crowds at the Zócalo. Unquestionably, the SME was the best represented. After over a decade of pushing, arguing that the nation's economic independence demanded a nationalized electric industry, the government had finally agreed. The union's treasurer, Antonio Ortiz Mena, crowned the 45-minute transfer ceremony by hoisting the national flag while the crowd sang the national anthem, released balloons, and whistled. *Lux* raved that the scene was a vigorous explosion of patriotism.[1]

Workers in the procession that followed carried placards with President Adolfo López Mateos's image, but nothing could compare with the massive banner that the SME had printed for the occasion (fig. 7.1). At the top, in

FIGURE 7.1. *Banner visualizing state ownership of electricity carried by workers. Cover of* Lux, *September 1960. Courtesy of the Sindicato Mexicano de Electricistas.*

boastful nationalistic tones, it cried, "Mexico for Mexicans!" The president's determined face below directed the viewer's attention to the foreground, where his fist was grasping an electrical lever. The enlarged fist (relatively the same size as the president's face) was the focus of the image. The raised switch, which is used to control the current of electricity in a circuit, was a metaphor for the industry's Mexicanization. The message could not be delivered without the help of those holding up the banner up for the rest to see and read. On one level, the SME's banner gave credit to the president, but the oversized hand, a recurrent visual trope in the union's lexicon, clearly alluded to the SME, an implicit claim that the "hand of the worker" deserved credit for the revolutionary feat.

As the president stepped onto the presidential palace balcony, the crowd greeted him with cheers, fireworks, *matracas* (noisemakers), and bells ringing from the Metropolitan Cathedral. López Mateos declared that by taking over Mexlight the people had concluded a long struggle to control the electric energy generated in the country.[2] Electricity belonged to Mexico, and nationalization was a fulfillment of the revolution's promise. He invited them to accelerate its expansion so that every home in the country could have electricity. The president's words were nearly identical to what the SME had been saying for years.

Luis Aguilar Palomino, secretary general of the SME, then spoke on behalf of the union, waxing eloquent on his hope that the transmission cables that crossed the nation's valleys and sierras would transmit the voice of progress. The government of López Mateos, he declared, was a revolutionary one. The industry's nationalization was on par with previous acts carried out by the heroes of the country's independence, liberal reform, and revolution. Speaking from the height of a crane, Palomino held that from the Zócalo, "the greatest soundbox of the pueblo's voice, the electricistas accompanied by hundreds of labor groups have come to . . . show our support for the president's transcendental step," "an indestructible column for Mexico's industrialization and economic progress."[3] As the president, cabinet members, and new governing board of the company returned to the presidential palace, the capitalist effigy that ridiculed the former foreign company went up in flames.

The celebrations lasted into the night. In another address, Palomino reaffirmed the SME's revolutionary pedigree and the electricistas' prime position within the industry. "Founded in 1914 in the crucible of the Revolution," the SME recognized the groundbreaking step taken with the industry's nationalization, he said, adding proudly that electricistas "were ready to assume the responsibility that comes with being the main element in the Mexican electrical industry."[4] Palomino lost no opportunity to highlight the electrical workers' place in the country's history and in its electrified future.

In justifying nationalization, López Mateos and other top federal officials underlined Mexlight's exploitation of the Mexican people. But the passivity that they emphasized stands in stark contrast with the reality of the historical actors found throughout this book. The Manichaean rhetoric of the speeches shows how the official history of electrification was being transcribed with errors and inaccuracies. The history of the electricscape—the interwoven actions and choices of electric producers, the government, and, most important, electric consumers—is one of growth, infiltration, and acceptance so complete that electricity became something like a fundamental right.

The references to the potential of electricity for the country's industrialization and material progress that peppered the speeches delivered throughout the day's ceremonies echoed sentiments first raised seven decades before by Porfirian education minister Justo Sierra in the speech that symbolically granted Mexican citizenship to electricity in 1886. Much had changed in the country, but hopes for an industrialized nation were still closely identified with the invisible current. Electricity as a technology was no longer the "beautiful stranger" that needed to be captured by students at the National Preparatory School, but a technology destined to cross valleys and sierras, expanding the capital's electricscape. Electricity had become a household name so common in the seventy-four years separating the two events that it could often be taken for granted.

Mexico had answered the call of the electric revolution taking the world by storm in the last quarter of the nineteenth century. Capitalinos constructed the electricscape, an expansion of the electrical possibilities of "modernity" that was debated, modified, and appropriated by real historical actors. Issues involving the definition of electricity and who could use it, under what terms, and its effects were not laid out for Mexicans. Rather, they defined these questions themselves and, in doing so, in many ways changed the technology itself. Unlike older technologies, electricity retained its association as a modern technology. That is, the arrival of electricity in new areas in the country was crafted in the language of modernity. Campaigns to electrify the countryside after the industry's nationalization presented electricity as a civilizing tool. A catalog of the Cámara Nacional de Manufacturas Eléctricas (CANAME; National Chamber of Electric Manufacturing), for example, is subtitled with the slogan "To electrify is to civilize."[5] In the lexicon of the period, the blueprints of the capital's electricscape were ready to combat the dark shadows and underdevelopment in rural Mexico.

The nationalization of the industry and the state's rural electrification efforts thereafter developed a particular discourse that not only was *presidencialista* (presidentialist), abstract, and disconnected from the imaginary future consumers but also failed to recognize the myriad ways in which capitalinos, and Mexicans at large, had shaped the electrification of their spaces and the intimate relationships that many had forged with electricity in general. Central to this erasure was the development of a visual language that placed López Mateos as the leading force behind the nationalization. Echoing the address by Aguilar Palomino, secretary general of the SME, various visual compositions placed both the president and the nationalization of the electric industry within the nation's revolutionary history. Electricity joined the agrarian reforms and the oil expropriation to form the trinity of economic independence: López Mateos joined the heroes of independence, the Reforma, and the revolution.[6]

A stream of animated cartoons served to amplify this discourse. In one of these, a cheerful caricatured López Mateos is bestowing a sombrero (per-

FIGURE 7.2. *President López Mateos crowns the nationalized electric industry.* Lux, *August 1960, 23. Courtesy of the Sindicato Mexicano de Electricistas.*

haps the most iconic part of the charro outfit) on an ecstatic anthropomorphized light bulb (fig. 7.2). In a sartorial world in which head coverings still carried a great deal of significance, the sombrero as a national emblem served as a heartfelt welcome to the nationalized electric industry.[7] The presence of a traditional Stratton felt campaign hat of the Royal Canadian Mounted Police in the image is a direct reference to Mexlight's Canadian ownership and reinforces the privileged position of the Mexican masculine symbol par excellence. The cartoonist's visual language left no room for misinterpretation. While the Canadian hat lies discarded on the floor, the radiant light bulb receives the sombrero adorned with the national flag with open arms. Like a monarch's crown, the Mexicanized light bulb's sombrero is a new symbol of economic sovereignty.

These pictorial devices collectively emphasized the president's role in the Mexicanization of the industry. They left no room for those who had debated, embraced, appropriated, rejected, and shaped the ways in which electricity would enter their lives and spaces for decades. The preceding chapters restore center stage to capitalinos who used electricity to reorder public space. Electrical lighting provided a rather easy medium to paint the city as modern and showcase its achievements abroad among modern sister countries. But accident reports revealed the violence of industrialized time, the resourcefulness and anxieties of those coping with the fears, excitement, and opportunities brought by a faster mode of travel. Users also actively conditioned the entrance of electricity into public, semiprivate, and private spaces, questioning its nature and appropriation. We see how the terms of consumption established by electric companies were probed and many times challenged by consumers.

The script on the proper use of the technology dictated by companies and authorities alike was tested even before the ink dried. Electrical knowledge could not be monopolized, opening the door for the free trade of skills and devices in a market critical of the rules of the game. The policing of consumption was not predetermined: it emerged as a response to those who contested scripted consumption. Not only consumption scripts but sales pitches for electricity as well were constantly altered. Social anxieties after the revolution permeated the sale and consumption of household articles. The rhetorical and commercial construction of women who were to inhabit modern spaces inside the safe haven of their homes was quick to exploit class and racial anxieties. Nonetheless, it could not be successfully imported wholesale from abroad. Mediators wove associations between technologies and spaces (social and physical) in order to domesticate exciting but foreign and unknown items. Home advisors and cookbook writers navigated murky waters in their efforts to make technologies desirable and palatable at home. To do so they had to rewrite the scripts, employing their own cultural baggage. The story of the consumption of appliances shows that not everything about the electricscape was bad and violent. People could imagine that they were modern in the same way the nation (or at least the city) was becoming modern: they fit in the city. Electricistas exploited and expanded the electricscape tangibly and symbolically. Like no other group, they embodied the ways in which electricity was appropriated to redefine the nation, the self, and the other.

Mexlight had conditioned the consumption of electric energy in the capital through its rates, limited investments, and service quality. The experience of users and nonusers of electricity had been to a certain extent influenced by the actions of this company, but their experience was never limited to their interaction with Mexlight. Their interactions took place in a much wider context of use, in which they themselves became electrifying agents. They wrestled with the challenges and opportunities presented by electrification, as their creativity paved different approaches to the construction of an electriscscape. Just as the human chain had made the Leyden jar demonstration possible in 1886, the hands holding the gigantic banner in 1960 boosted hopes for an electrified nation.

Notes

ABBREVIATIONS

AGN	Archivo General de la Nación
AHDF	Archivo Histórico del Distrito Federal
SCOP	Secretaría de Comunicaciones y Obras Públicas
SME	Sindicato Mexicano de Electricista
TSJDF	Tribunal Superior de Justicia del Distrito Federal

INTRODUCTION

1. Cuéllar, *Los fuereños*, 28–30. All translations are mine unless otherwise noted. Note that Doña Candelaria drops the "c" in her pronunciation ("elétrica"), a common representation of provincial speech patterns. For Cuéllar's treatment of Mexico City, see Zavala Díaz, "Espejismos de la modernidad."

2. For a history of power outages in the United States, see Nye, *When the Lights Went Out.*

3. Arnold, *Everyday Technology*, 5.

4. Misa, "The Compelling Tangle of Modernity and Technology," 2.

5. Basalla, "The Spread of Western Science"; also see Arnold, "Europe, Technology, and Colonialism." For diffusionism in general, see Blaut, "Diffusionism: A Uniformitarian Critique."

6. Arnold, *Everyday Technology*, 5. On technological determinism, see Smith and Marx, *Does Technology Drive History?*

7. Chatterjee, *The Nation and Its Fragments*, 5.

8. Misa, "The Compelling Tangle of Modernity and Technology," 9, 2–3.

9. Morton, "Reviewing the History," 60.

10. Hughes, *Networks of Power*. Large technical systems of infrastructural power

have been overlooked in Mexico, with the exception of studies on the railroad, as noted by Hill, "Circuits of State," 16.

11. Historians of technology and social scientists have taken the LTS approach to map out the particular subtleties of national styles as well as the significance of system builders and have followed the path dependencies and reverse salients that systems encountered as they developed. See, for example, Lagendijk, *Electrifying Europe*.

12. Bijker, Hughes, and Pinch, *The Social Construction of Technological Systems*, xvii.

13. My initial interest in this subject and my title were inspired by Nye, *Electrifying America*. Meiton, *Electrical Palestine*, 8 (quotation). Indeed, Meiton argues that the grid served as "the base plate of a future Jewish state" (2). For South Africa, see Christie, *Electricity, Industry and Class*. For India, see Kale, *Electrifying India*; and Dubash, Kale, and Bharvirkar, *Mapping Power*. For Palestine, see Meiton, *Electrical Palestine*; and Shamir, *Current Flow*. For Russia, see Coopersmith, *The Electrification of Russia*. For energy expansion as a state-led moral project, see Chatterjee, "The Asian Anthropocene"; and Coleman, *A Moral Technology*. For Latin America, see Tendler, *Electric Power in Brazil*; McDowall, *The Light*; Armstrong and Nelles, *Southern Exposure*; Campos Aragón, *El modelo británico* and *La electricidad en la Ciudad de México*; Saes and Lanciotti, "La regulación de los servicios"; García Lázaro, "Luces y sombras en la Ciudad de México"; Liehr and Torres, "Las 'Free-Standing Companies'"; and Gámez, *Electricidad: Recurso estratégico*. For an early assessment of US-financed rural electrification efforts in Latin America, see Ross, *Cooperative Rural Electrification*.

14. Both this shortcoming and Nye's fleeting treatment of anxieties and doubts about electricity being overshadowed by positive experiences with the new technology are underlined in Gooday, *Domesticating Electricity*, 13. For the worries and fears of the American public, see Simon, *Dark Light*.

15. Gooday, *Domesticating Electricity*, 13; Jones, *Routes of Power*, 2 (quotation). For a thorough study of how the United States became the largest energy consumer in history, see Nye, *Consuming Power*. For the need to consider rivals in gas supply when looking at early electrical systems in the United States, see Platt, *The Electric City*; and Rose, *Cities of Light and Heat*. For the British case, see Clendinning, *Demons of Domesticity*.

16. The few studies on the history of electrification in the non-Western world concentrate primarily on urban areas. Hasenöhrl, "Rural Electrification," 12. She notes that historians of technology, who are the "natural" candidates to undertake such studies, have remained parochial in practice, rarely venturing out of Europe and North America. Two panels dedicated to the electrification of the Global South in the annual meeting of the Society for the History of Technology in 2019 are promising in this regard.

17. Hofmänner, *Science & Technology Studies Elsewhere*; Chakrabarty, "Postcoloniality and the Artifice of History," 2.

18. Hofmänner, *Science & Technology Studies Elsewhere*, 270.

19. Ibid., 260 (quotation). For the conceptualization of multiple modernities, see

Eisenstadt, "Multiple Modernities." For the postcolonial critique of modernity in the field of science, technology, and society (STS), see Postcolonial Technoscience, special issue of *Social Studies of Science* 32, nos. 5–6 (2002). For a critique of modernity, see Mitchell, *Questions of Modernity.*

20. Randeria, "Entangled Histories," 287.

21. For global financers in the electric industry, see Hausman, Hertner, and Wilkins, *Global Electrification.* For Canadian capital in the electric market in Latin America, see Armstrong and Nelles, *Southern Exposure.*

22. Lemon and Medina, "Technology in an Expanded Field," 113.

23. Portuondo, "Constructing a Narrative," 500 (quotation); Medina, da Costa Marques, and Holmes, "Introduction," in *Beyond Imported Magic*, 2.

24. Beatty, Pineda, and Sáiz, "Technology in Latin America's Past," 139.

25. Matthews, *The Civilizing Machine*; Castro, *Radio in Revolution*; Robles, *Mexican Waves*; Tinajero and Freeman, *Technology and Culture*; Beatty, *Technology and the Search for Progress*, 6 (quotation). Beatty notes that the history of technology has received "little attention from historians," with even recent works on Mexico's history failing to "treat technology in any explicit manner, despite its centrality to . . . economic growth and development." Beatty, *Technology and the Search*, 20.

26. Tenorio-Trillo, *I Speak of the City*, 307; see also Alexander, *City on Fire*; Vitz, *A City on a Lake*; and Agostoni, *Monuments of Progress.*

27. Lemon and Medina, "Technology in an Expanded Field," 112.

28. Jasanoff defines sociotechnical imaginaries as "collectively held, institutionally stabilized, and publicly performed visions of desirable futures, animated by shared understandings of forms of social life and social order attainable through, and supportive of, advances in science and technology." Jasanoff, "Future Imperfect," 6.

29. *El Partido Liberal*, December 19, 1886.

30. Posadas are nocturnal gatherings centered on the reenactment of Mary and Joseph's search for shelter in Bethlehem, performed from December 16 to 24.

31. *El Partido Liberal*, December 19, 1886.

32. Limited access to sources of inanimate power before 1900 frustrated efforts to mechanize the country's main industries (mining, agriculture, and manufacturing). Both the scarcity and high cost of water, wood, and coal had long-lasting consequences, hindering the diffusion of mills, steam engines, and overall mechanization. Beatty, *Technology and the Search for Progress*, 41–45. Lack of demand (*falta de consumô*) for industrial products did not "justify investment in anything more than artisanal and home production" (19).

33. *El Economista Mexicano*, December 24, 1886.

34. *El Tiempo*, December 17, 1886.

35. For the complete program, see *El Tiempo*, December 14, 1886.

36. *El Partido Liberal*, December 24, 1886.

37. *El Economista Mexicano*, December 24, 1886.

38. Boast, *The Machine in the Ghost*, 26.

39. *El Partido Liberal*, December 24, 1886.

40. *El Economista Mexicano*, December 24, 1886.

41. *El Tiempo*, August 22, 1886.

42. The Ayuntamiento (Town Hall) repeatedly answered petitions from provincial cities and even other Latin American cities to provide copies of the municipal contract for electric public lighting.

43. See Kline and Pinch, "Users as Agents of Technological Change."

44. Oudshoorn and Pinch, "Introduction," in *How Users Matter*, 16. See Silverstone and Hirsch, *Consuming Technologies*; and Lie and Sørensen, *Making Technology Our Own?*

45. Gooday, *Domesticating Electricity*. Although Gooday's treatment centers on "the extent to which electricity was interpreted as sufficiently tamed to be safely, reliably and comfortably introduced to the home," I have employed it to examine both public and private spaces.

46. Garza Toledo, *Historia de la industria eléctrica*; Reséndiz Núñez, *El sector eléctrico*.

47. Marichal, *Las inversiones extranjeras*; Tafunell, "La revolución eléctrica"; Campos Aragón, *El modelo británico*; Tafunell points to J. Fred Rippy's work as a precedent for this type of regional studies. See Rippy, "German Investments in Latin America" and *British Investments in Latin America*.

48. Brown, *A History of Mining*, 110.

49. Galarza, *La industria eléctrica*, 9.

50. Steam-generated electricity was used in the Santa Ana mine in the Catorce district, the largest mine site in San Luis Potosí, for drainage works; El Boleo mines of Santa Rosalía, in Baja California, for pumping, ventilation, crushing mills, illumination, and grinding; Real del Monte, in Pachuca, for pumping the deepest shafts and grinding ore. The Guggenheim Corporation plant in Aguascalientes was the first to employ the new energy for smelting metal. Galarza, *La industria eléctrica*, 9–24.

51. *Mexican Mining Journal* (October 1907): 8.

52. Among these were Cananea, Santa Rosalía, Ostotopaquillo, San Rafael, Dos Estrellas, Río Verde, and Natividad. Galarza, *La industria eléctrica*, 10.

53. Henry George Ward detailed the physical state of some of the large mines at the end of the independence wars. Many mines had caved in due to the flooding of shafts and decayed timbering. Several remained unattractive for decades, awaiting investors with deep pockets and up-to-date technology to make them profitable. Ward, *Mexico in 1827*.

54. During the Mexican Revolution, rebels not only repeatedly cut transmission lines but stole the horses and mules owned by the company's freighters. The prompt repair of destroyed lines was thus doubly significant to prevent bringing the industry to a standstill. In the north, energy could be imported from the United States when no national sources were available. El Tigre mine, in Sonora, for instance, received electricity from the Copper Queen Mining Company in Douglas, Arizona. Electric companies and mines became interdependent in many mining districts, while mining companies generated their own electricity elsewhere. Residential and commercial consum-

ers availed themselves of surplus energy. "Conditions in Sonora," *Mining and Scientific Press* 111 (July 1915): 80–81. On surplus energy see Galarza, *La industria eléctrica*, 14–15.

55. *American Machinist* (August 18, 1904): 1109.

56. Mexican Light and Power Company did suffer substantial financial losses as large traditional consumers cut operations. During the Decena Trágica (Tragic Ten Days: the series of events that took place from February 9 to 19, 1913), a number of utility posts bearing power lines throughout the city came down. The military intervened and also managed the city's tramways company for a few years.

57. Wionczek, "Electric Power," 34–35.

58. *Outlook* (March 15, 1913): 594.

59. Cited in Bloch, *The Historian's Craft*, 55.

60. Arnold, *Everyday Technology*; Jasanoff and Kim, *Dreamscapes of Modernity.*

61. Piccato, *City of Suspects*, 47; see also Porter, "'And That It Is Custom Makes It Law'"; and Pulido, *¡A su salud!*

62. *El Tiempo*, December 22, 1886.

63. The clergyman and physicist Jean-Antoine Nollet, reputed to have coined the term "Leyden jar," is recognized for popularizing science in France. He performed a demonstration before King Louis XV, forming a human chain with 180 soldiers at Versailles. At the king's request, he repeated the impressive presentation in Paris with 700 monks. See Pyenson and Gauvin, *The Art of Teaching Physics*. For the use of spectacular demonstrations to exhibit electricity's potential, see Beauchamp, *Exhibiting Electricity*; Delbourgo, *A Most Amazing Scene*; Morus, *Frankenstein's Children*; and Adams, "The Promotion of New Technology."

64. *El Diario del Hogar*, December 19, 1886; *El Tiempo*, December 22, 1886 (quotation). The press applauded the experiment as a popularization of science.

CHAPTER 1. SENSING THE BEAUTIFUL STRANGER

1. Prantl and Groso, *La Ciudad de México*, xix, 687 (quotation).

2. Ibid., 688.

3. Ibid.

4. Otter, *The Victorian Eye*, 2, 19–21.

5. Few sources have emerged on the equipment used for these trials. De Vecchi and De Vecchi suggest in "Primeros ensayos públicos" that the lamp utilized for these trials must have been a voltaic arc lamp. Pedro Romero de Terreros hailed from an affluent family whose wealth could be traced back to the Pachuca mines. He authored a book on telegraphy and electric lighting: *La ciencia puesta al alcance.*

6. *El Siglo Diez y Nueve*, November 5, 1850.

7. *El Monitor Republicano*, November 20, 1850; *El Siglo Diez y Nueve*, November 19, 1850.

8. Bille and Sørensen, "An Anthropology of Luminosity," 264.

9. Ingold, *The Perception of the Environment*, 248, 259.

10. Bille and Sørensen, "An Anthropology of Luminosity," 265.

11. Ibid., 265 (quotation), 266–267.

12. Johns, *The City of Mexico*, 13.

13. Lira, *Comunidades indígenas*. Recent scholarship has shown that indigenous communities gained concessions from liberal governments through processes of negotiation, confrontation, and conditional cooperation with local, regional, and national leaders and with the help of intermediaries, ultimately altering the application of liberal reforms to accord with local contexts. See, for example, Ruiz Medrano, *Mexico's Indigenous Communities*. For the city's environmental transformation, see Vitz, *A City on a Lake*.

14. Rodríguez Kuri, *La experiencia olvidada*, 106–107, 110 (quotation), 112. As Rodriguez Kuri notes, until we have a study of the urban planning thought of liberals between 1850 and 1860 we do not know to what extent the destructive impulse responded to a political reprisal and to what extent it was about a desired city.

15. Gamboa, *Del natural*, 150. *Liberalotes* is a derogatory nickname for liberals.

16. Szuchman, "The City as Vision," 5.

17. Jiménez, *La traza del poder*, 66.

18. Agostoni, *Monuments of Progress*, 79–81 (quotation on 80).

19. Ibid., 46. Agostoni offers a thorough discussion of the city's expansion.

20. Garner, *Porfirio Díaz*, 188–189.

21. De Quevedo cited in Vitz, *A City on a Lake*, 47. For a discussion of these individuals, see Jiménez, *La traza del poder*.

22. Tenorio-Trillo, *I Speak of the City*, 14.

23. Aréchiga Córdoba, "Lucha de clases," 20–21, discusses these differences and the long-term consequences they would have on the urbanization process.

24. Agostoni, *Monuments of Progress*.

25. Vitz, *A City on a Lake*, 39.

26. "Actas de Cabildo," 1897, Archivo Histórico del Distrito Federal, Mexico City (hereafter cited as AHDF)), 121.

27. *La Patria*, May 20, 1889.

28. Rodríguez Kuri, *La experiencia olvidada*, 212.

29. Arizpe, *El alumbrado público*, 90, 59 (quotation).

30. Ibid., 57, 59 (quotation).

31. For a thorough study of the gas company, see García Lázaro, "Luces y sombras."

32. For the development of gas lighting and particularities of the contracts, see Arizpe, *El alumbrado público*.

33. Aréchiga Córdoba, "Lucha de clases," 282.

34. García Lázaro, "Luces y sombras," 187–188.

35. "Actas de Cabildo," 1882, AHDF, 353.

36. *Municipio Libre*, February 10, 1881. Outside the area serviced by gas lighting, an army of workers kept the oil and turpentine lamps lit. A major, four managers, a scribe, a reserve corporal, and station corporal along with sixty-one *encendedores* (indi-

viduals who lit the lamps) constituted the army charged with the immense operation. In 1878 encendedores were responsible for 1,406 lamps (210 oil and 1,196 turpentine) across the city. The lighting of the lamps began an hour earlier than in the contracted schedule because of the arduousness of the task. In 1881 the Alumbrado Commission deemed oil lamps useless and "suffering from a thousand defects." Among other problems, its dull and gaunt light could barely illuminate a space of twenty steps on calm nights. Having taken the commission's finding to heart, the Ayuntamiento granted the Aguirre Hermanos Company an exclusive concession for 1,500 turpentine lights later that year, officially eliminating oil lamps from the city's lightscape, although they remained in use unofficially for years after.

37. *El Telégrafo*, May 10, 1881.

38. *La Voz de México*, November 11, 1881.

39. *El Telégrafo*, July 12, 1881.

40. "Actas de Cabildo," 1882, AHDF, 319–321. In July 1882 the municipal budget set 7,109.37 pesos for gas and 6,518.36 and 4,181.21 pesos for public works and police, respectively.

41. *La Patria Ilustrada*, December 8, 1884.

42. "Actas de Cabildo," 1882, AHDF, 353 (quotation), 295. In the summer the Ayuntamiento moved to examine delays in the permanent installation of electric lighting and determine measures to oblige the company to uphold the contract's stipulations.

43. Contreras and Almaraz were assigned payment of 300 pesos for their work.

44. "Actas de Cabildo," 1882, AHDF, 416.

45. Arizpe, *El alumbrado público*, 53. Even turpentine lighting, which perhaps should have been more reliable, as a more rudimentary technology, also suffered delivery problems. When a cholera epidemic hit the capital in 1850, the indigenous people who supplied the tar that became turpentine refused to deliver it.

46. *La Patria Ilustrada*, January 5, 1885.

47. Paz, *Leyendas históricas*, 66–67.

48. *La Voz de México*, September 18, 1881.

49. Calhoun, *The Gilded Age*, 40.

50. *La Voz de México*, September 18, 1881.

51. *El Monitor Republicano*, September 18, 1881; Hoffmann, *Tales of Hoffmann*. A vara is a unit of measurement, just over thirty-three inches.

52. *La Patria Ilustrada*, May 19, 1884.

53. *El Monitor Republicano*, September 18, 1881.

54. *El Nacional*, November 8, 1881. The weekly *La Imprenta* repeatedly published advertisements in verse for this rice powder beauty product. See the issues for June 8, 1884, and June 13, 1884.*The Ballet of the Nuns* in the opera was partially created to exhibit the gas lighting installation at the Paris Opéra, which allowed the creation of ghastly effects. Kirstein, *Four Centuries of Ballet*, 142.

55. *El Nacional*, November 8, 1881. Najéra's remarks might be a general commentary on the public presence of women. Journalist and novelist Francisco Zarco had taken a different stand on artificial lighting three decades before, stating that tertulias,

dances, and concerts should be held at night because women needed artificial lighting to highlight their beauty. Zarco, "México de noche," 159.

56. *El Diario del Hogar*, November 6, 1881.

57. *El Telégrafo*, November 6, 1881.

58. Nájera in *El Nacional*, November 8, 1881.

59. *La Voz de México*, November 6, 1881.

60. *El Telégrafo*, November 6, 1881.

61. For the development of arc lighting, see Schivelbusch, *Disenchanted Night*, 52–57.

62. Jones, *Empires of Light*, 35.

63. Candlepower (CP) was the measurement unit for luminous intensity at the time, indicating how much light a source produced.

64. *Journal of Gas Lighting, Water Supply, & Sanitary Improvement* (August 20, 1878): 277.

65. Schivelbusch, *Disenchanted Night*, 118.

66. *La Voz de México*, November 6, 1881. José Zorrilla was a Spanish Romantic poet and dramatist.

67. *La Voz de México*, December 14, 1881.

68. Crary, *Techniques of the Observer*, 413. The physiology of the senses and the development of optical devices received paramount attention throughout the nineteenth century. A new kind of observer and an altogether new model of vision emerged.

69. *Journal of the Society of Arts*, 381, no. 8 (March 9, 1860): 274. Regnauld developed a treatise on electricity's influence over the muscles and binocular vision.

70. *Gaceta Médica de México* (November 1, 1883): 412–415.

71. Ibid., 414–415.

72. *Gaceta Médica de México* (November 15, 1883): 442–446.

73. Ibid., 446.

74. *Gaceta Médica de México* (November 15, 1883): 436–441.

75. Ahuehuete, also known as Montezuma bald cypress, held a special place in the city's history and culture, having been used as an ornamental tree since Pre-Columbian times. For the importance of gardens in the Pre-Columbian urban landscape, see Granziera, "Concept of the Garden."

76. *Gaceta Médica de México* (January 1, 1886): 21.

77. Gamboa, *Del natural*, 327.

78. For the significance of means and location, see chapter 5 of Sloan's *Death in the City*.

79. "Actas de Cabildo," 1885, AHDF, 365. It is not sure whether there was an installation for the occasion. Knight notified the Ayuntamiento that it was impossible to complete the installation by that night, offering instead to illuminate it with additional gaslights. A historical review a few years later underlined the solemn character of Múgica's performance, with the theater adorned and illuminated with electric lights paid by the Ayuntamiento. Olavarría y Ferrari, *Reseña histórica*, 435.

80. Briseño Senosiain, *Candile de la calle*, 110.

81. Gamboa, *Del natural*, 176.

82. *El Siglo XIX*, October 15, 1895 (emphasis in the original).

83. Schivelbusch, *Disenchanted Night*, 146.

84. Bunker, *Creating Mexican Consumer Culture*, 142–143.

85. *Municipio Libre*, June 20, 1885.

86. Zarco, "México de Noche," 158.

87. Gamboa, *Metamorfosis*, 466–467.

88. Ibid., 470, 472.

89. Unless otherwise cited, details of stores come from Doménech, *Guía general descriptiva*, vol. 2, 755–783.

90. Bunker, *Creating Mexican Consumer Culture*, 102–105. Bunker provides a thorough analysis of the inauguration, which he claims elucidates the confluence of commerce with the city's public rituals of domination.

91. Doménech, *Guía general descriptiva*, vol. 2, 776.

92. Gamboa, *Del natural*, 25–26.

93. Doménech, *Guía general descriptiva*, vol. 2, 782.

94. *Harper's Weekly*, October 20, 1883.

95. *El Tiempo Ilustrado*, May 30, 1909.

96. Ibid.

97. *El Centinela Español*, November 18, 1881.

98. For the environmentalist understanding of the disease, see Agostoni, *Monuments of Progress*.

99. *El Mundo Ilustrado*, February 13, 1898.

100. Altamirano, *Paisajes y leyendas*, 138–139.

101. *El Telégrafo*, July 12, 1881.

102. Glasco, *Constructing Mexico City*, 127.

103. Voekel, "Peeing on the Palace," 185.

104. Scardaville, "(Hapsburg) Law and (Bourbon) Order," 5. Serenos strolled the streets, announcing the hour and temperature.

105. Glasco, *Constructing Mexico City*, 4.

106. Pulido, "Policía: Del buen gobierno," 1606.

107. Voekel, "Peeing on the Palace," 190.

108. Pulido, "Policía: Del buen gobierno," 1607.

109. *El Foro*, February 19, 1898.

110. Ibid. *El Foro* pledged to circulate the new ideas of criminologists, whose theories were "masterpieces of the modern spirit." For a thorough discussion of criminology in Mexico, see Piccato, *City of Suspects*.

111. *La Patria*, March 6, 1881.

112. *El Diario del Hogar*, February 25, 1882.

113. Rodríguez Kuri, *La experiencia olvidada*, 211–213 (quotation on 211). Details of these requests come from this source.

114. "Actas de Cabildo," 1897, AHDF, 786.

115. "Actas de Cabildo," 1899, AHDF, 100.

116. Rodríguez Kuri, *La experiencia olvidada*, 213.

117. Ibid., 14. For the power of petitioning, see C. Jiménez, "Popular Organizing."

118. *La Voz de México*, November 11, 1881.

119. Schivelbusch, *Disenchanted Night*, 121–126 (quotation on 122). The earliest system of towers was proposed as early as 1799 in Paris.

120. "Actas de Cabildo," 1882, AHDF, 423–424. Alderman Manuel González Cosío argued that the company failed to deliver six hundred lights and ten towers because of lack of funds in the municipal coffers.

121. "Actas de Cabildo," 1887, AHDF, 444.

122. "Actas de Cabildo," 1886, AHDF, 119.

123. "Actas de Cabildo," 1885, AHDF, 519.

124. "Actas de Cabildo," 1887, AHDF, 441–446 (quotation on 444).

125. Ibid., 269.

126. Details for the following discussion come from ibid., 269–275.

127. This information comes from the May 24, 1887, council meeting minutes.

128. "Actas de Cabildo," 1887, AHDF, 414.

129. Arizpe, *El alumbrado público*, 59–60. In 1893 Genaro Vergara offered to replace turpentine lights with his petroleum-fueled lamp, but no contract materialized. Arizpe notes that lighting solarina lamps was faster than lighting turpentine lamps, a particular advantage given that the contract stipulated penalties for delays.

130. Ibid., 98.

131. *El Mundo Ilustrado*, February 13, 1898.

132. Arizpe, *El alumbrado público*, 149.

133. *El Imparcial*, February 18, 1898.

134. *El Imparcial*, June 2, 1897.

135. *Semana Mercantil*, June 7, 1897. *El Correo Español* also cited the advantage of eliminating the clashing posts on February 16, 1898.

136. Arizpe, *El alumbrado público*, 117–118.

137. *El Popular*, June 6, 1897.

138. *El Imparcial*, February 18, 1898.

139. Months before securing the street lighting contract, Brockman secured permission for aerial or underground lines in a contract granted by the federal government for the exploitation of waterfalls in the state of Mexico.

140. "Actas de Cabildo," 1884, AHDF, 49.

141. Ibid., 269.

142. *La Patria Ilustrada*, May 20, 1889.

143. "Actas de Cabildo," 1890, AHDF, 376.

144. Details of the 1896 contract come from its publication in "Actas de Cabildo," 1896, AHDF, 432–444. The unit for measuring electric current is the ampere (abbreviated amp).

145. *El Municipio Libre*, June 2, 1897.

146. "Actas de Cabildo," 1887, AHDF, 270.

147. *La Patria*, July 9, 1897.

148. *El Imparcial*, July 17, 1897.

149. *Electrical Engineer*, August 30, 1893, 211. For a thorough discussion of danger, see Gooday, *Domesticating Electricity*.

150. *El Imparcial*, July 15, 1897.

151. *El Imparcial*, July 20, 1897.

152. *El Imparcial*, July 15, 1897.

153. *El Universal*, October 2, 1897.

154. *El Universal*, July 21, 1897, published in its "Electricity Section."

155. Ibid.

156. Matthews, *The Civilizing Machine*, 145.

157. "Actas de Cabildo," 1898, AHDF, 13.

158. *El Nacional*, February 16, 1898.

159. "Actas de Cabildo," 1895, AHDF, 248–249.

160. *El Popular*, February 16, 1898.

161. "Actas de Cabildo," 1898, AHDF, 23.

162. Arizpe, *El alumbrado público*, 191.

163. Rodríguez Kuri, *La experiencia olvidada*, 195. For a longer discussion of this claim, see 189–198.

164. "Actas de Cabildo," 1897, AHDF, 409 (quotation), 515.

165. Ibid., 733, 735.

166. *El Nacional*, April 22, 1897.

167. Pond, *Calcium Carbide and Acetylene*, 10. Although known for much of the century, acetylene attained commercialization as an illuminant only in the 1890s with the arrival of the first factory producing calcium carbide. Among these was the development of a gas burner designed to provide adequate air to the flame in order to eliminate smoke and soot, which produced a white light ten to twelve times brighter than any commercial light source in use.

168. Rodríguez Kuri, *La experiencia olvidada*, 200.

169. Hausman, Hertner, and Wilkins, *Global Electrification*, 112. For a discussion of free-standing companies, see Liehr and Torres, "Las 'Free-Standing Companies.'"

170. Arizpe, *El alumbrado público*, 125–126.

CHAPTER 2. EXHIBITING THE ELECTRIC CITY

1. *El Siglo Diez y Nueve*, January 20, 1882.

2. For construction of the railroad line, see Pletcher, "The Building of the Mexican Railway."

3. The significant decline in overland transport thanks to railway development, a regime friendly to foreign investment, and an untapped market presented a welcoming

environment for those ready to transport knowledge and goods. Machinery and hardware imports led the way, with a third of the country's import bill from 1870 to 1910. Beatty, *Technology and the Search for Progress*, 60–61.

4. *El Monitor Republicano*, March 26, 1882.

5. Artificial incubation was indeed not new. Egyptians and Chinese had done without brooding hens many years before by subjecting the eggs to artificial heat. *La Patria*, April 16, 1882.

6. Circulation of the eggs began from above and moved downward. Thanks to a strong reflector, observers saw the spectacle of the incubation process. Knight, *Knight's New Mechanical Dictionary*, 494.

7. *Siglo XIX*, March 26, 1882. It seems that the fuel used was oil: an 1880 ad proclaimed that the incubator "burned less oil according to the number of eggs set." See Ferris, *Practical Artificial Incubation*, 111.

8. Olavarría y Ferrari, *Reseña histórica del teatro en México*, 344. The exhibition's success seems to have motivated José Trinidad Herrera: months after the egg-hatching exhibit he requested the municipal council's permission to set up an amusement automaton display at the spot where the hen had been. "Actas de Cabildo," 1882, Archivo Histórico del Distrito Federal (hereafter cite as AHDF), 352.

9. *The Two Republics*, March 26, 1882. Experts on incubators abroad also considered hens obsolete for hatching. See E. E. Bishop, "Hens for Hatching, a Thing of the Past," *Poultry World* 10, no. 3 (March 1881): 75–76.

10. Although Axford's display was a money-making success, we have no idea whether his hen had any success in the Mexican egg business.

11. Maxson, *A Practical Up-to-Date Guide to Mexico City*, 23. Over 200 feet tall, the bell towers offered an unparalleled view of the city and the surrounding valley. For a small gratuity, visitors could ascend and see the city spread out in front of them as a vast panorama, like a city in the Holy Land.

12. Terry, *Terry's México*, 273. Since the late colonial period, visitors have applauded the cathedral's massiveness and the majesty of its architecturally mixed baroque and neoclassical design. City guides advised tourists to pay a visit, making the cathedral a goal for sightseers.

13. *Illuminating Engineer* (November 1910).

14. For this dual significance, see Briseño Senosiain, "La fiesta de luz."

15. Bennett, *The Birth of the Museum*, 67.

16. Tenorio-Trillo, *Mexico at the World's Fairs*, 12. Tenorio-Trillo's masterful analysis of Mexico's participation at world's fairs argues that to examine the country's role is "to write the history of what modernity and progress meant for them."

17. Quoted in Bennett, *The Birth of the Museum*, 81–82. Benjamin underlines how "earlier progressivist taxonomy based on stages of production was subordinated to the dominating influences of principles of classifications based on nations and the supra-national constructs of empires and races." Nations thus became responsible for erecting their separate competing pavilions zoned into racial groups (81–82).

18. Morus, *Frankenstein's Children*, 5.

19. Nye, *Electrifying America*, 33–47. The intertwined history of electricity and international exhibitions is too broad to do it justice here. See Marvin, *When Old Technologies Were New*; Beauchamp, *Exhibiting Electricity*; Delbourgo, *A Most Amazing Scene*; and Morus, *Frankenstein's Children*.

20. Platt, *The Electric City*, 59.

21. Marvin, *When Old Technologies Were New*, 171. The World's Columbian Exhibition is also well known as the battlefield for proponents of alternating current and direct current. See Jones, *Empires of Light*.

22. Platt, *The Electric City*, 63. Beauchamp argues in *Exhibiting Electricity* that the impetus for the use of electricity for heating and cooking in the home can be traced back to the fairgrounds.

23. *El Veterinario y el Agricultor Prácticos*, April 1, 1882.

24. The article claimed that this was particularly beneficial for Mexican chickens, because they were "so tender and careful, suffering a true disease during the incubation that exhausted and slimmed them down to the point that it took some time before they recovered and were ready to lay again." Ibid.

25. AHDF, Collection: Ayuntamiento Gobierno del Distrito Federal, Section: Diversiones Públicas, 802/648/1882.

26. *El Monitor Republicano*, March 26, 1882.

27. *La Patria*, April 16, 1882.

28. Many questions were raised. Who would love the chicks? Who would tenderly and affectionately choose their nourishment? Who would shelter them? Why did sons and daughter have no mother? Why did the chicks have a glass mother?

29. Best, *Noticia sobre las aplicaciones*. For more on Mexican statisticians during the Porfiriato, see Tenorio-Trillo, *Mexico at the World's Fairs*, 127–131.

30. Best had lectured on telephony and the application of electricity in smelting metals during the 1886 Electricity Exposition. *El Economista Mexicano*, December 24, 1886.

31. The following details come from their report, the *Revista Telegráfica de México*, published in installments from January 16 to April 1, 1889.

32. *Revista Telegráfica de México*, January 16, 1889.

33. Best, *Noticia sobre las aplicaciones*, 6.

34. Yeager, "Porfirian Commercial Propaganda," 240.

35. Best, *Noticia sobre las aplicaciones*, 12.

36. Tenorio-Trillo, *Mexico at the World's Fairs*, 48.

37. Arizpe, *Estadística de las aplicaciones*, 13.

38. Ibid., 14.

39. Arizpe, *El alumbrado público*, 5. Both of Arizpe's publications were also distributed at the Pan-American Exposition of 1901.

40. While industrializing countries propelled the transformation of their energy regimes for much of the nineteenth century through the incorporation of steam engines, Mexico's leading industries reported a marginal consumption of steam technologies. The efficiency and incentives to adopt these methods were recognized but were

"not strong enough to overcome the substantial obstacles posed by the high cost of the machines themselves, furthered encumbered by the cost of overland transportation, technical expertise, and fuel." Thus, to be perceived as a potential consumer, it was necessary to eclipse the slow deployment of older technologies. Beatty, *Technology and the Search for Progress*, 44–45.

41. Sherratt, *Mexican Vistas*, 212–214.

42. Ruiz, *Americans in the Treasure House*, 3.

43. Carson, *Mexico: The Wonderland of the South*, v–vi.

44. *Mexican Financier*, February 11, 1898.

45. Winter, *Mexico and Her People*, 58–59.

46. For an analysis of the paradigm developed by foreigners, see Beezley, *Judas at the Jockey Club*.

47. Iglehart, *Face to Face*, 188–189, 396.

48. Blake and Sullivan, *Mexico: Picturesque*, 79.

49. Brown, *The Tour of the 400*, 140.

50. Outside the purview of foreign travelers was the way electrical power and electrical lighting in mines and factories altered the nature of labor as well as daily working, eating, and sleeping habits. These scenes were not as apparent on urban streets for travelers to comment on them.

51. Ruiz, *Americans in the Treasure House*, 3.

52. *Electrical World*, May 14, 1887.

53. Ibid. For a discussion of these images and symbolic inversions, see Beezley, *Judas at the Jockey Club*, 67–88.

54. *Electrical World*, May 21, 1887.

55. Arizpe, *Estadística de las aplicaciones*, 94, 157–158.

56. Ibid., 159.

57. Tenorio-Trillo, *Mexico at the World's Fairs*, 126.

58. The library of the Bureau of the American Republics noted the acquisition of Arizpe's *Estadística de las aplicaciones* and *El alumbrado público* a year later. See *Accessions to the Library*, no. 3 (July–December 1901).

59. For cartographic projects and surveying, see Craib, *Cartographic Mexico*.

60. Tenorio-Trillo, *Mexico at the World's Fairs*, 131.

61. For an analysis, see Vanderwood, "Los bandidos de Manuel Payno." This had become an integral part of the country's external reputation since the 1830s, with nearly all travelers remarking on the ubiquitous nature of rural banditry.

62. See, in particular, Carrera, *Traveling from New Spain*; Tenorio-Trillo, *Mexico at the World's Fairs*.

63. Mapmaking as a cartographic routine became a federal obsession in order to facilitate economic development and political administration. Craib, *Cartographic Mexico*, 56.

64. Tenorio-Trillo, *Mexico at the World's Fairs*, 3–4.

65. Arizpe, *El alumbrado público*, 203.

66. Arizpe, *Estadística de las aplicaciones*, 125.

67. Krauze, *Mexico: Biography of Power*, 6.

68. Tenenbaum, "Streetwise History," 128, 135.

69. For a discussion of the history of festive lighting, see Briseño Senosiain, "La fiesta de luz."

70. Curcio-Nagy, *The Great Festivals*, 71.

71. Matthews, *The Civilizing Machine.*

72. *Mexican Herald*, November 13, 1904.

73. *El Correo Español*, December 2, 1904.

74. *El Tiempo*, December 3, 1904.

75. *El Tiempo*, December 4, 1904.

76. Nava and Fernández, "La campana de Dolores"; Fernández and Nava, "Images of Independence."

77. "Letter from Thomas Edison to Porfirio Díaz, 9 October, 1889," Edison Papers Digital Edition, http://edison.rutgers.edu/digital/document/.

78. "Letter from D. J. Balleras to Thomas Alva Edison, 8 November, 1889," Edison Papers Digital Edition, http://edison.rutgers.edu/digital/document/.

79. Nava and Fernández, "La campana de Dolores," 141.

80. Three individuals reportedly died in the plaza that night. *El Siglo Diez y Nueve*, September 17, 1896.

81. Campbell, *Campbell's New Revised Complete Guide*, 89; *La Patria*, September 16, 1896.

82. *El Siglo Diez y Nueve*, October 28, 1881.

83. *La Patria*, February 20, 1898.

84. *Álbum de la coronación*, 99.

85. Colloquially known as "The House of Tiles," this building was an eighteenth-century palace built by the Count del Valle de Orizaba family.

86. *Álbum de la coronación*, 101.

87. *Illuminating Engineer*, November 1910, 478.

88. *El Imparcial*, July 3, 1910.

89. *El Imparcial*, July 10, 1910.

90. *Mexican Herald*, August 28, 1910.

91. Spectacular lighting reordered space and edited the city down to a few idealized locales, according to Nye, *Electrifying America*, 60.

92. *El Diario*, September 20, 1910.

93. Starr, *Mexico and the United States*, 65–66.

94. *El País*, September 15, 1910.

95. Starr, *Mexico and the United States*, 61.

96. Ibid., 65.

97. Frank Kahlo, "Mexlight Centennial Lighting," in "Las iluminaciones." Besides powering a million lights for the celebration through the electricity generated at the Necaxa, the Mexican Light and Power Company also promoted patriotism by of-

fering to lease Tungsten lamps free of charge to those interested in joining the celebration. From their—albeit self-serving—perspective, real Mexican patriots honored the nation with electric lights.

98. Ruiz, *Americans in the Treasure House*, 49.

99. De Orellana, "El poder de la memoria fugaz."

100. Mraz, *Looking for Mexico*, 39–42.

CHAPTER 3. TRAPPED UNDER THE WHEELS OF MODERNITY

1. *El Popular*, February 2, 1902.

2. Manta was the most widely produced fabric, favored by the lower classes throughout the country.

3. McKay, *Tramways and Trolleys*.

4. Travel writers formulated "informal hierarchies of cosmopolitanism" grounded on "physical and technological markers that distinguish the modern city." Rosenthal, "The Streetcar in the Urban Imaginary," 162.

5. Rodríguez Kuri, *La experiencia olvidada*.

6. *Electric Railway Journal* (May 1, 1909): 812–820.

7. Bijker and Law, *Shaping Technology*.

8. Oudshoorn and Pinch, "Introduction," in *How Users Matter*, 16. See Silverstone and Hirsch, *Consuming Technologies*; Lie and Sørensen, *Making Technology Our Own?*

9. Oudshoorn and Pinch, "Introduction," 14–16.

10. Since the late nineteenth century, technological innovations and the incorporation of technologies into everyday life have profoundly shaped these ideas. Arnold, *Everyday Technology*, 5.

11. For a social history of streetcars, see Leidenberger, *La historia viaja en tranvía*.

12. García, *Memorias íntimas*, 24–25.

13. *Electric Railway Journal* (May 1, 1909): 812–820.

14. The rapid tram and railway expansion in Mexico resembled the situation in the United States, where 28 percent of the railway kilometers were electrified by 1890 and 84 percent by the end of 1895. In contrast, the British railway development lagged. In its first year (1897) a mere 6 percent was electrified, and it took five years to reach 38 percent. McKay, *Tramways and Trolleys*, 71–73.

15. While animal-powered traction was generally being phased out, it carved out a unique niche in those markets where electric traction remained unprofitable and stayed in service for a remarkably long time. The last animal-powered line ran until 1932.

16. Brown, *The Tour of the 400*, 70.

17. Moses, *Today in the Land*, 68.

18. *El Imparcial*, June 1, 1901.

19. *Electric Railway Journal* (June 5, 1909).

20. Barton, *Impressions of Mexico*, 74. A tour guide warned visitors that second-

class cars were "usually stuffed with ill-smelling *pelados* [riffraff]." Terry, *Terry's México*, 236.

21. *El Universal*, January 16, 1900.

22. Ibid.

23. Ibid.

24. Rosenthal, "The Streetcar in the Urban Imaginary," 163.

25. Sloan, *Death in the City*, 14. Sloan provides a succinct discussion of liberalism and scientific politics.

26. Agostoni, *Monuments of Progress*, 27–28.

27. For statistics and suicide, see Sloan, *Death in the City.*

28. Tenorio-Trillo, *Mexico at the World's Fairs.*

29. Details come from the 1905 and 1907 editions of *Reseña histórica y estadística de los ferrocarriles.*

30. *El Imparcial*, January 18, 1901.

31. *El Imparcial*, May 7, 1903.

32. Gamboa, *La llaga*, 372.

33. *Gaceta Médica de México*, September 1, 1900.

34. *Gaceta Médica de México*, January 1, 1905. Although official data for yellow fever registered few deaths from 1900 to 1904 (5, 0, 7, 4, and 0, respectively), deaths attributed to tuberculosis were consistently over 1,100 (1,155, 1,345, 1,325, 1,138, 1,125, respectively). For a breakdown of premature deaths, see Agostoni, *Monuments of Progress*, 67.

35. Muñoz-Serrano, *Actualidades*, 49–50.

36. Nuñez was particularly equipped to do so, having served as medical inspector and director of the Juárez Hospital (1886–1891), where he established the orthopedic clinic and outpatient services. Cavazos Guzmán and Carrillo Arriaga, *Historia y evolución de la medicina*, 77.

37. *Gaceta Médica de México*, December 1, 1906.

38. *El Popular*, April 18, 1900. The newspaper also referred to a "human meatball." *El Popular*, April 1, 1901. For an analysis of salvavidas and the use of humor, see Montaño, "*Machucados* and *Salvavidas*."

39. Reference to blood was implicit in this description: the meatball soup recipe called for simmering the albóndigas in a seasoned tomato broth.

40. *El Popular*, May 13, 1900.

41. Ramos y Duarte, *Diccionario de mejicanismos*, 280; *El Popular*, August 27, 1903.

42. *El Popular*, August 27, 1903.

43. Pilcher, *The Sausage Rebellion*, 62. The verb *sacrificar* (to sacrifice or to slaughter) provided the colorful language to describe the sacrifice of capitalinos to "Electric Civilization." *El Popular*, June 18, 1903.

44. Del Castillo Troncoso, "El surgimiento de la prensa," 108.

45. Changes to content included new advertising strategies, reduced treatment of political topics along with the consolidation of the interview, and reporting as preferred techniques to cover social issues. The imagined audience was "nonspecialized

readers, passive in political terms and much more interested in conjugal tragedies and contemporary police reports than in well versed and sensible doctrinaire and political reflections." Ibid., 111.

46. Mexico participated in a sensational visual culture that commodified accidental death and other forms of violence. Public spheres were forged in both Mexico and Brazil predicated on a "collective consumption of spectacles of violence." Navitski, *Public Spectacles*, 12. This culture, she argues, resulted from the cross-fertilization of explicit and rousing descriptions of injury and bodily peril across media.

47. George Beard's work on nervousness and Max Nordau's on degeneration identified newspapers as part of the stimuli affecting individuals. Shaya, "Mayhem for Moderns," 55 (on the press and the crowd in France).

48. Wynne, "Unruly Technology," 149–150.

49. *El Imparcial*, August 14, 1899; *El Imparcial*, November 19, 1899.

50. *El Imparcial*, October 3, 1899.

51. Parra, *Pacotillas*, 17.

52. Moses, *Today in the Land*, 26–27.

53. Quoted in Borg, "The 'Chauffeur Problem,'" 800.

54. *El Mundo Ilustrado*, November 26, 1899.

55. *El Imparcial*, April 7, 1902.

56. *El Correo Español*, August 15, 1902.

57. *El Diario*, January 13, 1911. Francisco Amezcua decided to get off after a driver declined to stop in front of his house. He lost his balance and fell to the ground, suffering several injuries. Amezcua went on to file a complaint at the precinct against the driver.

58. *Cómico*, May 26, 1901.

59. Matthews, *The Civilizing Machine*, 219–220.

60. Borg, "The 'Chauffeur Problem,'" 798.

61. *Electrician*, October 16, 1903.

62. *Western Electrician*, June 8, 1901.

63. *Ki-ki-ri-kí*, September 13, 1903 (emphasis in the original).

64. Archivo General de la Nación, Secretaria de Comunicaciones y Obras Publicas (hereafter cited as AGN/SCOP), File 3/214-1, Year 1877.

65. *El País*, March 31, 1900.

66. *El País*, October 27, 1900.

67. *Mexican Herald*, March 4, 1900.

68. *El Imparcial*, May 7, 1902.

69. *El Imparcial*, January 16, 1901; August 1, 1901; September 22, 1901; November 18, 1902.

70. *El País*, February, 9, 1900; *El Imparcial*, May 7, 1903; *El País*, January 4, 1904.

71. Muñoz-Serrano, *Actualidades*, 63–64.

72. *Cómico*, May 5, 1901.

73. Muñoz-Serrano, *Actualidades*, 78–79. A 1910 article held that the first idea after boarding an eléctrico was death. *El País*, January 29, 1910.

74. *Cómico*, April 7, 1901.

75. *Cómico*, February 4, 1900.

76. Gamboa, *La llaga*, 377.

77. Shaya argues in "Mayhem for Moderns" that the formation of crowds in France legitimated the attention to grisly sights and causes célèbres, making them a prominent and profitable category of news.

78. *La Patria*, November 26, 1902.

79. *El Imparcial*, November 25, 1902.

80. *La Patria*, November 26, 1902.

81. *El País*, April 17, 1900.

82. *El Imparcial*, October 30, 1901.

83. Published by *El Imparcial*, March 14, 1901; and *La Voz de México*, March 15, 1901.

84. This was also true elsewhere. The *Electrical Review*, writing about the United States on December 22, 1893, commented: "[I]it seems inexplicable to the public at large that the mystery about electricity is not dispelled." I found no campaign to educate the population about the dangers of electrical wires in the archives or newspaper collections. Such a campaign was not unheard of. The *Atlanta Constitution*, for instance, ran the article "The Deadly Wires: How and Why You Must Avoid Them" to educate those who valued their lives about the dangers of telephone, telegraph, and electric wires around the city on March 4, 1889.

85. *Electrical Review*, December 22, 1893.

86. *El Popular*, June 14, 1897.

87. *La Patria*, July 2, 1897.

88. *El Tiempo*, November 4, 1900.

89. *El Imparcial*, November 3, 1900.

90. *El Popular*, September 26, 1900.

91. *Gaceta Médica*, January 1, 1905.

92. Barbosa, *El trabajo en las calles*, 57.

93. Godoy, *When I Was a Girl*, 35, 61.

94. *Electric Railway Journal* (May 1, 1909): 813–815.

95. Tweedie, *Mexico As I Saw It*, 208.

96. *Cómico*, April 16, 1899.

97. *Cómico*, April 15, 1900.

98. *Cómico*, April 7, 1901.

99. Piccato, "Urbanistas, Ambulantes, and Mendigos."

100. Barbosa, *El trabajo en las calles*, 78.

101. Porter, "'And That It Is Custom,'" 112.

102. Barbosa, *El trabajo en las calles*, 81.

103. Porter, "'And That It Is Custom,'" 115.

104. Díaz y de Ovando, "La Ciudad de México," 128–129.

105. *El Popular*, March 3, 1903.

106. Ibid.

107. This emphasis on distracted, undisciplined, and "unproductive" walking reflected new theories of culture and modern life. Attentiveness emerged as a scientific object and social problem in the 1870s, reflecting the belief that urban life required learning to exclude sensual stimuli. The German sociologist Georg Simmel argued in "The Metropolis and Mental Life" that learning to buffer oneself from a continually changing environment would produce more rational urban dwellers. The physician and social critic Max Nordau associated failure of attentiveness with sociopathic behavior. The problem of attention arose within a new capitalist economic system: "through the new imperatives of attentiveness . . . , the perceiving body [of urban dwellers] was deployed and made productive and orderly." Crary, *Suspensions of Perception*, 16.

108. *Gaceta de Policía*, November 12, 1905.

109. *El Tiempo*, November 5, 1905.

110. Cited in Barbosa, *El trabajo en las calles*, 54.

111. *Diario Oficial*, February 24, 1900.

112. *El Imparcial*, August 7, 1901.

113. Ibid.

114. For advertising technology. see Bunker, *Creating Mexican Consumer Culture*, 30–48.

115. Camacho Morfín, *Las historietas de El Buen Tono*, 65–55, which includes a good discussion of industrial lithography.

116. *Country Life*, December 7, 1901.

117. *El Mundo Ilustrado*, July 24, 1904.

118. Years earlier, an article titled "An Imitator of Don Tancredo" narrated the audacity of a man who was dared by friends to stand in the eléctrico's path to force it to stop. *El Imparcial*, April 9, 1901.

119. "Hacer el don Tancredo" or *dontancredismo* became a metaphor for immobility and was used colloquially to describe someone unable to face problems, unmoved by changes, or imperturbable in the face of a threat or problem. The connotation was always negative, underlining the individual's inability to react. In that period reaction time was one of the behaviors that attention lab studies identified as both measurable and externally observable.

120. *El Mundo Ilustrado*, October 9, 1904.

121. For rural society and technology, see Beezley, *Judas at the Jockey Club*, 67–88. For the theme of alertness among American cartoonists, see Bukatman, *The Poetics of Slumberland*.

122. Bunker, *Creating Mexican Consumer Culture*, 44–45.

123. Muñoz-Serrano, *Actualidades*, 19–20.

124. Justo Guerrero drank away his love troubles at a cantina before concluding that life was no longer worth living. He decided to jump in front of a Tacubaya eléctrico. The driver managed to stop the motorcar, so Guerrero incurred only minor injuries. *El Imparcial*, January 6, 1902. Another suicide attempt was frustrated when a young man, after discovering that his brother had poisoned himself, jumped on the rails as an eléctrico was approaching. *El Diario*, August 7, 1911.

125. Archivo General de la Nación/Tribunal Superior de Justicia del Distrito Federal (hereafter cited as AGN/TSJDF), Box 118, File 2-703.

126. Nervo, *El libro*, 145–146.

127. *El Imparcial*, July 3, 1903.

128. Schivelbusch analyzes the railroad's effect on passengers' bodies, particular how travel transformed perception, in *The Railway Journey*.

129. *El Mundo Ilustrado*, January 17, 1904.

130. Ibid.

131. For efforts of the working classes to distinguish themselves from other sectors of the lower classes, see Buffington, *A Sentimental Education*.

132. *El Imparcial*, February 15, 1903.

133. For the work of physicians, hygienists, and sanitary engineers, see Agostoni's chapter "The Control of the Environment," in *Monuments of Progress*. Carlos Roumagnac, a journalist-turned-police-officer-turned-criminologist, for instance, undertook empirical studies in the form of interviews with inmates in Porfirian jails. He went on to publish *Los criminales de México* and *Crímenes sexuales y pasionales*. Piccato, *City of Suspects*, 62.

134. Buffington, *Criminal and Citizen*, 75.

135. *El Imparcial*, May 18, 1902.

136. *El Imparcial*, December 3, 1902.

137. *El Diario*, January 20, 1908.

138. Antonio Vanegas Arroyo, *Canciones modernas 1906* (Mexico City: Editor Vanegas Arroyo, 1906), 13.

139. *El Imparcial*, December 20, 1905.

140. *El País*, October 28, 1901.

141. "Prohibición para que los empleados de los Ferrocarriles del Distrito Federal en servicio penetren a los expendios de bebidas embriagantes," *México*, May 15, 1904.

142. Dublán and Lozano, *Legislación mexicana*, 331. The decree (12,762) was issued on September 29, 1894.

143. Article 21, section II.

144. Pulido Esteva, *¡A su salud!*, 11.

145. Piccato, *City of Suspects*, 30.

146. *El Popular*, June 18, 1903.

147. *El Imparcial*, January 7, 1908.

148. AHDF, Fondo: Ayuntamiento, Sección: Gobierno del Distrito Federal, 1738/987/1910–11.

149. *El Imparcial*, September 27, 1902.

150. Ibid.

151. Cited in Rodríguez, *Los tranviarios*, 130.

152. *El Imparcial*, November 4, 1902.

153. AGN/SCOP, 3/250-1.

154. *El Universal*, October 5, 1901.

155. AGN/TSJDF, Box 987, File 172713, Year 1910.

156. Based on article 322.

157. AGN/TSJDF, Box 545, File 95393, Year 1906.

158. AGN/TSJDF, Box 1039, File 184510, Year 1911.

159. AGN/Compañía Luz y Fuerza del Centro, Box 1041, File 1/501–3a, March 19, 1907.

160. Ibid.

161. AGN/Compañía Luz y Fuerza del Centro, Box 1041, File 1/501-3c10, June 1907, 503-21-22.

162. *El Imparcial*, March 19, 1903.

163. *El Imparcial*, September 11, 1903 (emphasis added).

164. Ibid.

165. *Cómico*, August 18, 1901.

166. *El Imparcial*, October 30, 1901.

167. Ibid.

CHAPTER 4. *LADRONES DE LUZ*

1. *El Imparcial*, December 19, 1901.

2. Details from *Diario de Jurisprudencia del Distrito y Territorios Federales*, January 20 and 21, 1904.

3. Lipsett-Rivera underlines the value of these minutiae in *Gender and the Negotiation of Daily Life*.

4. Piccato, *City of Suspects*, 3.

5. Findlay, "Courtroom Tales of Sex and Honor," 201.

6. Levine, "Michel de Certeau and Latin America," 317.

7. Pretel, Inkster, and Wendt, "Technology in Latin American History," 1, 7.

8. Mexican cultural historians have explored elite visions of criminality and their efforts to police public behavior. For example, see Buffington, *Criminal and Citizen*; Piccato, *City of Suspects*; Speckman-Guerra, *Crimen y castigo*; Garza, *The Imagined Underworld*; Pérez Montfort, del Castillo Yurrita, and Piccato, *Hábitos, normas y escándalo*; and Pulido Esteva, *¡A su salud!* For the region, see Johnson, *The Problem of Order*; Salvatore and Aguirre, *The Birth of the Penitentiary*; Aguirre and Buffington, *Reconstructing Criminality*; Caufield, *In Defense of Honor*; Salvatore, Aguirre, and Joseph, *Crime and Punishment*; and Ruggiero, *Modernity in the Flesh*.

9. See, in particular, Lear, *Workers, Neighbors, and Citizens*; Porter, *Working Women in Mexico City*; and Piccato, *City of Suspects* and "Urbanistas, Ambulantes, and Mendigos."

10. For technology's impingement on banditry, see Vanderwood, *Disorder and Progress*, 88–90. For technology's role in the rise and policing of transnational crime, see Galeano, *Criminosos viajantes*.

11. Weiner, *Race, Nation, and Market*, 56–59 (quotation on 59).

12. *El Imparcial*, December 27, 1907.

13. This was a response to the immediate need to find industrial contracts for thousands of HP and to flood the market with cheaper electricity, to pressure independent companies into selling their properties and concessions. The unprecedented concession was valid for ninety-seven years. Galarza, *La industria eléctrica en México*, 121. For an ad, see *El Tiempo*, February 26, 1904.

14. *El Popular*, April 5, 1908.

15. *El Imparcial*, November 4, 1906.

16. For the increased frequency and intensity of fire hazards, see Alexander, *City on Fire*.

17. *El Imparcial*, December 19, 1901.

18. *Nueva Era*, June 6, 1912.

19. AGN/TSJDF, Box 1118, File 198038, Year 1912.

20. AGN/TSJDF, Box 1297, File 226252, Year 1915.

21. *El Diario*, August 24, 1913.

22. AGN/TSJDF, Box 1254, File 218874, Year 1914.

23. The company's lawyer withdrew the accusation, stating that the Compañía Molinera indeed had a "special contract" for several addresses and that someone had improperly documented the connection.

24. *Diario de Jurisprudencia*, August 13, 1907.

25. Five years later, Mexlight would further strengthen its legal position by adding a twelfth clause, stipulating "unconditional consent and authorization for company employees to have access to the house or places where the subscriber uses electric energy." The company understood how essential inspections would be to its profitability. AGN/TSJDF, Box 1143, File 203598, Year 1912.

26. Lipsett-Rivera, *Gender and the Negotiation of Daily Life*, 24–25.

27. See Kuznesof, "Gender, Ideology, Race"; and Arrom, *The Women of Mexico City*.

28. Lipsett-Rivera, *Gender and the Negotiation of Daily Life*, 13–14.

29. Lipsett-Rivera cites studies that recognize women as defenders of their honor and their households. In the case of *vendedoras* (market women) who occupied public spaces out of economic necessity, their conception of honor was not divorced from the binary spatial division but instead built on it, defying their characterization as immoral based on their publicness. See Porter, "'And That It Is Custom.'"

30. The angst experienced by women who received men into their homes without the patriarch's authorization was perfectly captured as late as 1949 in the film *Una familia de tantas* (A family like many others), in which a middle-class family allows a male salesperson to demonstrate the usefulness of an electric vacuum.

31. AGN/TSJDF, Box 1507, File 269076, Year 1918. Montaño denied entrance to the gendarmes and delayed the entry of the inspectors, while allegedly ordering a servant to cut the fraudulent installation.

32. Lipsett-Rivera, *Gender and the Negotiation of Daily Life*, 38. For hierarchies of housing in the nineteenth century, see 49–58.

33. AGN/TSJDF, Box 1591, File 285122, Year 1920.

34. *El Imparcial*, December 27, 1907.

35. As in the case of the Ambos Rumbos Hotel, business and residential owners were often unwilling to pay a company for installation. Instead, they installed the wires themselves. This was only illegal when tapping into a power line on the street without a contract. See AGN/TSJDF, Box 381, File 67615, Year 1905 and AGN/TSJDF, Box 1469, File 260451, Year 1918.

36. *El País*, March 4, 1914. In another case, a commercial house that closed at 10 P.M. installed a cable to prevent recording overnight usage. *Nueva Era*, January 31, 1913.

37. *El Imparcial*, April 27, 1906.

38. AGN/TSJDF, Box 1143, File 203598, Year 1912.

39. *El Tiempo*, July 4, 1912. Vázquez Suárez also claimed that he had no need for such a device because artificial lighting was used for half an hour every night when he and his wife returned home. A new inspection would prove that the building had no commercial equipment or any bathroom, kitchen, irons, or electric heaters that demanded electricity.

40. Lipsett-Rivera, *Gender and the Negotiation of Daily Life*, 38.

41. AGN/TSJDF, Box 495, File 087332, Year 1906.

42. The recognized penologist Demetrio Sodi asserted that "servants, as well as workers and artisans, are generally thieves." Sodi, *Nuestra ley penal*, 8.

43. Macedo, *La criminalidad en México*, 28–29.

44. *El Diario*, December 15, 1911.

45. *El Demócrata*, October 27, 1924.

46. Morgan, "Proletarians, Politicos, and Patriarchs." Water-powered textile factories, for instance, had remained dispersed on the outskirts of the Federal District in the towns of Tlalpan, Contreras, and San Ángel, Santa Fe, Chalco, and Tlalnepantla until the 1870s. Electric energy undercut former sources of energy, altering hydraulic power dependency. For hydraulic sources used for industrial purposes, see Galarza, *La industria eléctrica*, 56–60.

47. Ibid. Although the emergence of modern factories made inroads in some sectors, these remained numerically unimportant in terms of labor, representing a mere 4 percent of the city's workforce in 1910.

48. Ibid., 62–66.

49. Garza, *El proceso de industrialización*, 120. Six textile factories were located in the Federal District in 1877, employing approximately 900 men, 300 women, and 220 children. The workforce doubled by 1910, while production increased fivefold. Lear, *Workers, Neighbors, and Citizens*, 59.

50. Best, *Noticia sobre las aplicaciones.*

51. A 1904 regulation established that match factories were to use only incandescent electric lighting.

52. Becerril, *Las fábricas de San Antonio Abad*, 95. A report of a visit by students three years later noted that electricity was being introduced as motive power, but steam machinery was dedicated to whitewashing and printing.

53. *El Imparcial*, April 26, 1903.

54. *El Tiempo*, April 26, 1903; *El Imparcial*, April 26, 1903.

55. Lear, *Workers, Neighbors, and Citizens*, 55–58. Vendors, the second-largest census category after domestic service, increased by 58 percent, going from 19,000 in 1895 to 30,000 by the end of the regime. For foreign merchants, see Buchenau, *Tools of Progress*; Schell, *Integral Outsiders*; Collado, *La burguesía mexicana*; and Bunker, *Creating Mexican Consumer Culture*.

56. *Nueva Era*, November 21, 1911.

57. Código Penal para el Distrito Federal, article 368.

58. *El Imparcial*, July 4, 1904.

59. Pallares authored the study "El robo de electricidad," reproduced in Sodi, *Nuestra ley penal*, 19–38. Sources do not provide its exact date, but it was probably in 1903, when Pallares led Torres's defense, or shortly after that.

60. *El Imparcial*, July 4, 1904.

61. Reprinted in *Diario de Jurisprudencia*, January 20 and 21, 1904.

62. *Diario de Jurisprudencia del Distrito y Territorios Federales*, January 1, 1904.

63. Quoted in Sodi, "Nuestra ley penal," 28.

64. Ibid. 13–42.

65. Ruling reprinted in *Diario de Jurisprudencia del Distrito y Territorios Federales*, January 25 and 26, 1904.

66. *Diario de Jurisprudencia*, August 11, 1904.

67. *Diario de Jurisprudencia*, August 8, 1904.

68. *Diario de Jurisprudencia*, July 20, 1904 (emphasis added).

69. *Diario de Jurisprudencia*, August 1, 1904.

70. *Diario de Jurisprudencia*, August 10, 1904.

71. *Diario de Jurisprudencia*, September 17, 1904.

72. *Diario de Jurisprudencia*, September 1, 1904. *El País* on April 23 and September 19, 1906, welcomed the clarification, maintaining that in the past criminals had received light punishment or had escaped it entirely.

73. AGN-TSJDF, Box 565, File 102030, Year 1906.

74. *Boletín de la Secretaría de Fomento*, October 1907. At the time of the report, the tannery was working far below capacity, with about fifty cowhides and thirty dozen sheep, goat, and calf skins.

75. It was also known as the Fratelli Durio process after the Italian brothers who perfected it.

76. Haber, *Industry and Underdevelopment*, 79–80. Haber argues that "the merchant-financier-industrialist nexus excluded the Mexican-born from industry," as foreigners had historically dominated large-scale commerce.

77. *Industria é Invenciones*, October 8, 1898.

78. Davis, *The Manufacture of Leather*, 300.

79. Herrero Bervera, *Los empresarios mexicanos*, 145.

80. Dr. I. Maschke, "A Review of Technical Progress," *Leather Manufacturer*, March 1898, 46.

81. AGN/TSJDF, Box 565, File 102030, Year 1906.

82. Ibid.

83. Ibid.

84. Granados had likely contracted tuberculosis at the Belem jail. He was released on a 500-peso bond for the duration of his trial. The jail suffered from overpopulation, creating a tuberculosis epidemic.

85. *El Correo Español*, June 29, 1906. The note did not specify the cause of death.

86. AGN-TSJDF, Box 491, File 087193, Year 1906.

87. Such was the case of Damaso de la Concha, owner of La Villa de Arriendas and Las Montañas de Solórzano. AGN/TSJDF, Box 506, File 088733, Year 1906; *El País*, October 16, 1906.

88. *El Popular* followed del Villar's case closely: see April 23, 24, and 27, 1906.

89. *El País*, April 23, 1906.

90. AGN/TSJDF, Box 491, File 087193.

91. *Memoria de la Secretaría de Gobernación (1900–1904)*, 448.

92. AGN/TSJDF, Box 746, File 131627, Year 1908.

93. The court initiated an inquiry into the molinos in the area and found that three molinos had opened in the last two years.

94. Ibid., 295.

95. See Keremitsis, "Del metate al molino," 286.

96. Lear, *Workers, Neighbors, and Citizens*, 65. This trend continued as military hostilities came to an end in the next decade: by 1923 the company claimed 119 of the 147 mills in operation. Keremitsis, "Del metate al molino," 292–293. A handful of merchants also controlled flourmills. For a study of bread, see Weis, *Bakers and Basques*.

97. *El País*, November 4, 1908.

98. AGN/TSJDF, Box 746, File 131627, Year 1908.

99. AGN/TSJDF, Box 643, File 113625, Year 1907.

100. No additional information on the resolution of the case is available.

101. AGN/TSJDF, Box 643, File 113625, Year 1907.

102. *Nueva Era*, March 30, 1912.

103. AGN/TSJDF, Box 1401, File 247414, Year 1917.

104. AGN/TSJDF, Box 1370, File 242373, Year 1916 (emphasis added).

105. *El País*, July 4, 1903.

106. AGN/TSJDF, Box 1185, File 207147, Year 1913.

107. Emiliano Zapata and his troops arrived at the outskirts of Mexico City in February 1913.

108. Details from *Diario de Jurisprudencia del Distrito y Territorios Federales*, August 24, 1905.

109. Ibid.

110. Vera challenged the sentence, pointing out the court's failure to provide him with legal representation and that he did not confess to carrying out the installation; instead, he helped El Chato to do so. The judge commuted the sentence to three months of jail time.

111. *El País*, August 13, 1912.

112. AGN/TSJDF, Box 491, File 087193, Year 1906.
113. *El País*, October 16, 1906.
114. *El Demócrata*, October 27, 1924.
115. Piccato, *City of Suspects*, 3.
116. Findlay, "Courtroom Tales of Sex and Honor," 201.
117. Levine, "Michel de Certeau and Latin America," 317.

CHAPTER 5. BECOMING *ELECTRO-DOMÉSTICAS*

1. As this chapter reveals, the world of appliances in Mexico was gendered in a way that defies spelling. The word for electrical appliances in Spanish is *electrodomésticos* (male). The word for female maids in Spanish is *domésticas*. *Electro-domésticas* is a play on the fusion of these words.

2. Aguilar de Carbia, *Almanaque 1936*, 204–206.

3. Pilcher, *¡Que Vivan Los Tamales!*, 130. By 1934 *Marichu va a la cocina* was in its second edition.

4. A metate is a flat stone for grinding, while a *molcajete* is a grinding stone bowl.

5. Aguilar de Carbia, *Almanaque 1936*, 295-296.

6. Novo, *Cocina mexicana*, 141–142.

7. Lacking full citizenship rights, women could still participate in nation-building efforts and simultaneously uphold a modernized patriarchal society.

8. "The discourse of tradition . . . offered a moralistic response to modernity that included (yet was not limited to) the world it described as traditional." Rubenstein, *Bad Language*, 84.

9. As a location where class, gender, consumption, nationalism, and technology interact daily, the home provides a unique window into how individuals experience, define, interpret, and make sense of their lives and self-image. Strangely, however, the intersection between domestic space and modernity has not received due attention. Histories of household technology have demonstrated how transformations of domestic gadgetry followed broader technological advancements in lockstep, but histories of the physical and mechanical transformation of domestic spaces are still few. Cowan, *More Work for Mother*; Pilcher, *¡Que Vivan Los Tamales!* and *The Sausage Rebellion*. In the few studies of Latin American households, domestic technology has hardly been a priority. Owensby, *Intimate Ironies*; French and James, *The Gendered Worlds*; Hershfield, *Imagining La Chica Moderna* and "Domestic Technologies"; an exception is French-Fuller, "Gendered Invisibility." For commercial technology, see Pilcher, "Fajitas and the Failure." Scholarship on women has centered on the factory and the anxieties provoked by women joining the workforce, but some studies are now interrogating the changing ideals of womanhood. Sanders, *Gender and Welfare in Mexico*; Sluis, *Deco Body*; Blum, "Cleaning the Revolutionary Household"; Farnsworth-Alvear, *Dulcinea in the Factory*; James, *Doña María's Story*; Porter, *Working Women*; Olcott, Vaughan, and Cano, *Sex in Revolution*; Wolfe, *Working Women*.

10. *VIII censo general de población 1960* (Mexico City: Secretaría de Industria y Comercio, 1963).

11. *IX censo general de población 1970* (Mexico City: Secretaría de Industria y Comercio, 1973).

12. Previous censuses did not record this. Also, while smaller and cheaper items experienced faster diffusion across the city, expensive larger machines, which required access to reliable water and electric networks and were costly to run, remained out of reach for most until the late century. Another factor to take into account is desirability. French-Fuller, in "Gendered Invisibility," found that, in prioritizing purchases of consumer durables, families in Santiago favored items that improved the family's leisure time (radio, TV) over those that specifically facilitated the housewife's work.

13. The introduction of electric lighting into the elite's domestic spaces in the nineteenth century "scarcely succeed in dissipating the tinieblas . . . [but] among the poor, the darkness was more uniform." Briseño Senosiain, *Candil de la calle*, 186.

14. *El Imparcial*, October 7, 1910. Gas and oil companies marketed the advantages of their own illuminant fuels. El Águila Mexican Petroleum Company, for instance, ran an ad for its product: "Smokeless and odorless, it illuminates splendidly. Elegant families preferred it." *El Diario*, August 4, 1912.

15. Briseño Senosiain, *Candil de la calle*, 142. She examines the attention that novelists dedicated to the ways in which artificial lighting altered life and experience in domestic spaces. See her chapter "Oscuridad de su casa."

16. *El País*, May 8, 1910.

17. *El Imparcial*, October 7, 1910.

18. "We give expert service free": *Mexican Herald*, June 30, 1912.

19. *Mexican Herald*, September 1, 1912.

20. *El País*, September 4, 1910.

21. *El Imparcial*, October 2, 1913.

22. *El Imparcial*, January 12, 1911.

23. Francois, "La lavandería," 40. For the everyday economic and social significance of women's clothing, see Francois, *A Culture of Everyday Credit*.

24. *El País*, January 19, 1911.

25. *El País*, March 5, 1908. An article on washing and ironing advised owning between four and six irons.

26. *El Imparcial*, December 29, 1910.

27. *El País*, January 19, 1911.

28. *El Imparcial*, November 16, 1911.

29. Velázquez de León, *Manual práctico de cocina*, 127.

30. *El Diario*, July 21, 1912.

31. *El País*, December 29, 1910.

32. *El Imparcial*, January 19, 1911.

33. *Mexican Herald*, October 23, 1912.

34. *El Imparcial*, March 29, 1912.

35. *Mexican Herald*, February 8, 1914.

36. *Mexican Herald*, August 4, 1912. Indeed, the ad claimed that it was cheaper to heat by electricity than by any other method. A cent's worth of electricity, at a rate of ten cents per kilowatt, would operate an electric grill for six minutes, a radiator for eight minutes, an electric sewing machine motor for three hours, and an electric iron for five minutes.

37. *El Diario*, July 14, 1912. For example, a housewife could boil two liters of water, cook a dish in a pot, make four cups of coffee in an electric coffee percolator, and have a hot stove in four to fifteen minutes.

38. *Mexican Herald*, April 14, 1912.

39. For ads promoting this service, see D. M. V. de R., *Manual de cocina práctica* (unpaged).

40. *El Diario*, August 25, 1912.

41. *Idaho Daily Statesman*, June 16, 1910; *El Imparcial*, November 16, 1911.

42. Juárez López, *La cocina también es laberinto*, 12.

43. For Velázquez de León's background and her culinary patriotism, see Pilcher, "Josefina Velázquez de León."

44. Velázquez de León, *Manual práctico de cocina*.

45. Vaughan, "Modernizing Patriarchy," 194. Without fundamentally altering the society's patriarchal backbone, the state inserted the government as the paternal figure within the family. Sanders, *Gender and Welfare*. For similar findings in Brazil, see Besse, *Restructuring Patriarchy*; for an argument on the maternalistic nature of those arrangements, see Pite, "Entertaining Inequalities." In incorporating modern and traditional ideas of womanhood, "women simultaneously situated themselves as agents of progressive change and embodiments of cultural traditions." Olcott, *Revolutionary Women*, 24.

46. De Grazia, *How Fascism Ruled*.

47. Loyo, "De sierva a compañera," 161–162. For similar work by reformers in the Andes, see Larson, "Capturing Indian Bodies." For the reverence for motherhood of feminists in the region, see Lavrin, *Women, Feminism, and Social Change*; Stoner, *From the House to the Streets*.

48. Aguilar-Rodríguez, "Cooking Modernity."

49. Juárez López, *La cocina también es laberinto*, 17–41.

50. Hernández, *Libro social y familiar*. In 1935 she was part of a group of prominent activists who founded the Frente Único Pro Derechos de la Mujer (FUPDM; United Front for Women's Rights) to bring together active female members of the Mexican Communist Party and the Partido Nacional Revolucionario, the official party.

51. Peronist Argentina called upon housewives to revive the national economy through careful household consumption. See Milanesio, "The Guardian Angels."

52. Moreno, *Yankee Don't Go Home*, 146–150.

53. Porter, *From Angel to Office Worker*, 72–86, 199.

54. For the work of middle-class female reformers, see Sanders, *Gender and Welfare*.

55. For a thorough discussion of theaters and the destabilization of gender roles, see Sluis, *Deco Body*.

56. Cited in Loyo, "De sierva a compañera," 164.

57. Velázquez de León, *El Arte de Cocinar* 8 (March 1944): 1.

58. Juárez López, *La cocina también es laberinto*, 11.

59. Keratol is a waterproof material used in bookbinding.

60. Back cover, Velázquez de León, *La cocina de la recién casada*.

61. Velázquez de León, *Manual práctico de cocina y repostería*, 8.

62. Velázquez de León, *La cocina de la recién casada*, 98.

63. Ibid., 3.

64. Pilcher, "Josefina Velázquez de León," 202.

65. Velázquez de León, *El Arte de Cocinar* 17 (December 1944): back cover.

66. The Compañía Exportadora e Importadora Mexicana, SA (CEIMSA), the State Food Agency, collaborated with Velázquez de León to publish fourteen modest recipe pamphlets, each containing thirty recipes for favorite dishes. The agency later compiled the recipes into a larger publication titled *420 recetas de platillos populares mexicanos* (420 recipes for popular Mexican dishes). It claimed to have distributed 1.6 million copies free of charge in working-class neighborhoods. For an analysis of the State Food Agency, see Ochoa, *Feeding Mexico*.

67. Velázquez de León also held free demonstration classes (such as modern cake decoration, artistic gelatins, and novel pastries) every Wednesday. The items prepared during the class were raffled among attendees at the end of class. Invitations were made available to anyone who acquired an issue of *Arte de Cocinar* and those who spent one peso on Oro products bought in the school's store.

68. Velázquez de León, *Fuertomate en la cocina*.

69. Nestlé published cookbooks with a selection of the thousands of recipes submitted by women throughout the country, which were tested in the Sala Científica Nestlé (Nestlé Scientific Showroom) and judged by a committee of expert judges. Among those who participated in the several contests were Lupe H. de Muñoz de Cote, graduate of the Philadelphia School of Home Economics and director of the Instituto de Repostería Nestlé; Carmen Ramírez del Corro and Clementina Cerrillo, instructors at the Escuela Profesional de Trabajadoras Sociales y de Enseñanza Doméstica; Concepción G. de Kapmann, a General Electric home economics expert; and Miguel Garnica Espinosa, chef at Paolo restaurant, who had cooked for all the Mexican presidents. See, for example, Nestlé Mexico, *Postres Nestlé: Una maravilla de la repostería moderna* (Mexico City: Productos Nestlé, ca. 1930).

70. Velázquez de León, *Cocina*, 5. A similar cookbook was her *Cocina al minuto* (Cooking by the minute).

71. Velázquez de León, *Cocina instantánea*, dedication page, 7.

72. Although the first edition is not dated, in 1950 the second edition was published. Engineer and entrepreneur Luis Álvarez had been hard at work since the 1910s manufacturing and promoting his "Ideal Electrical Oven," via demonstrations in the Mercado de Toluca, in the state of Mexico. He later published two cookbooks. According to his grandson, Fernando Álvarez, although the books were published under his grandfather's name, his grandmother was the one who compiled and tested the recipes.

73. Velázquez de León, *Como cocinar en los aparatos modernos*, 5.

74. Ibid., dedication page.

75. Ibid., 32.

76. *El Arte de Cocinar* 46 (September 1949): 10.

77. Velázquez de León, *Como cocinar en los aparatos modernos*, 123–124.

78. Pilcher, "Josefina Velázquez de León," 206.

79. Velázquez de León, *Como cocinar en los aparatos modernos*, 272.

80. For culinary nationalism, see Pilcher, *¡Que Vivan Los Tamales!*; and Juárez López, *Nacionalismo culinario.*

81. For the widespread practice of pulque consumption, see Pulido Esteva, *¡A su salud!* For hygienic campaigns directed at the peasant family, see Vaughan, "Rural Women's Literacy." For those directed at the urban working-class family, see Sanders, *Gender and Welfare*; and Aguilar-Rodríguez, "Cooking Modernity."

82. Velázquez de León, *Como cocinar en los aparatos modernos*, 284.

83. López, *Crafting Mexico*, 9, 37.

84. An ad from Glasbake promoting crystal ovenware items as the "new monarchs of the kitchen and dining room" echoes this narrative. Various anthropomorphized crystal dishes energetically march into the kitchen, while chipped earthenware pots and teapots flee in the background. Their time has passed. This appealing crystal ovenware satisfies the female consumer's material needs through speed and economy and also her emotional needs: its beauty is a source of pride.

85. Heneghan, *Whitewashing America*, 7. Everyday household goods in the antebellum American imagination also served as a means of self-definition and national definition: "their ownership of [white things] and proper management constituted racial whiteness, and skin became simply another white thing" (7).

86. Silverstone and Haddon, "Design and the Domestication of Information," 60.

87. Oudshorn and Pinch, "Introduction," 14.

88. Cited in ibid.

89. Ibid.

90. Levenstein, "The 'Servant Problem,'" 133–134.

91. Frost, "Machine Liberation," 113.

92. Scott, *The Market Makers*, 2.

93. Frost in his article on home appliances in interwar France argues that the supposed French servant shortage, like its solution, "was probably as much a rhetorical device as a concrete reality." Frost, "Machine Liberation," 114.

94. Ironically, this remains an understudied area. Much of the work took place behind closed doors and its nature, patterns, and changes seldom entered traditional sources, which undermined its prominence as a field of study. Chaney and García Castro's *Muchachas No More: Household Workers in Latin America and the Caribbean* stands as an unsurpassed volume on the subject, despite having been written over two decades ago. Study of the Mexican case has greatly benefited from the creative mining of sources and the close inspection of traditional sources. See Arrom, *The Women of*

Mexico City; Blum, "Cleaning the Revolutionary Household" and *Domestic Economies*; and Francois, *A Culture of Everyday Credit* and "La lavandería."

95. Kuznesof, "A History of Domestic Service," 22.

96. Arrom, *The Women of Mexico City*, 187. Mexican constitutions and legislation in the 1800s legitimized the patriarchal family structure, further limiting women's economic and commercial opportunities. Housework was a construction, a product of the "simultaneous legal, economic, and cultural subordination of women and non-whites under patriarchal colonial and liberal regimes that devalued housework despite depending on it." Francois, "The Products of Consumption," 209. For a regional assessment of the significant changes during the nineteenth century, see Dore, "One Step Forward."

97. Arrom, *The Women of Mexico City*, 183–189, 69.

98. Arrom, *Containing the Poor*.

99. Blum, "Cleaning the Revolutionary Household," 71–72.

100. Blum, *Domestic Economies*, 48. Furthermore, the civil codes of 1870 and 1884 gave employers "the same rights as guardians in disciplining their underage domestic servants" (49).

101. *La Voz de México*, May 24, 1906.

102. For the conceptualization of housewives as angels during the Porfiriato, see French, "Prostitutes and Guardian Angels."

103. See Pilcher's chapter "The Tortilla Discourse" in *¡Que Vivan Los Tamales!*

104. González Navarro, "El porfiriato: la vida social," cited in Goldsmith, "Politics and Programs," 241.

105. *El Imparcial*, July 17, 1910.

106. Blum, "Cleaning the Revolutionary Household," 80. Goldsmith emphasizes in "Politics and Programs" that chapter 13 of the Federal Labor Law, which describes the rights and obligations of domestics, supports the employers' viewpoint instead of the workers'.

107. Blum, "Cleaning the Revolutionary Household," 76. As wards of the state, domestic workers would have been seen as being incapable of performing difficult work not only as "minors" per se but as indigenous people. The colonial legal code also designated indigenous peoples as minors.

108. Goldsmith, "Politics and Programs," n. 6, 240. The Supreme Court supported the position that those engaged in the domestic sector were not entitled to the same rights as other workers until 2019. This ruling, the social view of housework as menial unskilled labor mainly carried out as a "labor of love," and the peculiarities of their labor environment undermined the development of class consciousness and political involvement among domestic workers. Goldsmith, "Politics and Programs," 225–226. See the *Hispanic American Historical Review*'s special issue on reproductive labor in the United States and Latin America, particularly Olcott, "Introduction."

109. Lear, *Workers, Neighbors, and Citizens*, cited in Blum, "Cleaning the Revolutionary Household," 80.

110. Estados Unidos Mexicanos, *6º censo de población, 1940* (Mexico City: Se-

cretaría de la Economía Nacional, 1948). The census for 1950, which sets the population at 3,050,442, did not break down the population engaged in this sector. Unlike previous censuses, which documented those engaged in domestic service in the section "Servidumbre" (Servitude), the new census placed them in the category "Servicios" (Services), which included no breakdown by sex or labor activity. The census set the number employed in "Servicios" at 98,951; given the historic female predominance in this sector, however, the number of women employed likely continued to ascend.

111. Scholars recognize these difficulties. See, for example, Davis, *Discipline, and Development*; Walker, *Waking from the Dream*; Parker, *The Idea of the Middle Class*; and Parker and Walker, *Latin America's Middle Class.*

112. Cited in Walker, *Waking from the Dream*, 209–210.

113. Ibid., 6. For a similar conclusion about middle-class social workers and the consolidation of the PRI, see Sanders, *Gender and Welfare in Mexico.*

114. Coral García, "The Mexico City Middle Class," 3.

115. Rankin, *¡México, La Patria!*, 161.

116. Rosenberg, *Spreading the American Dream*, 59. Rosenberg offers an analysis of the pre-1945 US ideology of liberal developmentalism.

117. Blanco, "Clases medias," 95. He adds that "the dream is becoming a nightmare, but there's nothing else to dream."

118. Reproduced in *JWT News* 2, no. 31 (August 4, 1947).

119. Ibid.; *JWT News* 4, no. 4 (January 24, 1949).

120. D. M. V. de R., *Manual de cocina práctica*. For the "electrical idea" as a marketing concept, see Feldman, "Selling the 'Electrical Idea.'"

121. For analysis of conduct manuals and educational reform for women during the Porfiriato, see Eineigel, "Visualizing the Self."

122. Pardo, *Los 30 menús del mes*, 12. Few details of Pardo's life have emerged. He was a Spanish citizen who made a successful career in the gastronomic world of Mexico City in the early twentieth century. He also published *El verdadero práctico*, a three-volume cookbook.

123. Francois, *A Culture of Everyday Credit*, 107–108.

124. The need of the middle and upper classes to protect private property was responsible for the formal regulation of domestic servants. Ibid., 108.

125. Advice manuals emphasizing thriftiness and hygiene were not limited to a middle-class audience. Ana María Hernández wrote a few manuals for the working class and peasant women, including *Como mejorar* (1934). For hygienic campaigns directed at the peasant family, see Vaughan, "Rural Women's Literacy."

126. For government-sponsored classes, see Sanders, *Gender and Welfare*, 79.

127. *Universidad de Mexico* (May–June 1933): 3.

128. In her analysis of *chicas modernas* (modern girls), Hershfield notes that in the city's visual economy working women were "imagined . . . as models of femininity . . . [none of them] engaged in backbreaking manual labor, or working with heavy machinery." Rather, the physical effort that their labor entails "is minimized . . . their productive efforts absent, and instead . . . these women [become] exemplars of decorum

in their dress and their demeanor." Hershfield, *Imagining la Chica Moderna*, 113–115. This visual regime nicely complemented the prototypical representation of working-class masculinity "expressed with light-skinned mestizo features, Westernized dress, and a muscled, shirtless eroticism." Lear, *Picturing the Proletariat*, 124. Lear argues that this visual complementarity speaks to anxiety about women's skilled labor and union leadership. Union leaders responded to this threat by espousing middle-class domesticity and manufacturing the notion of the invisibility of women's work. For more on this see, Gauss, "Masculine Bonds."

129. Sluis, *Deco Body*, 3, 77. Sluis argues that this beauty ideal, predicated on white bodies, relied on form and not skin color, thus allowing mestizas and indigenous Mexican women to aspire to it. It gave the "transnational beauty industries . . . a new tool for selling merchandise in Mexico," including domestic technologies (72–77).

130. Solomon-Godeau, "The Other Side of Venus," 113.

131. D. M. V. de R., *Manual de cocina práctica*, no page.

132. Sluis, *Deco Body*, 92.

133. Hershfield, *Imagining la Chica Moderna*, 126–127. For technology as an extension of female bodies, see Hershfield, "Domestic Technologies"; for bodies and technology, see Dalton, *Mestizo Modernity*.

134. During the Porfiriato, Indian bodies "functioned in American travel discourse as sites of interest, desire, disgust, and sometimes, humor." Travelers and US citizens residing in Mexico used images of indigenous peoples to underline material and cultural differences. The representational politics at play turned them "into objects that [could] be symbolically possessed." Ruiz, *Americans in the Treasure House*, 38–39.

135. Cited in Moreno, *Yankee Don't Go Home!*, 125.

136. Gauss, *Made in Mexico*, 180–184.

137. Moreno, *Yankee Don't Go Home!*, 131, 135–136.

138. The word *lavadora* is female gendered and applicable to either the washing machine or the laundress.

139. *Almanaque de la felicidad: Con nuestros mejores deseas para 1955* (Mexico City: n.p., 1954), 118. The distancing and dry hands promised by ads was more wishful thinking than a reality. Advice literature emphasized taking care of hands and nails, which could imply someone's class and culture. The ad also promoted the company's two-year payment plan, which guaranteed its affordability: "less than the monthly payment for servants, saving food, shelter, lighting . . . and scolding." The time-saving inventions that were supposed to liberate women from domestic chores actually increased the workload. Cowan, *More Work for Mother*.

140. Oudshoorn and Pinch, "Introduction."

141. Arnold, *Everyday Technology*, 69.

142. Lewis, *The Children of Sánchez*, 207. One of the families even became the "Rockefellers of the Casa Grande" when the father began servicing refrigerators and the family renovated the home, first buying a gas stove, then a dining set, then a radio and even a *tele* (TV).

143. Juan Carlos Espinosa Cuock, "El cuarto de servicio," *Arquine*, February 26, 2019.

144. *JWT Weekly News* 1, no. 7 (July 15, 1946): 3–4.

145. For oral interviews as a possible source for lived experiences, see Aguilar-Rodríguez, "Cooking Technologies."

CHAPTER 6. THE PEOPLE, THEIR ELECTRICSCAPE, AND THE VANGUARD OF LABOR, 1930S–1960

1. *Lux*, February 1956, cover.

2. Ibid., 1.

3. In the case of Peronist Argentina, the state led a drive for a fairer distribution of national income, consumption, and industrialization. Milanesio, *Workers Go Shopping in Argentina*.

4. The role of light and power workers in Latin America is critical to understand the history of workers in the "modern sectors" in the postwar labor movements. Brennan, "Industrial Sectors," 41–42.

5. An analysis of real wage studies find that "most have to agree that Mexican economic growth through the mid-1950s accompanied falling wages and perhaps stagnating living standards." Bortz and Aguila, "Earning a Living," 129. In this context, even though installment plans had become standard among retailers, larger appliances still required monthly payments of over 100 pesos. See Ballent, "La publicidad de los ámbitos," 56.

6. Jeffrey Bortz, *Los salarios industriales en la Ciudad de México, 1939–1975* (1988), cited in Bortz and Aguila, "Earning a Living," 126. The provision of inexpensive foodstuffs for urban workers who were experiencing a real decline in their wages became a priority for the country's State Food Agency. See Ochoa, *Feeding Mexico*.

7. Alegre documents how the presidencies of Manuel Ávila Camacho and Miguel Alemán reintroduced policies to modernize the railroad industry "at the expense of workers' salaries and workplace control." Alegre, *Railroad Radicals*, 19.

8. Rankin, *¡México, La Patria!*, 258, 283.

9. The following examples come from ibid., 283–289.

10. *Foreign Relations of the United States*, Diplomatic Papers, 1944, American Republics, vol. 7, 812.50/5–2341, 1206 (https://history.state.gov/historicaldocuments/frus1944v07/d1187).

11. Gauss, *Made in Mexico*, 169.

12. For a comparison of the Brazilian and Argentinean cases, see Saes and Lanciotti, "La regulación de los servicios."

13. Hausman, Hertner, and Wilkins, *Global Electrification*, 213–226. Unless otherwise noted, details for the following discussion come from this source.

14. Enríquez, *Cuatro conferencias*, 37.

15. Hausman, Hertner, and Wilkins, *Global Electrification*, 234, 245.

16. Despite growing economic nationalism, the federal government recognized the value of having an assessment by an all-US team, particularly as international lending agencies required such studies. Under the leadership of Edward Falck, former director of the Office of War Utilities Administration, the consultants included Charles Collier, the former head of the Joint Chiefs of Staff and president of the Detroit Edison Company; Herbert Marks, a consultant to the State Department; and James Mamarchev, the chief electrical engineer at the Dow Chemical Plant Company. The group had led the general mobilization of the electric industry in the United States for the War Production Board during the war and then reestablished electrical services in several war-torn European countries. Together with knowledgeable Mexican engineers and technicians, the commission studied the nation's hydraulic works, hydrology, and fuel handling and supply in order to solve its electricity shortage. Enríquez, *Cuatro conferencias*, 37–40.

17. *Power and Engineer* 91, no. 1 (1947): 174. The commission offered three urgent resolutions to guarantee the continuity and even the expansion of power supply: (1) to buy or lease from the US Navy one or two power trains of 10,000 kW; (2) to buy or lease at least ten portable diesel power generators of 1,000 kW; and (3) to buy twenty-four pumps (1.5 m^3/sec) to secure water supply for the New Electric Chapala Company. Emphasis was placed on the acquisition of used equipment that could be operating within eighteen months. A lease arrangement with the US Navy department materialized within months.

18. Wionczek, "The State and the Electric-Power Industry," 539.

19. This preference was evident in the First Congress of the Electric Industries of Mexico held in late May 1948, which brought together representatives of both public and privately owned electric companies, mechanical and electrical manufacturing industries, and the general public to study the industry's most salient problems. As discussion unfolded, the congress was increasingly inclined to accept the recommendations of several prominent earlier studies. First and foremost, the production and distribution of power in the capital must be addressed. These studies also proposed the expansion of the Necaxa hydroelectric system, a diesel plant near Tacubaya, and the Nonoalco thermoelectrical station. In addition, one study examined the benefits of Mexlight's plan for yet another thermal or hydraulic plant to serve the city. Finally, in terms of distribution of power, the congress relied on a study that recommended a low-tension network for the capital. "Primer Congreso de la Industria Eléctrica Nacional, Actas de Asambleas Plenarias," Archivo General de la Nación/Compañía de Luz y Fuerza del Centro, Box 1192, File 1, Year 1948.

20. Reynolds, *The Mexican Economy*, 169.

21. Ibid., 169–170. Major political and population centers in Latin America mirrored this concentration. Capitals offered a mass market, easy access to labor, the ready availability of suppliers and credit facilities, and the proximity of governmental regulatory agencies.

22. Gauss, *Made in Mexico*, 171.

23. Ibid., 172.

24. Galarza, *La industria eléctrica*, 193.

25. Wionczek, "The State and the Electric-Power Industry," 540–541.

26. Galarza, *La industria eléctrica*, 93–106 (quotation on 95). Mexlight, for instance, charged 30 cents per kilowatt-hour (kWh) in the capital in 1905, with discounts for public entities and large customers. The rate for a load of 48 HP continuous service was 250 pesos, while at the nearby mining center of El Oro electricity was sold at 100 pesos for a continuous demand of 1,000 HP or more. The Mexico City Ayuntamiento paid for public lighting at a discount rate of 25.5 cents per kWh and only 3.5 cents per kWh for pumping water, with a minimum monthly consumption of 4,000 kWh. Rates for lighting in cities in the central region varied from 30 to 35 cents, while those in the north and south paid 40 cents per kWh. The Mexico City Ayuntamiento paid 400 pesos per horsepower by 1923, compared to the 250 pesos from 1910 to 1920. But it was not the discounts for the Ayuntamiento that caused concern, but those for large, private enterprises. The differential rates that electric companies like Mexlight could charge varied but could be so low that it hindered the development of small shops that paid higher rates for power.

27. For an analysis of the Cardenista administration relationship with railway workers, see Alegre, *Railroad Radicals*.

28. Ibid., 97. Halving the rates responded to the immediate need to find industrial contracts for thousands of HP and to flood the market to pressure independent companies into selling their properties and concessions.

29. Wionczek, "The State and the Electric-Power Industry," 534–535. The 1929 National Water Resources Law prohibited private companies from acquiring a concession for "future use" and set other limitations extending government control over water resources.

30. Galarza, *Industria eléctrica*, 93–106.

31. Ibid., 102. Besides affecting their financial well-being, residential users complained of the company workers' arbitrary methods of settling accounts and the troubles caused by inspections of electric installations.

32. Violent episodes resulted in a payment strike that used slogans such as "[U]sted no pague. Si le cortan, nosotros lo reconectamos" (Don't pay. If they cut it off, we'll reconnect it) and "¡Fusilamiento a los culpables!" (Death to the guilty!). In response, twenty-one companies called for military intervention. Galarza, *La industria eléctrica*, 104–105.

33. Ibid., 104.

34. Wionczek, "The State and the Electric-Power Industry," 536. The US-controlled company (American and Foreign Power Company) came onto the Mexican scene in 1928 with the biggest financial operation in the history of the country's public utilities until the industry's nationalization.

35. Galarza, *Industria eléctrica*, 118. Galarza concluded that the Mexican case mirrored the situation in countries with limited industrial development.

36. Campos Aragón, *Electricidad en la Ciudad de México*, 138. These divisions

were Alameda, Cuernavaca, the Federal District, El Oro, Ixtapantongo, Juandó, Necaxa, Pachuca, San Ildefonso, Temascaltepec, Tepuxtepec, and Toluca.

37. Wionczek, "The State and the Electric-Power Industry," 541.

38. Gauss, *Made in Mexico*, 178.

39. For labor politics in the 1930s and the Communist Party, see ibid., chapter 5.

40. Lear, *Picturing the Proletariat*, 165.

41. "El porque de la huelga eléctrica," *Futuro* (August 1936): 8.

42. For the exploitation of the Lerma-Chapala Basin, see Wester, "Capturing the Waters."

43. Lear, *Picturing the Proletariat*, 218–219.

44. Details of the strike come from Walter, "Lázaro Cárdenas."

45. Thompson, "The Development of Unionism," 179.

46. Ibid., 98.

47. Walter, "Lázaro Cárdenas," 100.

48. *Futuro* (August 1936): 4.

49. Asamblea General, November 19, 1936, SME, "Libro de Actas #9," 3.1.

50. Benjamin, *La Revolución*, 19.

51. Knight, "The Myth of the Revolution," 228–229. This myth included institutionalized government policies (land and labor reform, *indigenismo* [indigenism], education, and economic nationalism) and the broader diffusion of cultural coinage (images, icons, heroes, stories, slogans, songs, and anniversaries) minted and exchanged via widespread media (books, newspapers, film, radio, murals, monuments, and public buildings) by various organizations (schools, parties, trade unions, and ejidos).

52. Alegre, *Railroad Radicals*, 30–31. Although peasant mobilization dwarfed the participation of rieleros, the railroad workers wrote themselves into the heart of the revolutionary narrative by stressing that the armed conflict "was made on the rails." Alegre, *Railroad Radicals*, 31.

53. Mexlight suffered substantial financial losses, as large traditional consumers paralyzed operations. Wionczek, "The State and the Electric-Power Industry," 34–35. In early 1913, at a formal banquet organized at the Necaxa hydroelectric plant, President Francisco Madero assured attendees that Mexico welcomed foreign capital and that his government would protect their property from rebel bands. *Outlook* (March 15, 1913): 594.

54. Thompson, "The Development of Unionism," 35–36. The electric industry housed one of the country's most technically advanced operations by 1914. Electricistas were among the most highly trained manual workers. Their relative skill and affluence made them good candidates for leadership and financial support in the eyes of other labor unions.

55. For a thorough discussion of the capital's working-class political culture and mobilization, see Lear, *Workers, Neighbors, and Citizens*. Right after establishing their union, electricistas joined the anarchist Casa del Obrero Mundial (House of the World Worker) but soon demonstrated uneasiness. The Casa, founded by former members of the Luz cultural group and craftworkers and radical intellectuals who had left the

Partido Liberal, earned a record for triggering unionism and working-class militancy. Thompson shows that the electricistas were divided on whether to support the Conventionalists (led by Emiliano Zapata and Francisco "Pancho" Villa) or the Constitutionalists (headed by Venustiano Carranza and Álvaro Obregón). The Casa committed to providing military units to the Constitutionalists in 1915. These units came to be known as the Red Battalions. Carranza tilted the balance by promising direct aid (buildings), free rein to organize workers, and political support when it came to "just claims" against employers. Initially divided on whether to back the Casa's pact, the SME ultimately rejected it, arguing that labor groups should concentrate on activities related to unionism. Thompson, "The Development of Unionism," 41–43.

56. Lear, *Picturing the Proletariat*, 244–245.

57. Asamblea General, March 24, 1938, SME, "Libro de Actas #10," 128–129.

58. *Lux* (July 1941): 36–37.

59. Lear, *Picturing the Proletariat*, 213, 219. Lear provides an in-depth analysis of the magazine from 1920 to 1940 and its use of art from 1920 to 1940 in chapter 6. The magazine became part of the visual culture that emerged in the aftermath of the revolution. As Lear demonstrates, the "postrevolutionary state and society were made by artists with brushes and workers with tools as much as by campesinos [peasants] and middle-class generals with guns" (3).

60. *Lux* (September 1936).

61. For a discussion of allegorical representations of enemies, see Bonnell, *Iconography of Power*, 194.

62. Bonnell views chains and other popular symbols of labor iconography as debris of the old order in the Soviet Union, in *Iconography of Power*, 26.

63. *Lux* (October 1941): 34.

64. *Lux* (April 1941).

65. *Lux* (January 1953).

66. Thompson, "The Development of Unionism," 200.

67. Gustafson, "'He Loves the Little Ones,'" 186.

68. Alegre, *Railroad Radicals*, 5.

69. *Lux* (November 1949): 33.

70. *Lux* (May 1941): back cover.

71. *Lux* (September 1950): 4.

72. Iconographies that depicted traditional enemies (employers, capitalists, and foreigners) developed throughout this period served as models for future generations of artists in the postrevolutionary period across the board, from murals to the avant-garde Estridentista movement and even early modernist photography. For the portrayal of the working class prior to the revolution, see Lear, *Picturing the Proletariat*, chapter 1.

73. Buffington, *A Sentimental Education*, 35. Buffington argues not only that Posada's caricatures depict the ruling class and foreigners but that the richness of his visual vocabulary as well as the vocabulary of the penny press in general responded not merely to a desire to embellish texts but to the needs of its semiliterate audience, with images as necessary as the text, if not more so.

74. *Lux* (April 1952); Rose, *Giants, Monsters, and Dragons*, 100.

75. *Lux* (February 1953).

76. Hands are omnipresent in working-class literature as "reductive identifiers and lucid maps to the geography of human complicity." Zandy, *Hands*, 1. For the use of hands during the 1930s in the United States, see Fraser, "Hands Off the Machine."

77. An example is famed muralist Diego Rivera, who repeatedly capitalized on the revolutionary symbolism of hands in his mural schemes. His 1951 *The Hands of Nature Offering Water* mural crowns one of the tunnels with the hands of Tlaloc, the Aztec deity associated with rain and fertility.

78. A photomontage of the 1936 strike, which doubled as a flyer, showed a composite landscape peopled by electricistas, the Necaxa hydroelectric plant, and transmission towers crisscrossed by a web of electrical wires. In what is otherwise an image of the city's interconnectivity, a set of oversized hands grasping industrial shears literally cuts the electrical lines that until then had kept the city in motion. For a longer discussion of the strike's art, see chapter 6 of Lear, *Picturing the Proletariat*.

79. *Lux* (January 1959): 23. The hand of the SME was not just a celebration of labor but could reach out to others and build political alliances. One image in *Lux* in 1950 shows an assemblage of hands, each grasping a tool. The accompanying text explained that the hands of workers who "activated an electric plant or a telephone network, moved a loom or a printing press" could also build a fair society and could "rise . . . demanding a right, or protesting tyranny and injustice." *Lux* (August 1950): inside cover.

80. *Lux* (October 1954).

81. *Lux* (January 1956): 10.

82. *Lux* (December 1954). Mexican audiences had grown accustomed to melodramatic visualizations of poverty that sentimentalized the poor by the 1950s. Street children in Mexican cinema were no longer exclusively unromanticized victims. Luis Buñuel had introduced a new narrative in *The Young and the Damned*. Mora, *Mexican Cinema*.

83. Bortz and Aguila, "Earning a Living," 125–129.

84. *Lux* (February 1958): 19.

85. Gauss, *Made in Mexico*, 204.

86. *Lux* (October 1948).

87. By mid-century, artists had a well-developed "pictorial cartography of types" to codify difference. Dress in particular served as a signifier of place. Mexico as a nation is embodied here by the china poblana and the charro. Zanardi and Klich, *Visual Typologies*, 1–2.

88. López, *Crafting Mexico*, 36.

89. *Lux* (October 1955).

90. Bonnell, *Iconography of Power*, 9.

91. López-Maya, "The Change in the Discourse of US-Latin American Relations," 136.

92. Bethell and Roxborough, “Latin America between the Second World War,” 168.

93. See Alegre, *Railroad Radicals*, chapter 1.

94. Ibid., 186–187. Not in terms of salary, as they occupied 5th place, but in terms of administration.

95. Asamblea General, April 14, 1950, SME, “Libro de Actas #14,” 183. Both of these were direct references to the country’s revolution. *La Adelita* had become one of the most famous corridos of the armed struggle.

96. Asamblea General, March 8, 1950, SME, “Libro de Actas #14,” 161–165.

97. Ibid., 163. Setting this precedent would have immediate influence: the main industrial unions, including the mine and metal workers, were set to enter contract revisions later that year.

98. Asamblea General, March 8, 1950, SME, “Libro de Actas #14,” 163.

99. MSS 0109, George S. Messersmith papers, University of Delaware (digitized and available online at https://udspace.udel.edu/handle/19716/5973), May 11, 1948.

100. Ibid.

101. Gauss, *Made in Mexico*, 179–180.

102. MSS 0109, George S. Messersmith papers, May 11, 1948, 5.

103. Mexlight produced and supplied almost all the power in the region except for the power produced by a few “very small private plants,” which was also distributed by Mexlight.

104. MSS 0109, George S. Messersmith papers, May 11, 1948, 10.

105. Mason and Asher, *The World Bank since Bretton Woods*, 151 (quotation), 152, 158 (quotation), 237. The precedent for this came in 1949, when the bank “received assurances from the Mexican government in connection with its first power loan that the government would grant reasonable rates to the [Mexlight].”

106. Ellis and Wallich, *Economic Development*, 210.

107. Export-Import Bank Archives, EXIM Bank Press Release, December 7, 1947, EXIM Export-Import Bank of the United States Digital Archives/Bank Press Release collection.

108. Ellis and Wallich, *Economic Development*, 213.

109. MSS 0109, George S. Messersmith papers, May 11, 1948.

110. Ibid.

111. George S. Messersmith papers, November 18, 1947.

112. Niblo, *Mexico in the 1940s*, 297.

113. Hausman, Hertner, and Wilkins, *Global Electrification*, 100, 133 (quotation), 153.

114. Ibid., 155.

115. George S. Messersmith papers, November 6, 1947.

116. Ibid.

117. George S. Messersmith papers, November 17, 1947. President Juan Perón seemed to want to collaborate with private enterprises to address the Argentine power

problem. He did, nonetheless, hold a similar assessment of the public. Public opinion "is not sufficiently aware of the rightness of such views and we shall have to fight a fierce struggle in our respective fields . . . to have them admitted by the man in the street."

118. MSS 0109, George S. Messersmith papers, May 11, 1948 (emphasis added).

119. Niblo, *Mexico in the 1940s*.

120. AGN/Presidencias/Alemán/545.22/94, July 14, 1947.

121. Asamblea General, August 13, 1947, SME, "Libro de Actas #13," 585.

122. AGN/Presidencias/Alemán/545.22/94–31328, July 17, 1947.

123. AGN/Presidencias/Alemán/545.22/94–31973, August 1, 1947.

124. AGN/Presidencias/Alemán/545.22/94–30114, July 17, 1947. Smaller unions representing individual factories requested the president's intervention on behalf of workers to guarantee full salaries. See telegrams under AGN, Presidencias/Alemán/545.22/94–32542, August 6, 1947. Not every proposal to reach the president's desk was so angry in tone. Ulrich Richter, the director general of the Amacuzac Hydroelectric Company, wrote in a spirit of benevolent commerce: "The alarming news of the restrictions awakens the company's most sincere Mexican sentiment to collaborate as it has done in the past." Addressing the severe problem of energy shortage "as an act of loyalty," the director offered to distribute 1,600,000 kWh of electricity in the Federal District for domestic services and assured the president that this action would not affect service for Amacuzac's other customers. Mexlight lines could distribute the energy. Perhaps in an effort to capture some of Mexlight's market, Amacuzac would only charge a rate of 6 cents per kWh—significantly lower than Mexlight's 20 cents. AGN/Presidencias/Alemán/545.22/94–29641, July 12, 1947. Carlos Villaverde, a man with a keen interest in military affairs, had a wildly different proposal. The power crisis arose from a lack of rain in the Necaxa basin, so he came up with a way to make the rains return: cannons. Villaverde, who had witnessed heavy rains after artillery practices, proposed that an artillery unit fire cannons with blank grenades throughout the day in Necaxa. Even if the attempt was unsuccessful, he added, the effort could serve as training for the army. AGN/Presidencias/Alemán/545.22/94–31432, July 28, 1947.

125. *Lux* (January–February 1950): 4.

126. *Excélsior*, April 21, 1950.

127. Alanís Patiño, "La energía," 174.

128. *Excélsior*, September 23, 1949.

129. MSS 0109, George S. Messersmith papers, May 9, 1949.

130. *Excélsior*, September 24, 1949.

131. *Excélsior*, November 1, 1949. The retraction was also published the same day in *El Nacional*, *La Prensa*, *El Universal*, and *Novedades*.

132. *Lux* (October 1949): 4–6.

133. Calls for nationalization had emerged in the collective bargaining period in 1948. Lopez Mendiola, "La huelga de 1950."

134. *Excélsior*, April 22, 1950. The SME also presented evidence that the water shortage problem was a trick to increase electric rates. It claimed that Necaxa was not

the only water receptacle for hydroelectric power plants and that the system included a series of five dams and four plants: Necaxa, Tepexic, Tezcapa, and Nexapa. Together with plants found in other divisions, the company had twelve hydroelectric power plants in optimal operation levels. Yet the company continued to evoke a fictitious water shortage at Necaxa—even while it was raining torrentially in the Vasos Basin. The SME also denounced the systematic rationing imposed by the company to justify the water shortage at Necaxa, affecting national industries and leaving countless workers idle. López Mendiola, "La huelga de 1950," 127.

135. López Mendiola, "La huelga de 1950," 130.

136. *Lux* (March, April, and May 1950).

137. Asamblea General, May 10, 1950, SME, "Libro de Actas #14," 208–209.

138. Asamblea General, April 23, 1950, SME, "Libro de Actas #14," 200–201.

139. López Mendiola, "La huelga de 1950," 139.

140. *Lux* (March, April, and May 1950): 50–52.

141. Rivera Rojas went on to claim that the SME had also saved the animating principles of the nation by acting like a "barrier against foreign capital and attacks against the interests of the Revolution." Asamblea General, May 10, 1950, SME, "Libro de Actas #14," 219.

142. *Lux* (November 30, 1950): 4.

143. Asamblea General, May 10, 1950, SME, "Libro de Actas #14," 226.

144. Hausman, Hertner, and Wilkins, *Global Electrification*, 244, 237. For nationalization in the region, see 239–242.

145. Kiddle, *Mexico's Relations*, 109.

146. Asamblea General, March 31, 1938, SME, "Libro de Actas #10," 137–147.

147. Regardless of the comparable significance of nationalization, Paulín judged it vital for workers to back the Cardenista government. Workers had received the greatest benefits from the Cárdenas administration and were thus obliged to support him in the economic battle. The assembly voted to provide 100,000 pesos to the cause.

148. Asamblea General, April 23, 1950, SME, "Libro de Actas #13," 201–203.

149. *Lux* (April 1958): 17–20.

150. *Lux* (January–February 1950): 26–27.

151. *Lux* (November 1950): 4–6.

152. *Lux* (December 1952): 4–5.

153. Ibid., 5.

154. *Lux* (March 1953).

155. *Lux* (March 1955): 8. For a thorough analysis of the oil industry, see Meyer, *Mexico and the United States*.

156. Carranza, *Architecture as Revolution*, 177.

157. Benjamin, *La Revolución*.

158. Ibid., 130.

159. *Lux* (April 1958): 4–5.

160. Claiming economic independence as the basis for political independence served the dual purpose of redrawing the country's patriotic timeline and underlining

the next step to achieve full independence: the nationalization of the electric industry. The September 1956 issue set the revolutionary tone with a representation of the nation's greatest independence leader, Father Miguel Hidalgo, with fist in the air leading a heterogeneous group of individuals with their arms in the air. The editorial piece titled "National Independence Remains a Goal" reaffirmed this. *Lux* (September 1956).

161. Wionczek, "The State and the Electric-Power Industry," 546–547.

162. *Lux* (November 1958): 24–25.

163. *Lux* (February 1955). For a discussion of the coup and the context of US intervention in the region, see Rabe, *The Killing Zone.*

164. *Lux* (November 1958): 50–51.

165. *Lux* (December 1958): 2.

166. Alegre, *Railroad Radicals*, 172–173.

167. Ibid., 187.

168. Servín, "Los 'enemigos del progreso': Critica y resistencia al desarrollismo del medio signo," in *Del nacionalismo al neoliberalismo*, 107.

169. Ibid., 120–121. For a thorough discussion of Cárdenas's "resurrection," see Zolov, *The Last Good Neighbor*, chapter 1.

170. Ibid., 122. These two ideological camps held opposing views of what a revolutionary program looked like, with Cuba becoming one of the most critical topics of confrontation. Hernández Rodríguez, *Presidencialismo*, 159.

171. Servín, "Propaganda," 13.

172. Gómez Tagle, *Insurgencia*, 57.

173. Zolov, *The Last Good Neighbor*, 30.

174. Ibid.

175. *Lux* (January 1959): 4–9.

176. Zolov, *The Last Good Neighbor*, 31.

177. Alegre, *Railroad Radicals*, 191–193.

178. *Lux* (January 1960): 4–9, 21–22.

179. Gómez Tagle, *Insurgencia*, 57.

180. Wionczek, "Electric Power," 96, 98.

181. Hernández Rodríguez, *Presidencialismo*, 173–174.

182. Zolov, *The Last Good Neighbor*, 23, 106.

CONCLUSION

1. *Lux* (September 1960): 4–5.

2. *Excélsior*, September 28, 1960.

3. *Lux* (September 1960): 14–15 (quotation), 20.

4. Ibid., 18.

5. *Catálogo de productos: CANAME: Electrificar es civilizar.*

6. Two profile images of Father Hidalgo and López Mateos succinctly illustrate this. A black-and-white engraving of Father Hidalgo with a lit torch in his extended

right hand faces López Mateos's jubilant countenance and extended right arm firmly gripping a power lever. The torch and lever are symbols of the nation's future. While Hidalgo's struggle would bring forth a republic, López Mateos's revolutionary feat of Mexicanizing the industry would create an industrial modern nation. *El Universal*, September 3 and 14, 1960.

7. *Lux* (August 1960): 23.

Bibliography

ARCHIVES

Archivo General de la Nación (AGN), Mexico City
Compañía de Luz y Fuerza del Centro
Patentes
Presidencias
Secretaría de Comunicaciones y Obras Públicas (SCOP)
Tribunal Superior de Justicia del Distrito Federal (TSJDF)
Archivo Histórico del Distrito Federal (AHDF), Mexico City
Ayuntamiento
Biblioteca de la Gastronomía de la Fundación Herdez, Mexico City
Biblioteca Miguel Lerdo de Tejada, Mexico City
Siemens Historical Institute, Munich and Berlin, Germany
Sindicato Mexicano de Electricista (SME), Mexico City
Universidad Autónoma de México (UNAM), Mexico City
Acervo Histórico del Palacio de Minería
Biblioteca Nacional
Fondo Reservado de la Hemeroteca Nacional de México
Hemeroteca Nacional Digital de México
University of New Mexico, Center for Southwest Research and Special Collections, Albuquerque

SOURCES

"Actas de Cabildo." 1882–1899. Archivo Histórico del Distrito Federal (AHDF), Mexico City.

Adams, Judith. "The Promotion of New Technology through Fun and Spectacle: Electricity at the World's Columbian Exposition." *Journal of American Culture* 18, no. 2 (1995): 45–55.

Agostoni, Claudia. *Monuments of Progress: Modernization and Public Health in Mexico City, 1876–1910*. Calgary, CA: University of Calgary Press, 2003.

Aguilar de Carbia, María (Marichu). *Almanaque 1936: "La familia."* Mexico City: Editorial Sayrols, 1936.

Aguilar-Rodríguez, Sandra. "Cooking Modernity: Nutrition Policies, Class, and Gender in 1940s and 1950s Mexico City." *Americas* 64, no. 2 (2007): 177–205.

———. "Cooking Technologies and Electrical Appliances in 1940s and 1950s Mexico." In *Technology and Culture in Twentieth-Century Mexico*, edited by Araceli Tinajero and J. Brian Freeman, 43–54. Tuscaloosa: University of Alabama Press, 2013.

Aguirre, Carlos, and Robert Buffington. *Reconstructing Criminality in Latin America*. Wilmington, DE: SR Books, 2000.

Alanís Patiño, Emilio. "La energía en México." *Investigación Económica* 14, no. 2 (1954): 163–229.

Álbum de la coronación de la Sma. Virgen de Guadalupe. Mexico City: Imp. de "El Tiempo," 1895.

Alegre, Robert. *Railroad Radicals in Cold War Mexico Gender, Class, and Memory*. Lincoln: University of Nebraska Press, 2014.

Alexander, Anna. *City on Fire: Technology, Social Change, and the Hazards of Progress in Mexico City, 1860–1910*. Pittsburgh: University of Pittsburgh Press, 2016.

Altamirano, Ignacio. *Paisajes y leyendas: Tradiciones y costumbres de México*. Mexico City: Impr. y Litogr. Española, 1884.

Aréchiga Córdoba, Ernesto. "Lucha de clases en la ciudad ca. 1890–1930." In *Los trabajadores de la Ciudad de México, 1860–1950*, edited by Mario Barbosa and Carlos Illades, 19–50. Mexico City: ColMex, 2013.

Arizpe, Rafael. *El alumbrado público en la Ciudad de México*. Mexico City: Tip. La Europea, 1900.

———. *Estadística de las aplicaciones de la electricidad en la República Mexicana*. Mexico City: La Europea, 1900.

Armstrong, Christopher, and H. V. Nelles. *Southern Exposure: Canadian Promoters in Latin America and the Caribbean, 1896–1930*. Toronto: University of Toronto Press, 1988.

Arnold, David. "Europe, Technology, and Colonialism in the 20th Century." *History and Technology* 21 (2005): 85–106.

———. *Everyday Technology: Machines and the Making of India's Modernity*. Chicago: University of Chicago Press, 2015.

Arrom, Silvia Marina. *Containing the Poor: The Mexico City Poor House, 1774–1871*. Durham, NC: Duke University Press, 2000.

———. *The Women of Mexico City: 1790–1857*. Stanford, CA: Stanford University Press, 1992.

Ballent, Anahi. "La publicidad de los ámbitos de la vida privada: Representaciones de la modernización del hogar en la prensa de los años cuarenta y cincuenta en México." *Alteridades* 6, no. 11 (1996): 53–74.

Barbosa, Mario. *El trabajo en las calles: Subsistencia y negociación política en la Ciudad de México a comienzos del siglo XX*. Mexico City: ColMex, 2008.

Barton, Mary. *Impressions of Mexico: With Brush and Pen*. London: Methuen, 1911.

Basalla, George. "The Spread of Western Science." *Science* 156, no. 3775 (1967): 611–622.

Beatty, Edward. *Technology and the Search for Progress in Modern Mexico*. Berkeley: University of California Press, 2015.

Beatty, Edward, Yovanna Pineda, and Patricio Sáiz. "Technology in Latin America's Past and Present: New Evidence from the Patent Records." *Latin American Research Review* 52, no. 1 (2017): 138–149.

Beauchamp, Kenneth. *Exhibiting Electricity*. London: Institution of Electrical Engineers, 1997.

Becerril, José. *Las fábricas de San Antonio Abad y San Ildefonso (1842–1910)*. Mexico City: INAH, 2011.

Beezley, William. *Judas at the Jockey Club and Other Episodes of Porfirian Mexico*. Lincoln: University of Nebraska Press, 1989.

Benjamin, Thomas. *La Revolución: Mexico's Great Revolution as Memory, Myth, and History*. Austin: University of Texas Press, 2010.

Bennett, Tony. *The Birth of the Museum: History, Theory, Politics*. London: Routledge, 2013.

Besse, Susan. *Restructuring Patriarchy: The Modernization of Gender Inequality in Brazil, 1914–1940*. Chapel Hill: University of North Carolina, 1996.

Best, Alberto. *Noticia sobre las aplicaciones de la electricidad en la República Mexicana*. Mexico City: Imp. de la Secretaría de Fomento, 1889.

Bethell, Leslie, and Ian Roxborough. "Latin America between the Second World War and the Cold War: Some Reflections on the 1945–8 Conjuncture." *Journal of Latin American Studies* 20, no. 1 (1988): 167–189.

Bijker, Wiebe E., Thomas P. Hughes, and Trevor J. Pinch (eds.). *The Social Construction of Technological Systems: New Directions in the Sociology and History of Technology*. Cambridge, MA: MIT Press, 1987.

Bijker, Wiebe E., and John Law. *Shaping Technology/Building Society: Studies in Sociotechnical Change*. Cambridge, MA: MIT Press, 1992.

Bille, Mikkel, and Tim Flohr Sørensen. "An Anthropology of Luminosity: The Agency of Light." *Journal of Material Culture* 12, no. 3 (2007): 263–284.

Blake, Mary E., and Margaret F. Sullivan. *Mexico: Picturesque, Political, Progressive*. Boston: Lee and Shepard, 1888.

Blanco, José Joaquín. "Clases medias y cultura nacional." In *Las clases medias en la coyuntura actual*, edited by Soledad Loaeza and Claudio Stern, 89–95. Mexico City: El ColMex, 1990.

Blaut, J. M. "Diffusionism: A Uniformitarian Critique." *Annals of the Association of American Geographers* 77, no. 1 (1987): 30–47.

Bloch, Marc. *The Historian's Craft*. Manchester: Manchester University Press, 2012.

Blum, Ann. "Cleaning the Revolutionary Household: Domestic Servants and Public Welfare in Mexico City, 1900–1935." *Journal of Women's History* 15, no. 4 (2004): 67–90.

———. *Domestic Economies: Family, Work, and Welfare in Mexico City, 1884–1943*. Lincoln: University of Nebraska Press, 2009.

Boast, Robin. *The Machine in the Ghost: Digitality and Its Consequences*. London: Reaktion Books, 2017.

Bonnell, Victoria E. *Iconography of Power: Soviet Political Posters under Lenin and Stalin*. Berkeley: University of California Press, 1999.

Borg, Kevin. "The 'Chauffeur Problem' in the Early Auto Era: Structuration Theory and the Users of Technology." *Technology and Culture* 40 (1999): 797–832.

Bortz, Jeffrey, and Marcos Aguila. "Earning a Living: A History of Real Wage Studies in Twentieth-Century Mexico." *Latin American Research Review* 41, no. 2 (2006): 112–138.

Brennan, James. "Industrial Sectors and Union Politics in Latin American Labor Movements: Light and Power Workers in Argentina and Mexico." *Latin America Research Review* 30, no. 1 (1995): 39–68.

Briseño Senosiain, Lillian. *Candil de la calle, oscuridad de su casa la iluminación en la Ciudad de México durante el porfiriato*. Mexico City: Porrúa, 2008.

———. "La fiesta de luz en la Ciudad de Mexico: El alumbrado eléctrico en el Centenario." *Secuencia* 60 (2004): 90–109.

Brown, Grace Owen. *The Tour of the 400 to Mexico*. Chicago: Hartzell, Lord, 1907.

Brown, Jonathan. "Introduction: What Is Worker's Control?" In *Workers' Control in Latin America, 1930–1979*, edited by Jonathan C. Brown, 1–15. Chapel Hill: University of North Carolina Press, 1997.

———. *Oil and Revolution in Mexico*. Berkeley: University of California Press, 1993.

Brown, Kendall. *A History of Mining in Latin America: From the Colonial Era to the Present*. Albuquerque: University of New Mexico Press, 2012.

Buchenau, Jürgen. *Tools of Progress: A German Merchant Family in Mexico City, 1865–Present*. Albuquerque: University of New Mexico Press, 2004.

Buffington, Robert. *Criminal and Citizen in Modern Mexico*. Lincoln: University of Nebraska Press, 2000.

———. *A Sentimental Education for the Working Man: The Mexico City Penny Press, 1900–1910*. Durham, NC: Duke University Press, 2015.

Bukatman, Scott. *The Poetics of Slumberland: Animated Spirits and the Animating Spirit*. Oakland: University of California Press, 2012.

Bunker, Steven B. *Creating Mexican Consumer Culture in the Age of Porfirio Díaz*. Albuquerque: University of New Mexico Press, 2012.

Calhoun, Charles W. *The Gilded Age: Perspectives on the Origins of Modern America*. Lanham, MD: Rowman and Littlefield, 2007.

Camacho Morfín, Thelma. *Las historietas de El Buen Tono (1904–1922): La litografía al servicio de la industria*. Mexico City: UNAM, 2013.

Cámara Nacional de Manufacturas Eléctricas. *Catálogo de productos, CANAME: Electrificar es civilizar*. Mexico City: La Cámara, 1963.

Campbell, Reau. *Campbell's New Revised Complete Guide and Descriptive Book of Mexico*. Chicago: Rogers and Smith, 1909.

Campos Aragón, Leticia. *El modelo británico en la industria eléctrica mexicana*. Mexico City: Siglo XXI Ediciones, 2009.

———. *La electricidad en la Ciudad de México y área conurbada: Historia, problemas y perspectivas*. Mexico City: Siglo XXI Editores, 2005.

Carranza, Luis. *Architecture as Revolution: Episodes in the History of Modern Mexico*. Austin: University of Texas Press, 2010.

Carrera, Magali Marie. *Traveling from New Spain to Mexico: Mapping Practices of Nineteenth-Century Mexico*. Durham, NC: Duke University Press, 2011.

Carson, W. E. *Mexico: The Wonderland of the South*. New York: Macmillan, 1909.

Castro, Justin. *Radio in Revolution: Wireless Technology and State Power in Mexico, 1897–1938*. Lincoln: University of Nebraska Press, 2016.

Caufield, Sueann. *In Defense of Honor: Sexual Morality, Modernity, and Nation in Early-Twentieth-Century Brazil*. Durham, NC: Duke University Press, 2000.

Cavazos Guzmán, Luis, and José Carrillo Arriaga. *Historia y evolución de la medicina*. Mexico City: Editorial El Manual Moderno, 2000.

Chakrabarty, Dipesh. "Postcoloniality and the Artifice of History: Who Speaks for 'Indian' Pasts?" *Representations* 37 (Winter 1992): 1–26.

Chaney, Elsa M., and Mary García Castro. *Muchachas No More: Household Workers in Latin America and the Caribbean*. Philadelphia: Temple University Press, 1989.

Chatterjee, Elizabeth. "The Asian Anthropocene: Electricity and Fossil Developmentalism." *Journal of Asian Studies* 79, no: 1 (2019): 3–24.

Chatterjee, Partha. *The Nation and Its Fragments: Colonial and Postcolonial Histories*. Princeton, NJ: Princeton University Press, 1994.

Christie, Renfrew. *Electricity, Industry and Class in South Africa*. Albany: State University of New York Press, 1984.

Clendinning, Anne. *Demons of Domesticity: Women and the English Gas Industry, 1889–1939*. London: Routledge, 2017.

Código penal para el Distrito Federal y territorio de la Baja-California. Mexico City: Imp. del Gobierno, 1871.

Coleman, Leo. *A Moral Technology: Electrification as Political Ritual in New Delhi*. Ithaca, NY: Cornell University Press, 2017.

Collado, María del Carmen. *La burguesía mexicana: El emporio Braniff y su participación política: 1865–1920*. Mexico City: Siglo XXI, 1987.

Coopersmith, Jonathan. *The Electrification of Russia, 1880–1926*. Ithaca, New York: Cornell University Press, 2017.

Coral García, Emilio Mario. "The Mexico City Middle Class, 1940–1970: Between Tradition, the State, and the United States." PhD diss., Georgetown University, 2011.

Cowan, Ruth Schwartz. *More Work for Mother: The Ironies of Household Technology from the Open Hearth to the Microwave*. New York: Basic Books, 1983.

Craib, Raymond B. *Cartographic Mexico: A History of State Fixations and Fugitive Landscapes*. Durham, NC: Duke University Press, 2012.

Crary, Jonathan. *Suspensions of Perception: Attention, Spectacle, and Modern Culture*. Cambridge, MA: MIT Press, 2001.

———. *Techniques of the Observer: On Vision and Modernity in the Nineteenth Century*. Cambridge, MA: MIT Press, 1992.

Cuéllar, José Tomás de. *Los fuereños y La Noche Buena*. Santander, Mexico: Imprenta y Litografía de El Atlántico, 1890.

Curcio-Nagy, Linda. *The Great Festivals of Colonial Mexico City: Performing Power and Identity*. Albuquerque: University of New Mexico Press, 2004.

Dalton, David S. *Mestizo Modernity: Race, Technology, and the Body in Postrevolutionary Mexico*. Gainesville: University Press of Florida, 2018.

Davis, Charles T. *The Manufacture of Leather*. Philadelphia: H. C. Baird, 1885.

Davis, Diane. *Discipline and Development: Middle Classes and Prosperity in East Asia and Latin America*. New York: Cambridge University Press, 2004.

De Grazia, Victoria. *How Fascism Ruled Women: Italy, 1922–1945*. Berkeley: University of California Press, 1992.

Delbourgo, James. *A Most Amazing Scene of Wonders: Electricity and Enlightenment in Early America*. Cambridge, MA: Harvard University Press, 2006.

Del Castillo Troncoso, Alberto. "El surgimiento de la prensa moderna en México." In *La republica de las letras: Asomos a la cultura escrita del México decimonónico*, vol. 2, edited by Belem Clark de Lara and Elisa Speckman-Guerra, 105–118. Mexico City: UNAM, 2005.

De Orellana, Margarita. "El poder de la memoria fugaz." *Artes de México* 48 (1999): 6–7.

De Vecchi, Bruno, and Carmen de De Vecchi. "Primeros ensayos públicos de la luz eléctrica en la República Mexicana." *Quipu* 4 (1987): 71–80.

Díaz y de Ovando, Clementina. "La Ciudad de México en 1904." *Historia Mexicana* 24, no. 1 (1974): 122–144.

D. M. V. de R. *Manual de cocina práctica: Adaptada a paladares delicados y de buen gusto*. Mexico City: Papelería "El 5 de Mayo," 1929.

Doménech, J. Figueroa. *Guía general descriptiva de la República Mexicana*. 2 vols. Mexico City: Ramón de S. N. Araluce, 1899.

Dore, Elizabeth. "One Step Forward, Two Steps Back Gender and the State in the Long Nineteenth Century." In *Hidden Histories of Gender and the State in Latin America*, edited by Elizabeth Dore and Maxine Molyneux, 3–32. Durham, NC: Duke University Press, 2009.

Dubash, Navroz K., Sunila S. Kale, and Ranjit Bharvirkar. *Mapping Power The Political Economy of Electricity in India's States*. Oxford: Oxford University Press, 2018.

Dublán, Manuel, and José María Lozano. *Legislación mexicana*. Vol. 24. Mexico City: Imprenta de Eduardo Dublán, 1898.

Eineigel, Susanne. "Visualizing the Self: Modernity, Identity, and the Gente Decente in Porfirian Mexico." MA thesis. Vancouver: University of British Columbia, 2003.

Eisenstadt, S. N. "Multiple Modernities," *Daedalus* 129, no. 1 (2000): 1–29.

Ellis, Howard S., and Henry C. Wallich. *Economic Development for Latin America.* New York: St. Martin's, 1961.

Enríquez, Oscar. *Cuatro conferencias sobre la industria eléctrica mexicana.* Mexico City: O. R. Enríquez, 1960.

Farnsworth-Alvear, Ann. *Dulcinea in the Factory: Myths, Morals, Men, and Women in Colombia's Industrial Experiment, 1905–1960.* Durham, NC: Duke University Press, 2009.

Feldman, Andrew. "Selling the 'Electrical Idea': The Campaign to Electrify America in the 1920s." *Argumentation* 8, no. 4 (1994): 377–389.

Fernández, Isabel and Carmen Nava. "Images of Independence in the Nineteenth Century: The Grito of Dolores, History and Myth." In *¡Viva Mexico! ¡Viva La Independencia! Celebrations of September 16*, edited by William H. Beezley and David E. Lorey, 1–41. Wilmington: SR Books, 2001.

Ferris, J. F. *Practical Artificial Incubation.* Albany, NY: Ferris, 1880.

Findlay, Eileen J. "Courtroom Tales of Sex and Honor: 'Rapto' and Rape in Late-Nineteenth-Century Puerto Rico." In *Honor, Status, and Law in Modern Latin America*, edited by Sueann Caulfield and Sarah C. Chambers, 201–222. Durham, NC: Duke University Press, 2005.

Francois, Marie. *A Culture of Everyday Credit: Housekeeping, Pawnbroking, and Governance in Mexico City, 1750–1920.* Lincoln: University of Nebraska Press, 2007.

———. "La lavandería, la producción cultural y la economía política en la Ciudad de México." In *Género en la encrucijada de la historia social y cultural de México*, edited by Susie Porter and María Teresa Fernández Aceves, 33–66. Morelia, Mexico: ColMich, 2015.

———. "The Products of Consumption: Housework in Latin American Political Economies and Cultures." *History Compass* 6, no. 1 (2008): 207–242.

Fraser, M. "Hands Off the Machine: Workers' Hands and Revolutionary Symbolism in the Visual Culture of 1930s America." *American Art* 27, no. 2 (2013): 95–117.

French, John, and Daniel James. *The Gendered Worlds of Latin American Women Workers: From Household and Factory to the Union Hall and Ballot Box.* Durham, NC: Duke University Press, 2012.

French, William E. "Prostitutes and Guardian Angels: Women, Work, and the Family in Porfirian Mexico." *Hispanic American Historical Review* 72 (1992): 529–553.

French-Fuller, Katharine. "Gendered Invisibility, Respectable Cleanliness: The Impact of the Washing Machine on Daily Living in Post-1950 Santiago, Chile." *Journal of Women's History* 18, no. 4 (2006): 79–100.

Frost, Robert. "Machine Liberation: Inventing Housewives and Home Appliances in Interwar France." *French Historical Studies* 18 (1993): 109–130.

Galarza, Ernesto. *La industria eléctrica en México.* Mexico City: FCE, 1941.

Galeano, Diego. *Criminosos viajantes: Circulações transnacionais entre Rio de Janeiro e Buenos Aires (1890–1930).* Rio de Janeiro: Ministério da Justiça, Arquivo Nacional, 2016.

Galindo, Alejandro. *Una familia de tantas.* 1948.

Gamboa, Federico. *Del natural: Esbozos contemporáneos.* Guatemala: Tip. "La Unión," 1889.

———. *La llaga.* Mexico City: Eusebio Gómez de la Fuente, 1912.

———. *Metamorfosis.* Mexico City: Centro Mercantil, 1899.

Gámez, Moisés. *Electricidad: Recurso estratégico y actividades productivas, procesos de electrificación en el norte de México, siglos XIX–XX.* San Luis Potosí, Mexico: El Colegio de San Luis, 2013.

García, Emilio. "Mexico City Middle Class, 1940–1970: Between Tradition, the State and the United States." PhD diss., Georgetown University, 2011.

García, Jacinto S. *Memorias íntimas de México: A propósito de la revolución mexicana.* Lima: Fondo Editorial Universidad Nacional Mayor de San Marcos, 2005.

García Lázaro, Andrés. "Luces y sombras en la Ciudad de México: La compañía mexicana de gas y luz eléctrica y el servicio de alumbrado, 1868–1898." In *Investigaciones y enfoques contemporáneos en los estudios urbanos: Historia, organización social y estructura espacial en ciudades*, edited by Alejandra Trejo Nieto, Adolfo Sánchez Almanza, and Edith Pacheco, 175–215. Mexico City: ColMex, 2017.

Garner, Paul. *Porfirio Díaz.* London: Routledge, 2016.

Garza, Gustavo. *El proceso de industrialización en la Ciudad de México (1821–1970).* Mexico City: ColMex, 1985.

Garza, James Alex. *The Imagined Underworld: Sex, Crime, and Vice in Porfirian Mexico City.* Lincoln: University of Nebraska Press, 2009.

Garza Toledo, Enrique de la. *Historia de la industria eléctrica en México.* Mexico City: UAM-Unidad Iztapalapa, 1994.

Gauss, Susan M. *Made in Mexico: Regions, Nation, and the State in the Rise of Mexican Industrialism, 1920s–1940s.* University Park, PA: Penn State University Press, 2015.

———. "Masculine Bonds and Modern Mothers: The Rationalization of Gender in the Textile Industry in Puebla, 1940–1952." *International Labor and Working-Class History* 63 (2003): 63–80.

Glasco, Sharon Bailey. *Constructing Mexico City: Colonial Conflicts over Culture, Space, and Authority.* New York: Palgrave Macmillan, 2014.

Godoy, Mercedes. *When I Was a Girl in Mexico.* Boston: Lothrop, Lee and Shepard, 1919.

Goldsmith, Mary. "Politics and Programs of Domestic Workers." In *Muchachas No More: Household Workers in Latin America and the Caribbean*, edited by Elsa M. Chaney and María García Castro, 221–243. Philadelphia: Temple University Press, 1989.

Gómez Tagle, Silvia. *Insurgencia y democracia en los sindicatos electricistas.* Mexico City: ColeMex, 1980.

González Navarro, Moisés. *El porfiriato: La vida social.* Mexico City: Editorial Hermes, 1957.

Gooday, Graeme. *Domesticating Electricity: Technology, Uncertainty and Gender, 1880–1914.* London: Pickering and Chatto, 2008.

Granziera, Patrizia. "Concept of the Garden in Pre-Hispanic Mexico." *Garden History* 29, no. 2 (2001): 185–213.

Guerrero, Javier. *Narcosubmarines: Outlaw Innovation and Maritime Interdiction in the War on Drugs*. Singapore: Palgrave, 2020.

Gurza, Francisco X. *Guía de pasajeros a bordo de los tranvías*. Mexico City: Imprenta del Gobierno Federal, 1913.

Gustafson, Reid. "'He Loves the Little Ones and Doesn't Beat Them': Working Class Masculinity in Mexico City, 1917–1929." PhD diss., University of Maryland, 2014.

Haber, Stephen. *Industry and Underdevelopment: The Industrialization of Mexico, 1890–1940*. Stanford, CA: Stanford University Press, 1995.

Hasenöhrl, Ute. "Rural Electrification in the British Empire," *History of Retailing and Consumption* 4, no. 1 (2018): 10–27.

Hausman, William, Peter Hertner, and Mira Wilkins. *Global Electrification: Multinational Enterprise and International Finance in the History of Light and Power, 1878–2007*. Cambridge: Cambridge University Press, 2008.

Heneghan, Bridget T. *Whitewashing America: Material Culture and Race in the Antebellum Imagination*. Jackson: University Press of Mississippi, 2003.

Hernández, Ana María. *Como mejorar la alimentación del obrero y campesino: Libro social y familiar para la mujer obrera y campesina mexicana*. Mexico City: Tip. Moderna, 1934.

Hernández Rodríguez, Rogelio. *Presidencialismo y hombres fuertes en México: La sucesión presidencial de 1958*. Mexico City: ColMex, 2015.

Herrero Bervera, Carlos. *Los empresarios mexicanos de origen Vasco y el desarrollo del capitalismo en México, 1880–1950*. Mexico City: UAM, 2004.

Hershfield, Joanne. "Domestic Technologies: Gender, Technology, and Mexican Housewives, 1930–1950." In *Technology and Culture in Twentieth-Century Mexico*, edited by Araceli Tinajero and J. Brian Freeman, 59–69. Tuscaloosa: University of Alabama Press, 2013.

———. *Imagining La Chica Moderna: Women, Nation, and Visual Culture in Mexico, 1917–1936*. Durham, NC: Duke University Press, 2008.

Hill, Jonathan. "Circuits of State: Water, Electricity, and Power in Chihuahua, 1905–1936." *Radical History Review* 127 (2017): 13–38.

Hoffmann, E.T.A. *Tales of Hoffmann*. London: Penguin Books, 1982.

Hofmänner, Alexandra. *Science & Technology Studies Elsewhere*. Basel: Schwabe Verlag, 2020.

Hughes, Thomas P. *Networks of Power: Electrification in Western Society, 1880–1930*. Baltimore: Johns Hopkins University Press, 1983.

Iglehart, Fanny Chambers Gooch. *Face to Face with the Mexicans: Domestic Life, Educational, Social and Business Ways*. New York: Fords, Howard and Hulbert, 1887.

Ingold, Tim. *The Perception of the Environment: Essays on Livelihood, Dwelling and Skill*. London: Routledge, 2011.

James, Daniel. *Doña María's Story: Life History, Memory, and Political Identity*. Durham, NC: Duke University Press, 2007.

Jasanoff, Sheila. "Future Imperfect: Science, Technology, and the Imaginations of Modernity." In *Dreamscapes of Modernity: Sociotechnical Imaginaries and the Fabrication of Power*, edited by Sheila Jasanoff and Sang-Hyun Kim, 1–33. Chicago: University of Chicago Press, 2016.

Jasanoff, Sheila, and Sang-Hyun Kim. *Dreamscapes of Modernity: Sociotechnical Imaginaries and the Fabrication of Power.* Chicago: University of Chicago Press, 2016.

Jiménez, Christina M. "Popular Organizing for Public Services: Residents Modernize Morelia, Mexico, 1880–1920." *Journal of Urban History* 30, no. 4 (2004): 495–518.

Jiménez, Jorge. *La traza del poder: Historia de la política y los negocios urbanos en el Distrito Federal de sus orígenes a la desaparición del Ayuntamiento (1824–1928).* Mexico City: Secretaría de Cultura del Gobierno del Distrito Federal, 2012.

Johns, Michael. *The City of Mexico in the Age of Díaz.* Austin: University of Texas Press, 1997.

Johnson, Lyman. *The Problem of Order in Changing Societies: Essays on Crime and Policing in Argentina and Uruguay.* Albuquerque: University of Mexico Press, 1990.

Jones, Christopher. *Routes of Power: Energy and Modern America.* Cambridge, MA: MIT Press, 2016.

Jones, Jill. *Empires of Light: Edison, Tesla, Westinghouse, and the Race to Electrify the World.* New York: Random House, 2004.

Juárez López, José Luis. *La cocina también es laberinto: Mujeres: Génesis de una historia.* Mexico City: Ediciones Navarra, 2016.

———. *Nacionalismo culinario: La cocina mexicana en el siglo XX.* Mexico City: Conaculta, 2013.

Kale, Sunila S. *Electrifying India: Regional Political Economies of Development.* Stanford, CA: Stanford University Press, 2014.

Keremitsis, Dawn. "Del metate al molino: La mujer mexicana de 1910 a 1940." *Historia Mexicana* 33 (1983): 285–303.

Kiddle, Amelia M. *Mexico's Relations with Latin America during the Cárdenas Era.* Albuquerque: University of New Mexico Press, 2016.

Kirstein, Lincoln. *Four Centuries of Ballet.* New York: Dover, 1984.

Kline, Ronald, and Trevor Pinch. "Users as Agents of Technological Change: The Social Construction of the Automobile in the Rural United States." *Technology and Culture* 37, no. 4 (October 1996): 763–795.

Knight, Alan. "The Myth of the Mexican Revolution." *Past and Present* 209, no. 1 (2010): 223–273.

Knight, Edward Henry. *Knight's New Mechanical Dictionary: A Description of Tools, Instruments, Machines, Processes, and Engineering, with Indexical References to Technical Journals (1876–1880).* Boston: Houghton, Mifflin, 1884.

Krauze, Enrique. *Mexico: Biography of Power.* New York: HarperCollins World, 1998.

Kuznesof, Elizabeth. "Gender, Ideology, Race, and Female Headed Households, 1750–1850." In *State and Society in Spanish America during the Age of Revolution*, edited by Victor Uribe-Uran, 149–172. Wilmington, DE: SR Books, 2001.

———. "A History of Domestic Service in Spanish America, 1492–1980." In *Muchachas No More: Household Workers in Latin America and the Caribbean*, edited by Elsa M. Chaney and Maria García Castro, 17–36. Philadelphia: Temple University Press, 1989.

Lagendijk, Vincent. *Electrifying Europe: The Power of Europe in the Construction of Electricity Networks*. Amsterdam: Aksant, 2008.

Larson, Brooke. "Capturing Indian Bodies, Hearths, and Minds: The Gendered Politics of Rural School Reform in Bolivia, 1920s–1940s." In *Natives Making Nation: Gender, Indigeneity, and the State in the Andes*, edited by Andrew Canessa, 32–59. Tucson: University of Arizona Press, 2011.

Lavrin, Asunción. *Women, Feminism, and Social Change in Argentina, Chile, and Uruguay, 1890–1940*. Lincoln: University of Nebraska Press, 1995.

Lear, John. *Picturing the Proletariat: Artists and Labor in Revolutionary Mexico, 1908–1940*. Austin: University of Texas Press, 2017.

———. *Workers, Neighbors, and Citizens: The Revolution in Mexico City*. Lincoln: University of Nebraska Press, 2001.

Leidenberger, Georg. *La historia viaja en tranvía el transporte público y la cultura política de la Ciudad de México*. Mexico City: UAM, 2011.

Lemon, Michael, and Eden Medina. "Technology in an Expanded Field: A Review of History of Technology Scholarship on Latin America in Selected English-Language Journals." In *Beyond Imported Magic: Essays on Science, Technology, and Society in Latin America*, edited by Eden Medina, Ivan da Costa Marques, and Christina Holmes, 111–136. Cambridge, MA: MIT Press, 2014.

Levenstein, Harvey. "The 'Servant Problem' and American Cookery." *Revue Française d'Études Américaines* 27–28 (1986): 127–137.

Levine, Robert M. "Michel de Certeau and Latin America." In *The Other Mirror: Grand Theory through the Lens of Latin America*, edited by Miguel Angel Centeno and Fernando López-Alves, 309–328. Princeton, NJ: Princeton University Press, 2001.

Lewis, Oscar. *The Children of Sánchez: Autobiography of a Mexican Family*. New York: Random House, 1961.

Lie, Merete, and Knut H. Sørensen. *Making Technology Our Own? Domesticating Technology into Everyday Life*. Oslo: Scandinavian University Press, 2002.

Liehr, Reinhard, and Mariano Torres. "Las 'Free-Standing Companies' británicas en el Mexico del porfiriato, 1884–1911." *Historia Mexicana* 187 (1998): 605–653.

Lipsett-Rivera, Sonya. *Gender and the Negotiation of Daily Life in Mexico, 1750–1856*. Lincoln: University of Nebraska Press, 2012.

Lira, Andrés. *Comunidades indígenas frente a la Ciudad de México: Tenochtitlan, Tlatelolco, sus pueblos y barrios, 1812–1919*. Mexico City: ColMex, 1995.

López, Rick. *Crafting Mexico: Intellectuals, Artisans, and the State after the Revolution*. Durham, NC: Duke University Press, 2010.

López-Maya, Margarita. "The Change in the Discourse of US-Latin American Relations from the End of the Second World War to the Beginning of the Cold War." *Review of International Political Economy* 2, no. 1 (1995): 135–149.

López Mendiola, Rubén. "La huelga de 1950: Los trabajadores electricistas de la Cia. Mexicana de luz y fuerza Motriz S.A., y subsidiarias." PhD diss., Escuela Nacional de Antropología e Historia, 2010.

Loyo, Engracia. "De sierva a compañera: La imagen de la mujer en textos y publicaciones oficiales (1920–1940)." In *Persistencia y cambio acercamientos a la historia de las mujeres en México*, edited by Lucía Melgar-Palacios, 159–184. Mexico City: ColMex, 2008.

Macedo, Miguel. *La criminalidad en México*. Mexico City: Secretaria de Fomento, 1897.

Marichal, Carlos. *Las inversiones extranjeras en América Latina, 1850–1930: Nuevos debates y problemas en historia económica comparada*. Mexico City: ColMex, 1996.

Marvin, Carolyn. *When Old Technologies Were New: Thinking about Electric Communication in the Late Nineteenth Century*. New York: Oxford University Press, 2007.

Mason, Edward S., and Robert E. Asher. *The World Bank since Bretton Woods*. Washington, DC: Brookings Institution, 1973.

Matthews, Michael. *The Civilizing Machine*. Lincoln: University of Nebraska Press, 2014.

Maxson, Harold R. *A Practical Up-to-Date Guide to Mexico City and Vicinity*. Mexico City: American Book and Print, 1907.

McDowall, Duncan. *The Light: Brazilian Traction, Light, and Power Company Limited, 1899–1945*. Toronto: University of Toronto Press, 1988.

McKay, John. *Tramways and Trolleys: The Rise of Urban Mass Transport in Europe*. Princeton, NJ: Princeton University Press, 1976.

Medina, Eden, Ivan da Costa Marques, and Christina Holmes. *Beyond Imported Magic: Essays on Science, Technology, and Society in Latin America*. Cambridge, MA: MIT Press, 2014.

Meiton, Fredrik. *Electrical Palestine: Capital and Technology from Empire to Nation*. Oakland: University of California Press, 2019.

Memoria de la Secretaría de Gobernación (1900–1904). Mexico City: Imp. del Gobierno, 1906.

Meyer, Lorenzo. *Mexico and the United States in the Oil Controversy, 1917–1942*. Austin: University of Texas Press, 2014.

Milanesio, Natalia. "'The Guardian Angels of the Domestic Economy': Housewives' Responsible Consumption in Peronist Argentina." *Journal of Women's History* 19, no. 3 (2006): 91–117.

———. *Workers Go Shopping in Argentina: The Rise of Popular Consumer Culture*. Albuquerque: University of New Mexico Press, 2015.

Misa, Thomas. "The Compelling Tangle of Modernity and Technology." In *Modernity and Technology*, edited by Thomas J. Misa, Phillip Brey, and Andrew Feenberg, 1–30. Cambridge, MA: MIT Press, 2003.

Mitchell, Timothy, ed. *Questions of Modernity*. Minneapolis: University of Minnesota Press, 2000.

Montaño, Diana J. "*Machucados* and *Salvavidas*: Patented Humor in the Technified Spaces Of Everyday Life in Mexico City, 1900–1910." *History of Technology* 34 (2019): 43–64.

Mora, Carl J. *Mexican Cinema: Reflections of a Society, 1896–2004*. Jefferson, NC: McFarland, 2012.

Moreno, Julio. *Yankee Don't Go Home! Mexican Nationalism, American Business Culture, and the Shaping of Modern Mexico, 1920–1950*. Chapel Hill: University of North Carolina Press, 2007.

Morgan, Tony. "Proletarians, Politicos, and Patriarchs: The Use and Abuse of Cultural Customs in the Early Industrialization of Mexico City, 1880–1910." In *Rituals of Rule, Rituals of Resistance: Public Celebrations and Popular Culture in Mexico*, edited by William Beezley, Cheryl English Martin, and William E. French, 151–171. Wilmington, DE: SR Books, 1994.

Morton, David L. "Reviewing the History of Electric Power and Electrification." *Endeavour* 26, no. 2 (June 2002): 60–63.

Morus, Iwan Rhys. *Frankenstein's Children: Electricity, Exhibition, and Experiment in Early-Nineteenth-Century London*. Princeton, NJ: Princeton University Press, 2016.

Moses, Jasper Turney. *Today in the Land of Tomorrow: Sketches of Life in Mexico*. Indianapolis: Christian Woman's Board of Missions, 1909.

Mraz, John. *Looking for Mexico: Modern Visual Culture and National Identity*. Durham, NC: Duke University Press, 2010.

Muñoz-Serrano, Benito. *Actualidades: Recopilación de humoradas en verso, correspondientes a los asuntos del día*. Mexico City: Católica, 1902.

Nava, Carmen and Isabel Fernández, "La campana de Dolores en el imaginario patriótico." *Boletín del AGN* series 6, no. 2 (November–December 2003): 123–142.

Navitski, Rielle. *Public Spectacles of Violence Sensational Cinema and Journalism in Early Twentieth-Century Mexico and Brazil*. Durham, NC: Duke University Press, 2017.

Nervo, Amado. *El libro que la vida no me dejó escribir: Una antología general*. Mexico City: Fondo de Cultura Económica, 2006.

Niblo, Stephen R. *Mexico in the 1940s: Modernity, Politics, and Corruption*. Wilmington, DE: SR Books, 2001.

Novo, Salvador. *Cocina mexicana: Historia gastronómica de la Ciudad de México*. Mexico City: Editorial Porrúa, 2010.

Nye, David E. *Consuming Power: A Social History of American Energies*. Cambridge, MA: MIT Press, 2001.

———. *Electrifying America: Social Meanings of a New Technology, 1880–1940*. Cambridge, MA: MIT Press, 1992.

———. *When the Lights Went Out: A History of Blackouts in America*. Cambridge, MA: MIT Press, 2013.

Ochoa, Enrique. *Feeding Mexico: The Political Uses of Food since 1910*. Wilmington, DE: SR Books, 2002.

Olavarría y Ferrari, Enrique de. *Reseña histórica del teatro en México*, 2nd ed. Vol. 3. Mexico City: La Europea, 1895.

Olcott, Jocelyn. "Introduction: Researching and Rethinking the Labors of Love." *Hispanic American Historical Review* 91, no. 1 (2011): 1–27.

———. *Revolutionary Women in Post-revolutionary Mexico*. Durham, NC: Duke University Press, 2005.

Olcott, Jocelyn, Mary K. Vaughan, and Gabriela Cano. *Sex in Revolution: Gender, Politics, and Power in Modern Mexico*. Durham, NC: Duke University Press, 2008.

Otter, Chris. *The Victorian Eye: A Political History of Light and Vision in Britain, 1800–1910*. Chicago: University of Chicago Press, 2008.

Oudshoorn, Nelly, and Trevor Pinch. "Introduction: How Users and Non-Users Matter." In *How Users Matter: The Co-Construction of Users and Technology*, edited by Nelly Oudshoorn and Trevor Pinch, 1–25. Cambridge, MA: MIT Press, 2005.

Owensby, Brian. *Intimate Ironies: Modernity and the Making of Middle-Class Lives in Brazil*. Stanford, CA: Stanford University Press, 2001.

Pardo Landa, Alejandro. *El verdadero práctico: Con recetas enteramente nuevas y ensayadas por su autor*. Madrid: Imprenta de Juan Pueyo, 1914.

———. *Los 30 menús del mes: Manual de la cocina casera*. Mexico City: Noriega Editores, 1917.

Parker, David. *The Idea of the Middle Class: White-Collar Workers and Peruvian Society, 1990–1950*. University Park, PA: Penn State University Press, 1998.

Parker, David, and Louise E. Walker. *Latin America's Middle Class: Unsettled Debates and New Histories*. Lanham, MD: Lexington, 2013.

Parra, Porfirio. *Pacotillas*. Barcelona: Salvat é Hijo, 1900.

Paz, Ireneo. *Leyendas históricas de la independencia*. Mexico City: Impr. de I. Paz, 1887.

Pérez Montfort, Ricardo, Alberto del Castillo Yurrita, and Pablo Piccato. *Hábitos, normas y escándalo: Prensa, criminalidad y drogas durante el porfiriato*. Mexico City: Ciesas, Plaza y Valdes Editores, 1997.

Piccato, Pablo. *City of Suspects: Crime in Mexico City, 1900–1931*. Durham, NC: Duke University Press, 2001.

———. "Urbanistas, Ambulantes, and Mendigos: The Dispute for Urban Space in Mexico City, 1890–1930." *Reconstructing Criminality in Latin America*, edited by Carlos Aguirre and Robert Buffington, 113–148. Wilmington, DE: SR Books, 2000.

Pilcher, Jeffrey M. "Fajitas and the Failure of Refrigerated Meatpacking in Mexico: Consumer Culture and Porfirian Capitalism." *Americas* 60, no. 3 (2004): 411–429.

———. "Josefina Velázquez de León, Apostle of the Enchilada." In *The Human Tradition in Mexico*, edited by Jeffrey M. Pilcher, 199–209. Wilmington, DE: SR Books, 2003.

———. *¡Que Vivan Los Tamales! Food and the Making of Mexican Identity*. Albuquerque: University of New Mexico Press, 2008.

———. *The Sausage Rebellion: Public Health, Private Enterprise, and Meat in Mexico City, 1890–1917*. Albuquerque: University of New Mexico Press, 2006.

Pite, Rebekah. "Entertaining Inequalities: Doña Petrona, Juanita Bordoy, and Domestic Work in Mid-Twentieth-Century Argentina." *Hispanic American Historical Review* 91 (2011): 97–128.

Platt, Harold. *The Electric City: Energy and the Growth of the Chicago Area, 1880–1930*. Chicago: University of Chicago Press, 1991.

Pletcher, David M. "The Building of the Mexican Railway." *Hispanic American Historical Review* 30, no. 1 (1950): 26–62.

Pond, George. *Calcium Carbide and Acetylene*. State College: Pennsylvania State College, 1908.

Porter, Susie. "'And That It Is Custom Makes It Law': Class Conflict and Gender Ideology in the Public Sphere, Mexico City, 1880–1910." *Social Science History* 24, no. 1 (2000): 111–148.

———. *From Angel to Office Worker: Middle-Class Identity and Female Consciousness in Mexico, 1890–1950*. Lincoln: University of Nebraska Press, 2018.

———. *Working Women in Mexico City: Public Discourses and Material Conditions, 1879–1931*. Tucson: University of Arizona Press, 2003.

Portuondo, María. "Constructing a Narrative: The History of Science and Technology in Latin America." *History Compass* 7, no. 2 (2009): 500–522.

Prantl, Adolfo, and José L. Groso. *La Ciudad de México: Novísima guía universal de la capital de la República Mexicana*. Mexico City: Juan Buxó, 1901.

Pretel, David, Ian Inkster, and Helge Wendt. "Technology in Latin American History: Perspectives, Scales, and Comparisons." *History of Technology* 34 (2019): 1–22.

Pulido Esteva, Diego. *¡A su salud!: Sociabilidades, libaciones y prácticas populares en la ciudad de México a principios del siglo XX*. Mexico City: ColMex, 2016.

———. "Policía: Del buen gobierno a la seguridad, 1750–1850." *Historia Mexicana* 60, no. 3 (2011): 1595–1642.

Pyenson, Lewis, and Jean-François Gauvin. *The Art of Teaching Physics: The Eighteenth-Century Demonstration Apparatus of Jean Antoine Nollet*. Sillery, Québec: Septentrion, 2002.

Rabe, Stephen G. *The Killing Zone: The United States Wages Cold War in Latin America*. New York: Oxford University Press, 2016.

Ramos y Duarte, Feliz. *Diccionario de mejicanismos*. Mexico City: Imp. de Eduardo Dublan, 1895.

Randeria, Shalini. "Entangled Histories of Uneven Modernities: Civil Society, Caste Solidarities and the Post-Colonial State in India." In *Unravelling Ties: From Social Cohesion to New Practices of Connectedness*, edited by Y. Elkana, I. Krastev, E. Macamo, and S, Randeria, 284–311. Frankfurt: Campus St. Martin's Press, 2002.

Rankin, Monica A. *¡México, La Patria! Propaganda and Production during World War II*. Lincoln: University of Nebraska Press, 2010.

Reseña histórica y estadística de los ferrocarriles de jurisdicción federal. Mexico City: Telégrafos Federales, 1905.

Reseña histórica y estadística de los ferrocarriles de jurisdicción federal. Mexico City: Telégrafos Federales, 1907.

Reséndiz Núñez, Daniel. *El sector eléctrico de México*. Mexico City: CFE, 1994.

Reynolds, Clark Winton. *The Mexican Economy: Twentieth-Century Structure and Growth*. New Haven, CT: Yale University Press, 1970.

Rippy, J. Fred. *British Investments in Latin America, 1822–1949: A Case Study in the Operations of Private Enterprise in Retarded Regions*. Hamden, CT: Archon Books, 1977.

———. "German Investments in Latin America." *Journal of Business of the University of Chicago* 21, no. 2 (1948): 63–73.

Robles, Sonia. *Mexican Waves: Radio Broadcasting along Mexico's Northern Border, 1930–1950*. Tucson: University of Arizona Press, 2019.

Rodríguez, Miguel. *Los tranviarios y el anarquismo en México, 1920–1925*. Puebla, Mexico: Editorial Universidad Autónoma de Puebla, 1980.

Rodríguez Kuri, Ariel. *Historia del desasosiego: La revolución en la Ciudad de México, 1911–1922*. Mexico City: ColMex, 2010.

———. *La experiencia olvidada: El Ayuntamiento de México: Política y gobierno, 1876–1912*. Mexico City: ColMex, 1996.

Romero de Terreros, Pedro. *La ciencia puesta al alcance de todas las inteligencias: Opúsculos sobre la luz eléctrica y telegrafía eléctrica*. Mexico City: Imprenta de Cumplido, 1850.

Rose, Carol. *Giants, Monsters, and Dragons: An Encyclopedia of Folklore, Legend, and Myth*. New York: Norton, 2001.

Rose, Mark H. *Cities of Light and Heat: Domesticating Gas and Electricity in Urban America*. University Park: Penn State University Press, 1995.

Rosenberg, Emily. *Spreading the American Dream: American Economic and Cultural Expansion, 1890–1945*. New York: Farrar, Straus and Giroux, 2013.

Rosenthal, Anton. "The Streetcar in the Urban Imaginary of Latin America." *Journal of Urban History* 42, no. 1 (2016): 162–179.

Ross, James E. *Cooperative Rural Electrification: Case Studies of Pilot Projects in Latin America*. New York: Praeger, 1972.

Roumagnac, Carlos. *Crímenes sexuales y pasionales*. Mexico City: Libreria de Ch. Bouret, 1906.

———. *Los criminales en México*. Mexico City: Tipogr. "El Fénix," 1904.

Rubenstein, Anne. *Bad Language, Naked Ladies, and Other Threats to the Nation: A Political History of Comic Books in Mexico*. Durham, NC: Duke University Press, 1998.

Ruggiero, Kristin. *Modernity in the Flesh: Medicine, Law, and Society in Turn-of-the-Century Argentina*. Stanford, CA: Stanford University Press, 2004.

Ruiz, Jason. *Americans in the Treasure House: Travel to Porfirian Mexico and the Cultural Politics of Empire*. Austin: University of Texas Press, 2014.

Ruiz Medrano, Ethelia. *Mexico's Indigenous Communities: Their Lands and Histories, 1500–2010*. Boulder: University Press of Colorado, 2011.

Saes, Alexandre M., and Norma S. Lanciotti. "La regulación de los servicios de electricidad en Argentina y Brasil (1890–1962)." *Economia e Sociedade* 21, no. 2 (2012): 409–447.

Salvatore, Ricardo, and Carlos Aguirre, eds. *The Birth of the Penitentiary in Latin Amer-*

ica: Essays on Criminology, Prison Reform, and Social Control, 1830–1940. Austin: University of Texas Press, 1996.

Salvatore, Ricardo, Carlos Aguirre, and Gilbert M. Joseph. *Crime and Punishment in Latin America: Law and Society since Late Colonial Times*. Durham, NC: Duke University Press, 2002.

Sanders, Nichole. *Gender and Welfare in Mexico: The Consolidation of a Postrevolutionary State*. University Park, PA: Penn State University Press, 2011.

Scardaville, Michael C. 2000. "(Hapsburg) Law and (Bourbon) Order: State Authority, Popular Unrest, and the Criminal Justice System in Bourbon Mexico City." In *Reconstructing Criminality in Latin America*, edited by Carlos Aguirre and Robert Buffington, 1–17. Wilmington, DE: SR Books, 2000.

Schell, William. *Integral Outsiders: The American Colony in Mexico City, 1876–1911*. Wilmington, DE: SR Books, 2001.

Schivelbusch, Wolfgang. *Disenchanted Night: The Industrialization of Light in the Nineteenth Century*. Berkeley: University of California Press, 1995.

———. *The Railway Journey: The Industrialization of Time and Space in the Nineteenth Century*. Berkeley: University of California Press, 2014.

Scott, Peter. *The Market Makers: Creating Mass Markets for Consumer Durables in Inter-War Britain*. Oxford: Oxford University Press, 2017.

Servín, Elisa. *Del nacionalismo al neoliberalismo, 1940–1994*. Mexico City: FCE, 2010.

———. "Propaganda y Guerra Fría: La campana anticomunista en la prensa mexicana del medio siglo." *Signos Históricos* 11 (January–June 2004): 9–39.

Shamir, Ronen. *Current Flow: The Electrification of Palestine*. Stanford, CA: Stanford University Press, 2013.

Shaya, Gregory. "Mayhem for Moderns: The Culture of Sensationalism in France, c. 1900." PhD diss., University of Michigan, 2000.

Sherratt, Harriott Wight. *Mexican Vistas Seen from Highways and By-ways of Travel*. Chicago: Rand, McNally, 1899.

Silverstone, Roger, and Leslie Haddon. "Design and the Domestication of Information and Communication Technologies: Technical Change and Everyday Life." In *Communication by Design: The Politics of Information and Communication Technologies*, edited by Robin Mansell and Roger Silverstone, 44–74. Oxford: Oxford University Press, 1996.

Silverstone, Roger, and Eric Hirsch, *Consuming Technologies: Media and Information in Domestic Spaces*. London: Routledge, 2005.

Simmel, Georg. "The Metropolis and Mental Life." In *The Blackwell City Reader*, edited by Gary Bridge and Sophie Watson, 11–19. Oxford, UK: Wiley-Blackwell, 2002.

Simon, Linda. *Dark Light: Electricity and Anxiety from the Telegraph to the X-Ray*. Orlando, FL: Harcourt Books, 2004.

Sloan, Kathryn. *Death in the City: Suicide and the Social Imaginary in Modern Mexico*. Oakland: University of California Press, 2018.

Sluis, Ageeth. *Deco Body, Deco City: Female Spectacle and Modernity in Mexico City, 1900–1939*. Lincoln: University of Nebraska Press, 2016.

Smith, Merritt Roe, and Leo Marx, eds. *Does Technology Drive History? The Dilemma of Technological Determinism.* Cambridge, MA: MIT Press, 1994.

Sodi, Demetrio. *Nuestra ley penal.* Mexico City: Carranza y Comp. Impresores, 1905.

Solomon-Godeau, Abigail. "The Other Side of Venus: The Visual Economy of Feminine Display." In *The Sex of Things: Gender and Consumption in History,* edited by Victoria de Grazia and Ellen Furlough, 113–150. Berkeley: University of California Press, 1996.

Sørensen, Knut H., Margarethe Aune, and Morten Hatling, "Against Linearity: On the Cultural Appropriation of Science and Technology." In *Between Understanding and Trust: The Public, Science and Technology,* edited by Meinolf Dierkes and Claudia von Grote, 237–257. Amsterdam: Harwood Academic, 2000.

Speckman-Guerra, Elisa. *Crimen y castigo: Legislación penal, interpretaciones de la criminalidad y administración de justicia (Ciudad de México, 1872–1910).* Mexico City: ColMex, 2002.

Starr, Frederick. *Mexico and the United States: A Story of Revolution, Intervention and War.* Chicago: Bible House, 1914.

Stoner, Lynn K. *From the House to the Streets: The Cuban Woman's Movement for Legal Reform, 1898–1940.* Durham, NC: Duke University Press, 1991.

Szuchman, Mark. "The City as Vision: The Development of Urban Culture in Latin America." In *I Saw a City Invincible: Urban Portraits of Latin America,* edited by Gilbert M. Joseph and Mark Szuchman, 1–31. Lanham: SR Books, 1996.

Tafunell, Xavier. "La revolución eléctrica en América Latina: Una reconstrucción cuantitativa del proceso de electrificación hasta 1930." *Revista de Historia Económica/ Journal of Iberian and Latin American Economic History* 29, no. 3 (2011): 327–359.

Tendler, Judith. *Electric Power in Brazil: Entrepreneurship in the Public Sector.* Cambridge, MA: Harvard University Press, 1968.

Tenenbaum, Barbara. "Streetwise History: The Paseo de la Reforma and the Porfirian State, 1876–1910." In *Rituals of Rule, Rituals of Resistance: Public Celebrations and Popular Culture in Mexico,* edited by William Beezley, Cheryl English Martin, and William E. French, 127–150. Wilmington, DE: SR Books, 1994.

Tenorio-Trillo, Mauricio. *I Speak of the City: Mexico City at the Turn of the Twentieth Century.* Chicago: University of Chicago Press, 2015.

———. *Mexico at the World's Fairs: Crafting a Modern Nation.* Berkeley: University of California Press, 1996.

Terry, T. Philip. *Terry's México: Handbook for Travellers.* 2nd ed. London: Gay and Hancock, 1911.

Thompson, Mark. "The Development of Unionism among Mexican Electrical Workers." PhD diss., Cornell University, 1966.

Tinajero, Araceli, and J. Brian Freeman, eds. *Technology and Culture in Twentieth-Century Mexico.* Tuscaloosa: University of Alabama Press, 2013.

Tweedie, Ethel B. *Mexico As I Saw It.* 2nd ed. New York: Macmillan, 1902.

Vanderwood, Paul J. *Disorder and Progress: Bandits, Police, and Mexican Development.* Wilmington, DE: SR Books, 1992.

———. "Los bandidos de Manuel Payno." *Historia Mexicana* 44, no. 1 (1994): 107–139.

Vaughan, Mary Kay. "Modernizing Patriarchy: State Policies, Rural Households, and Women in Mexico, 1930–1940." In *Hidden Histories of Gender and the State in Latin America*, edited by Elizabeth Dore and Maxine Molyneux, 194–214. Durham, NC: Duke University Press, 2000.

———. "Rural Women's Literacy and Education during the Mexican Revolution: Subverting a Patriarchal Event?" In *Women of the Mexican Countryside, 1850–1990*, edited by Heather Fowler-Salamini and Mary Kay Vaughan, 106–124. Tucson: University of Arizona Press, 1994.

Velázquez de León, Josefina. *Cocina al minuto*. Mexico City: Ediciones J. Velázquez de León, 1956.

———. *Cocina instantánea*. Mexico City: Ediciones J. Velázquez de León, 1956.

———. *Cocina rápida para la mujer moderna*. Mexico City: Ediciones J. Velázquez de León, 1950.

———. *Cómo cocinar en los aparatos modernos*. Mexico City: Eds. J. Velázquez de León, 1950.

———. *Fuertomate en la cocina*. Mexico City: Distribuidora del Fuerte, n.d.

———. *La cocina de la recién casada*. Mexico City: Eds. J. Velázquez de León, 1951.

———. *La cocina económica*. Mexico City: Eds. J. Velázquez de León, 1945.

———. *Los 30 menus*. Mexico City: Eds. J. Velázquez de León, 1942.

———. *Manual práctico de cocina y repostería*. Mexico City: Academia de Cocina y Repostería Velázquez de León, 1943.

Vitz, Matthew. *A City on a Lake: Urban Political Ecology and the Growth of Mexico City*. Durham, NC: Duke University Press, 2018.

Voekel, Pamela. "Peeing on the Palace: Bodily Resistance to Bourbon Reforms in Mexico City." *Journal of Historical Sociology* 5, no. 2 (1992): 183–208.

Walker, Louise E. *Waking from the Dream: Mexico's Middle Classes after 1968*. Stanford, CA: Stanford University Press, 2013.

Walter, Jane. "Lázaro Cárdenas y la fuerza del trabajo: Tres huelgas en 1936." *Historias* 5 (1984): 67–107.

Ward, Henry George. *Mexico in 1827*. London: Colburn, 1828.

Weiner, Richard. *Race, Nation, and Market: Economic Culture in Porfirian Mexico*. Tucson: University of Arizona Press, 2004.

Weis, Robert. *Bakers and Basques: A Social History of Bread in Mexico*. Albuquerque: University of New Mexico Press, 2012.

Wester, Philippus. "Capturing the Waters: The Hydraulic Mission in the Lerma–Chapala Basin, Mexico (1876–1976)." *Water History* 1, no. 1 (2009): 9–29.

Winter, Nevin Otto. *Mexico and Her People of To-day*. London: Cassell, 1913.

Wionczek, Miguel S. "Electric Power: The Uneasy Partnership." In *Public Policy and Private Enterprise in Mexico*, edited by Raymond Vernon, 19–110. Cambridge, MA: Harvard University Press, 2013.

———. "The State and the Electric-Power Industry in Mexico, 1895–1965." *Business History Review* 39, no. 4: 527–566.

Wolfe, Joel. *Working Women, Working Men: São Paulo and the Rise of Brazil's Industrial Working Class, 1900–1955*. Durham, NC: Duke University Press, 1993.

Wynne, Brian. "Unruly Technology: Practical Rules, Impractical Discourses and Public Understanding." *Social Studies of Science* 18, no. 1 (1988): 147–167.

Yeager, Gene. "Porfirian Commercial Propaganda: Mexico in the World Industrial Expositions." *Americas* 34, no. 2 (1977): 230–243.

Zanardi, Tara, and Lynda Klich. *Visual Typologies from the Early Modern to the Contemporary: Local Contexts and Global Practices*. New York: Routledge, 2019.

Zandy, Janet. *Hands: Physical Labor, Class, and Cultural Work*. New Brunswick, NJ: Rutgers University Press, 2004.

Zarco, Francisco. "México de Noche." In *La ilustración mexicana*, vol. 3, edited by Ignacio Cumplido, 157–160. Mexico City: I. Cumplido, 1851.

Zavala Díaz, Ana Laura. "Espejismos de la modernidad: La Ciudad de México en *Los fuereños* de José Tomás de Cuéllar." *Literatura Mexicana* 27, no. 2 (2016): 9–30.

Zolov, Eric. *The Last Good Neighbor: Mexico in the Global Sixties*. Durham, NC: Duke University Press, 2020.

Index

Note: Page numbers in italic type indicate information contained in images or image captions.